Classics in Mathematics

N.P. Bhatia • G.P. Szegő

Stability Theory of Dynamical Systems

Springer
Berlin
Heidelberg
New York
Barcelona
Hong Kong
London
Milan
Paris
Tokyo

N.P. Bhatia • G.P. Szegő

Stability Theory of Dynamical Systems

Reprint of the 1970 Edition

Nam Parshad Bhatia
University of Maryland
Baltimore, MD 21250
USA

George Philip Szegő
University of Rome "La Sapienza"
00184 Rome
Italy

Originally published as Vol. 161 of the
Grundlehren der mathematischen Wissenschaften

Cataloging-in-Publication Data applied for

Die Deutsche Bibliothek - CIP-Einheitsaufnahme
Bhatia, Nam Parshad:
Stability theory of dynamical systems / N. P. Bhatia; G. P. Szegö. - Reprint of the 1970 ed.. - Berlin; Heidelberg; New York; Barcelona; Hong Kong; London; Milan; Paris; Tokyo: Springer, 2002
(Classics in mathematics)
ISBN 3-540-42748-1

Mathematics Subject Classification (2000): 37Cxx, 34Dxx

ISSN 1431-0821
ISBN 3-540-42748-1 Springer-Verlag Berlin Heidelberg New York

Springer-Verlag Berlin Heidelberg New York
a member of BertelsmannSpringer Science+Business Media GmbH

http://www.springer.de

Printed in Germany

Printed on acid-free paper SPIN 10855130 41/3142ck-5 4 3 2 1 0

N. P. Bhatia · G. P. Szegö

Stability Theory of Dynamical Systems

Springer-Verlag Berlin · Heidelberg · New York 1970

Professor Nam Parshad Bhatia
University of Maryland
Division of Mathematics, Baltimore
and Institute for Fluid Dynamics
and Applied Mathematics, College Park
Maryland, U.S.A.

Professor George Philip Szegö
University of Milan
Milan, Italy

Library of Congress Catalog Card Number 70-126892
Title No. 5144

N. P. Bhatia · G. P. Szegö

Stability Theory of Dynamical Systems

Springer-Verlag New York · Heidelberg · Berlin 1970

Professor Nam Parshad Bhatia

University of Maryland
Division of Mathematics, Baltimore
and Institute for Fluid Dynamics
and Applied Mathematics, College Park
Maryland, U.S.A.

Professor George Philip Szegö

University of Milan
Milan, Italy

Library of Congress Catalog Card Number 70-126892
Title No. 5144

To Sushiela and Emilia

Preface

This book contains a systematic exposition of the elements of the theory of dynamical systems in metric spaces with emphasis on the stability theory and its application and extension for ordinary autonomous differential equations.

In our opinion, the book should serve as a suitable text for courses and seminars in the theory of dynamical systems at the advanced undergraduate and beginning graduate level, in mathematics, physics and engineering.

It was never our intention to write a treatise containing all known results on the subject; but we have endeavored to include most of the important new results and developments of the past 20 years. The extensive bibliography at the end should enhance the usefulness of the book to those interested in the further exploration of the subject.

Students should have completed an elementary course in ordinary differential equations and have some knowledge of metric space theory, which is usually covered in undergraduate courses in analysis and topology.

Each author strongly feels that any mistakes left in the book are attributable to the other author, but each would appreciate receiving any comments from the scientific community.

We are obliged to Professor Aaron Strauss for reading the entire typescript and pointing out several corrections. We would also like to thank Doctors Florencio Castillo, Lawrence Franklin, Cesław Olech and Giulio Treccani for help in proofreading the galleys.

June 1970

N. P. Bhatia · G. P. Szegö

Contents

Notation

Set Theoretic Notation

Throughout the book standard set theoretic notations are used. Thus $\subset$, $\cup$, $\cap$, stand for set inclusion, set union and set intersections, respectively. For a given set M, ∂M, $\mathcal{J}M$, $\overline{M}$, $\mathcal{C}(M)$ denote the boundary, interior, closure, and complement of the set M, respectively. For given sets A, B, the set $A-B$ is the set difference.

Other standardly used set theoretic notations are:

X	a metric space with metric ϱ.
2^X	family of all subsets of X.
R	set of real numbers.
R^n	real n-dimensional euclidean space.
R^+	set of non-negative reals.
R^-	set of non-positive reals.
$\emptyset$	the empty set.
$\lvert\cdot\rvert$	absolute value of a real number.
$\lVert\cdot\rVert$	euclidean distance norm.
$\langle x, y\rangle$	the scalar product of vectors x, y in R^n.
$S(x, \alpha)$	for given $x \in X$ and $\varepsilon > 0$ is the open ball of radius $\alpha > 0$ centered at x, i.e., the set $\{y\colon \varrho(x, y) < \alpha\}$.
$S(M, \alpha)$	the set $\{y\colon \varrho(y, M) < \alpha\}$, where $M \subset X$ and $\alpha > 0$ are given.
$S[x, \alpha]$	the closed ball of radius $\alpha \geqq 0$ centered at x, i.e., the set $\{y\colon \varrho(x, y) \leqq \alpha\}$.
$S[M, \alpha]$	the set $\{y\colon \varrho(y, M) \leqq \alpha\}$.
$H(x, \alpha)$	the spherical hypersurface of radius $\alpha \geqq 0$ centered at x, i.e., the set $\{y\colon \varrho(x, y) = \alpha\}$.
$H(M, \alpha)$	the set $\{y\colon \varrho(y, M) = \alpha\}$.
$\{x_n\}$ or $\{x^n\}$	a sequence.
$x_n \to x$	sequence $\{x_n\}$ converges to x.
$\mathcal{C}^1$	family of continuously differentiable functions.
$\mathcal{C}^2$	family of twice continuously differentiable functions.

Notation Pertaining to Dynamical Systems

(X, R, π)	dynamical system on a space X (I, 1.1, p. 5).
π	phase map of a given dynamical system (I, 1.1, p. 5).
π^t	transition corresponding to a given $t \in R$ (p. 6).
π_x	motion through x (p. 6).
$\gamma(x)$	trajectory through x (II, 1.9, p. 14).
$\gamma^+(x)$ $(\gamma^-(x))$	positive (negative) semi-trajectory through x (II, 1.9, p. 14).

$\Lambda^+(x)$ $(\Lambda^-(x))$	positive (negative) limit set of x (II, 3.1, p. 19).
$D^+(x)$ $(D^-(x))$	positive (negative) prolongation of x (II, 4.1, p. 24).
$J^+(x)$ $(J^-(x))$	positive (negative) prolongational limit set of x (II, 4.1, p. 25).
$D^+_\alpha(x)$ $(D^-_\alpha(x))$	for given ordinal number α, the α-th positive (negative) prolongation of x (VII, 1.12, p. 123).
$J^+_\alpha(x)$ $(J^-_\alpha(x))$	for given ordinal number α, the α-th positive (negative) prolongational limit set of x (VII, 3.1, p. 129).
$D^+_u(M)$	uniform positive prolongation of a set M (VII, 2.11, p. 128).
$D^+(M, U)$	for given M, U in X, the positive prolongation of M, relative to U (II, 4.10, p. 29).
$A_\omega(M)$	region of weak attraction of $M \subset X$ (V, 1.1, p. 56).
$A(M)$	region of attraction of $M \subset X$ (V, 1.1, p. 56).
$A_u(M)$	region of uniform attraction of $M \subset X$ (V, 1.1, p. 56).
$\mathcal{R}$	the generalized recurrent set (VII, 3.6, p. 131).
$\mathcal{D}$	operator used in the definition of higher prolongation (VII, 1.1, p. 120).
$\mathcal{S}$	operator used in the definition of higher prolongation (VII, 1.1, p. 120).

Introduction

The theory of dynamical systems may be said to have begun as a special topic in the theory of ordinary differential equations with the pioneering work of HENRI POINCARÉ in the late 19th century. POINCARÉ, followed by IVAR BENDIXSON, studied topological properties of solutions of autonomous ordinary differential equations in the plane. The Poincaré-Bendixson theory is now a standard topic of discussion in courses on ordinary differential equations, and is adequately covered in all its details in the books of, say, CODDINGTON and LEVINSON [1], LEFSCHETZ [1], HARTMAN [1], SANSONE and CONTI [1], and NEMYTSKII and STEPANOV [1]. Of these, HARTMAN's book contains the most detailed and recent exposition.

Almost simultaneously with POINCARÉ, A. M. LIAPUNOV developed his theory of stability of a motion (solution) for a system of n first order ordinary differential equations. He defined in a precise form the concept of stability, asymptotic stability, and instability; and gave a "method" (the second or direct method of Liapunov) for the analysis of the stability properties of a given solution of an ordinary differential equation. Both his definition and his "method" characterize, in a strictly local setting, the stability properties of a solution of the differential equation. As such, the Liapunov theory is strikingly different from the Poincaré theory, in which, on the contrary, the study of the global properties of differential equations in the plane play a major role.

One of the main aspects of the Poincaré theory is the introduction of the concept of a trajectory, i.e., a curve in the x, $\dot{x}$ plane, parametrized by the time variable t, which can be found by eliminating the variable t from the given equations, thus reducing it to a first order differential equation connecting x and $\dot{x}$. In this way, POINCARÉ set up a convenient geometric framework in which to study qualitative behavior of planar differential equations. POINCARÉ was not interested in the integration of particular types of equations, but in classifying all possible behaviors of the class of all second order differential equations. By introducing this concept of trajectory, POINCARÉ was able to formulate and solve, as topological problems, problems in the theory of differential equations.

In the above fashion POINCARÉ paved the way for the formulation of the abstract notion of a dynamical system, which can be essentially attributed to A.A. MARKÓV and H. WHITNEY. These two authors separately noticed that one could study the qualitative theory of families of curves (trajectories) in a suitable space X, provided that these families are somehow restricted in their possible behavior, e.g., if they are defined, as having been generated by a general one-parameter topological transformation group acting on X.

Great impetus to the theory of dynamical systems came from the work of G.D. BIRKHOFF, who may truly be considered as the founder of the theory. His celebrated 1927 monograph on Dynamical Systems (BIRKHOFF [1]) is the basis of much of the research which came in the 1930's and 1940's and even today it is not outdated. BIRKHOFF established the two main streams of work on the theory of dynamical systems, namely, the topological theory and the ergodic theory.

In 1947, V.V. NEMYTSKII and V.V. STEPANOV [1] completed their "Qualitative Theory of Differential Equations" which to this day has served as a standard reference for all the major development in the theory of dynamical systems up to the middle 1940's. In 1949, NEMYTSKII [10] wrote a survey paper on the topological problems in the theory of dynamical systems, which sums up almost all the research into the topological theory to the end of 1940's.

During the 1950's a relatively large effort went into the generalization of the concept of a dynamical system to topological transformation groups. Thus in 1955 the book of W.H. GOTTSCHALK and G.A. HEDLUND [1] appeared, and a large body of research has appeared since in print. In this connection the work of R. ELLIS, H. FURSTENBERG, J. AUSLANDER, H. CHU, F. HAHN, S. KAKUTANI, besides GOTTSCHALK and HEDLUND, is noteworthy. On the other hand problems on structural stability in ordinary differential equations have led to the efforts at introducing the concepts and methods of differential topology and the theories of S. SMALE, D.V. ANOSOV, J. MOSER, M. PEIXOTO, and L. MARKUS. In this connection ANOSOV's monograph [1] is noteworthy.

More recently the basic theory has been extended by bringing in, the stability problems à la Liapunov, characteristically absent in earlier works on dynamical systems and topological transformation groups. In this connection, TARO URA's work, in particular his theory of prolongations, and its connections with stability has clearly shown that a significant portion of stability theory is topological in nature and hence belongs to the main stream of the theory of dynamical systems. An attempt in bringing in the direct method of Liapunov was made by V. I. Zubov [1]. However, Zubov mainly carried over to flows in metric spaces, results and methods previously known in differential equations without attempting to develop an independent theory.

The present volume was thus conceived to present in an easy and readable fashion the recent research in the theory of dynamical systems on metric spaces, and especially the stability theory together with its concrete applications to the theory of differential equations.

This book does not introduce several interesting areas of modern day research, such as the theory of structural stability (which requires some knowledge of differential topology), ergodic theory, and the general theory of topological transformation groups. To keep the presentation at a level easily accessible to undergraduate students in their junior or senior years (who have had some exposure to metric spaces and differential equations), we have not introduced local dynamical systems, although most results are true for such systems. Local semi-dynamical systems (N. P. Bhatia and O. Hajek [1]), flows without uniqueness (G. P. Szegö and G. Treccani [1]), are other major areas of development today. The present book should be helpful to all those desiring to work in any one of the above mentioned fields of study and research, which are not covered in this volume.

As to material covered in this volume, Chapters I—VII contain the basic theory of dynamical systems in metric spaces and Chapters VIII and IX contain applications and extensions of the stability theory (Chapter V) to dynamical systems defined by ordinary differential equations. Specifically, Chapter I contains the definition of a dynamical system and some examples to indicate various fields of application. Chapter II contains elementary notions which remain invariant under certain topological transformations of dynamical systems. Chapter III deals mainly with minimal sets and their structure. Chapter IV is devoted to the study of dispersive and parallelizable dynamical systems and concludes the part of the book devoted to the basic theory. Chapter V develops the main theme of the book, i.e., the stability and attraction theory. The theory presented here differs rather strongly from the one developed by Zubov, being essentially based on the concept of weak attraction (absent in Zubov's work). Chapter VI is devoted to a more specific problem: the classification of flows near a compact invariant set. Some results are given, but many problems in this are still open. Chapter VII contains the theory of higher prolongations originated by T. Ura with applications to absolute stability and generalized recurrence. Chapter VIII deals with the geometrical theory of stability for ordinary autonomous differential equations including various extensions of Liapunov's direct method. Chapter IX is again devoted to a more specific problem of characterizing stability and attraction concepts via non-continuous Liapunov functions; these are concepts like the weak attractor, which are not characterizable by continuous Liapunov functions.

Regarding the formal structure of the book, each chapter is divided into sections, followed by an un-numbered section of notes and references. In each section, individual items (Definitions, Theorems, etc.) are numbered consecutively. Each item may be subdivided into consecutively numbered subitems. References to the same chapter do not mention the chapter number. Thus, for example, reference 2.5.3 indicates, section 2, item 5, subitem 3 of the same chapter. References to other chapters contain an indication of the chapter number. Thus II, 3.17 denotes item 17 of section 3 in Chapter II. References to the bibliography are given by the author's name followed, if necessary, by an item number between brackets.

Chapter I

Dynamical Systems

In this chapter we introduce the definition of a dynamical system or what is also called a continuous flow. Several general examples are given to motivate and prepare the reader for the study of the theory of dynamical systems. Throughout the book the symbol X denotes a metric space with metric ϱ and R stands for the set of real numbers.

1. Definition and Related Notation

1.1 Definition. A *dynamical system* on X is the triplet (X, R, π), where π is a map from the product space $X \times R$ into the space X satisfying the following axioms:

1.1.1 $\pi(x, 0) = x$ for every $x \in X$ (identity axiom),

1.1.2 $\pi(\pi(x, t_1), t_2) = \pi(x, t_1 + t_2)$ for every $x \in X$ and t_1, t_2 in R (group axiom),

1.1.3 π is continuous (continuity axiom).

Given a dynamical system on X, the space X and the map π are respectively called the *phase space* and the *phase map* (of the dynamical system). Unless otherwise stated a dynamical system on X is always assumed given.

In the sequel we shall generally delete the symbol π. Thus the image $\pi(x, t)$ of a point (x, t) in $X \times R$ will be written simply as xt. The identity and the group axioms then read

1.1.4 $x0 = x$ for every $x \in X$,

1.1.5 $xt_1(t_2) = x(t_1 + t_2)$ for every $x \in X$ and t_1, t_2 in R.

In line with this notation, if $M \subset X$ and $A \subset R$, then MA is the set $\{xt\colon x \in M \text{ and } t \in A\}$. If either M or A is a singleton, i.e., $M = \{x\}$ or

$A = \{t\}$, then we simply write xA and Mt for $\{x\} A$ and $M\{t\}$, respectively. For any $x \in X$, the set xR is called the trajectory through x (see II, 1.9).

The phase map determines two other maps when one of the variables x or t is fixed. Thus for fixed $t \in R$, the map π^t: $X \to X$ defined by $\pi^t(x) = xt$ is called a *transition*, and for a fixed $x \in X$, the map π_x: $R \to X$ defined by $\pi_x(t) = xt$ is called a *motion* (through x). Note that π_x maps R onto xR.

The following theorem expresses an important property of the transitions.

1.2 **Theorem.** For each $t \in R$, π^t is a homeomorphism of X onto itself.

Proof. For any $t \in R$ the transition π^t is continuous as π is such. To see that π^t is one-to-one and onto observe that if $yt = zt$, then $y = z$ follows from $y = y\,0 = y(t - t) = yt(-t) = zt(-t) = z(t - t) = z\,0 = z$. Again, if $y \in X$, then $\pi^t(x) = y$ for $x = y(-t)$ is easily verified. Finally to see that π^t has a continuous inverse we need only show that the transition π^{-t} is the inverse of π^t. To see this note that for any two transitions π^t and π^s, the composition $\pi^t \circ \pi^s$ is the transition π^{t+s}, because for any $x \in X$,

$$\pi^t \circ \pi^s(x) = \pi^t(\pi^s(x)) = \pi^t(xs) = xs(t) = x(t + s) = \pi^{t+s}(x).$$

Note also that the transition π^0 is the identity, since for any $x \in X$, $\pi^0(x) = x\,0 = x$. Since now $\pi^{-t} \circ \pi^t = \pi^{t-t} = \pi^0$, the transition π^{-t} is the inverse of π^t.

1.3 *Exercises.*

1.3.1 Show that the transitions π^t, $t \in R$, form a commutative group with the group operation being the composition of transitions.

1.3.2 For any $x \in X$ and $[a, b] \subset R$, the set $x[a, b]$ is compact and connected.

2. Examples of Dynamical Systems

2.1 *Ordinary Autonomous Differential Systems.* Consider the autonomous differential system

2.1.1 $$\frac{dx}{dt} = \dot{x} = f(x),$$

where f: $R^n \to R^n$ (R^n is the real n-dimensional euclidean space) is continuous and moreover assume that for each $x \in R^n$ a unique solution

$\varphi(t, x)$ exists which is defined on R and satisfies $\varphi(0, x) = x$. Then it is well known (see for example CODDINGTON and LEVINSON [1], chapters 1 and 2) that the uniqueness of solutions implies

2.1.2 $$\varphi(t_1, \varphi(t_2, x)) = \varphi(t_1 + t_2, x) \quad \text{for } t_1, t_2 \text{ in } R$$

and considered as a function from $R \times R^n$ into R^n, φ is continuous in its arguments (section 4, chapter 2 in CODDINGTON and LEVINSON [1]). It is clear that the map $\pi: R^n \times R \to R^n$ such that $\pi(x, t) = \varphi(t, x)$ defines a dynamical system on R^n. We remark that the conditions on solutions of 2.1.1, as required above, are obtained, for example, if the function f satisfies a global Lipschitz condition, i.e., there is a positive number k such that

2.1.3 $$\|f(x) - f(y)\| \leqq k \|x - y\| \quad \text{for all } x, y \text{ in } R^n.$$

2.2 *Ordinary Autonomous Differential Systems (Continued).* To illustrate that the theory of dynamical systems as defined in this chapter is applicable to a much larger class of ordinary autonomous differential systems we consider a system

2.2.1 $$\frac{dx}{dt} = \dot{x} = f(x), \quad x \in R^n$$

where $f: D \to R^n$ is a continuous function on some open set $D \subset R^n$, and for each $x \in D$, 2.2.1 has a unique solution $\varphi(t, x)$, $\varphi(0, x) = x$ defined on a maximal interval (a_x, b_x), $-\infty \leqq a_x < 0 < b_x \leqq +\infty$. For each $x \in D$ define $\gamma^+(x) = \{\varphi(t, x): \; 0 \leqq t < b_x\}$, and $\gamma^-(x) = \{\varphi(t, x): a_x < t \leqq 0\}$. $\gamma^+(x)$ and $\gamma^-(x)$ are respectively called the positive and negative trajectory through the point $x \in D$. We will show that to each system 2.2.1, there corresponds a system

2.2.2 $$\frac{dx}{dt} = \dot{x} = g(x), \quad x \in R^n,$$

where $g: D \to R^n$, such that 2.2.2 defines a dynamical system on D with the property that for each $x \in D$ the systems 2.2.1 and 2.2.2 have the same positive and the same negative trajectories. Thus in general it is sufficient to consider 2.2.2 instead of 2.2.1.

If $D = R^n$, then given 2.2.1, we set

2.2.3 $$\frac{dx}{dt} = \dot{x} = g(x) = \frac{f(x)}{1 + \|f(x)\|},$$

where $\|\cdot\|$ is the euclidean-distance norm. If $D \neq R^n$, then $\partial D \neq \emptyset$ and is closed. In this case, given 2.2.1, we set

2.2.4 $$\frac{dx}{dt} = \dot{x} = g(x) = \frac{f(x)}{1 + \|f(x)\|} \, \frac{\varrho(x, \partial D)}{1 + \varrho(x, \partial D)},$$

where $\varrho(x, \partial D) = \inf\{\|x - y\|: y \in \partial D\}$. One can now show that 2.2.3 or 2.2.4 defines a dynamical system on R^n or D, whenever the system 2.2.1 has unique solutions. Moreover, 2.2.3 and 2.2.4 have the same positive and negative trajectories as 2.2.1.

2.3 *Ordinary Autonomous Differential Systems Containing Parameters.* Consider the differential system

2.3.1 $\dot{x} = f(x, \mu), \quad x \in R^n,$

where μ is a parameter in a metric space M, and $f: R^n \times M \to R^n$ is continuous. Assume that for each $(x, \mu) \in R^n \times M$ there is a unique solution $\varphi(t, (x, \mu))$ defined on R and satisfying $\varphi(0, (x, \mu)) = x$. Set $X = R^n \times M$, and define $\pi: X \times R \to X$ by setting $\pi((x, \mu), t) = (\varphi(t, (x, \mu)), \mu)$. Then it is easily verified that π defines a dynamical system on $R^n \times M = X$.

2.4 *Ordinary Non-autonomous Differential Systems.* Consider the differential system

2.4.1 $\dot{x} = f(t, x), \quad x \in R^n, t \in R,$

where $f: R \times R^n \to R^n$ is continuous. Assume that for each $(t_0, x_0) \in R \times R^n$, 2.4.1 possesses a unique solution $\varphi(t, t_0, x_0)$, $\varphi(t_0, t_0, x_0) = x_0$, defined for all $t \in R$. Consider the map $\pi: X \times R \to X$, where $X = R \times R^n$, defined by $\pi((t, x), s) = (s + t, \varphi(s + t, t, x))$. Then π defines a dynamical system on X. This dynamical system in fact corresponds to the equivalent autonomous differential system in R^{n+1}

2.4.2 $\dot{y} = g(y),$

where $y = (t, x)$, and $g(y) = (1, f(y))$.

2.5 *The Bebutov Dynamical System.* Consider the set X of all continuous functions $f: R \to R^n$. Now define $\varrho: X \times X \to [0, +\infty)$ as follows: For any integer m, let $I_m = [-m, m]$, and for $f, g \in X$ set

2.5.1 $\sigma_m(f, g) = \max\{\|f(t) - g(t)\|: t \in I_m\},$

2.5.2 $\varrho_m(f, g) = \dfrac{\sigma_m(f, g)}{1 + \sigma_m(f, g)},$

and finally

2.5.3 $\varrho(f, g) = \sum_{m=1}^{\infty} \dfrac{1}{2^m} \varrho_m(f, g).$

One may now verify that ϱ defines a metric on X (i.e., X is a metric space with metric ϱ given by 2.5.3) and that with this metric the space X is complete. One can verify in fact that if $\{f_n\}$ is any sequence in X, then $\varrho(f_n, f) \to 0$ as $n \to \infty$ if and only if $\{f_n(t)\}$ converges to $f(t)$ uniformly on every compact subset of R.

On the metric space X defined above we define a dynamical system as follows: $\pi\colon X \times R \to X$ is given by

$$\pi(f, t) = g$$

where g is given by $g(s) = f(t + s)$ for all $s \in R$. To verify that π is a dynamical system, note first that the identity axiom is trivially satisfied. To see the group axiom, let $\pi(\pi(f, t), s) = h$ and $\pi(f, t) = g$. Then $h(\tau) = g(s + \tau) = f(t + s + \tau)$ by definition. Hence $h = \pi(f, t + s)$. Finally, to see that π is continuous, let f_n be a sequence in X, $\{t_n\}$ a sequence in R, such that $f_n \to f$ and $t_n \to t$. To show is $\pi(f_n, t_n) \to \pi(f, t)$. Let $K \subset R$ be compact. Let $\pi(f_n, t_n) = g_n$, and $\pi(f, t) = g$. We must show that given $\varepsilon > 0$ there is an integer N such that $\|g_n(\tau) - g(\tau)\| < \varepsilon$ for $n \geqq N$ and $\tau \in K$. Now $g_n(\tau) - g(\tau) = f_n(t_n + \tau) - f(t + \tau) = f_n(t_n + \tau) - f_n(t + \tau) + f_n(t + \tau) - f(t + \tau)$. Thus

$$\|g_n(\tau) - g(\tau)\| \leqq \|f_n(t_n + \tau) - f_n(t + \tau)\| + \|f_n(t + \tau) - f(t + \tau)\|.$$

Note now that $f_n(t) \to f(t)$ uniformly on any compact subset of R so that the sequence $\{f_n(t)\}$ is equi-continuous. The result is now easily seen from the last inequality. The dynamical system π defined above has many interesting properties. This has been included to show that the theories expounded in this book are applicable to situations which may not necessarily arise out of ordinary differential equations. However, we shall not study such examples in any detail.

2.6 *Dynamical Systems on a Torus.* Consider a differential system in the plane

2.6.1 $$\frac{dx}{dt} = f(x), \qquad x \in R^2,$$

where $f\colon R^2 \to R^2$ is continuous and the conditions of example 2.1 are satisfied. Set $x = (\varphi, \theta)$, $f = (f_1, f_2)$ and assume that the functions f_1 and f_2 are periodic of period 1 in each of the variables θ and φ. Thus for all real φ and θ

2.6.2 $$f_i(\varphi, \theta) = f_i(\varphi + 1, \theta) = f_i(\varphi, \theta + 1), \qquad i = 1, 2.$$

It is customary to think of the equation 2.6.1 satisfying condition 2.6.2 as representing a dynamical system on the torus, instead of in the plane R^2, the immediate advantage being the fact that the torus is a compact 2-manifold. To see this clearly one takes as a convenient representation of the torus, the square

$$\{(\varphi, \theta)\colon 0 \leqq \varphi < 1, 0 \leqq \theta < 1\}$$

in which the pairs of opposite sides $\varphi = 0$, $\varphi = 1$ and $\theta = 0$, $\theta = 1$ are identified. This yields, in particular, that the vertices $(0, 0)$, $(0, 1)$, $(1, 0)$, $(1, 1)$ are all identified. The plane R^2 is now projected on the torus by identifying any point (φ, θ) in the plane with a point $(\tilde{\varphi}, \tilde{\theta})$ of the square for which $\varphi - \tilde{\varphi}$, and $\theta - \tilde{\theta}$ are integers. One usually expresses this by $\varphi = \tilde{\varphi}$ (mod 1), and $\theta = \tilde{\theta}$ (mod 1). The projection of trajectories of 2.6.1 in the plane onto the torus then yields the trajectories of a dynamical system on the torus. Thus 2.6.1 is said to define a dynamical system on the torus when condition 2.6.2 holds.

Notes and References

Section 1. The introduction of the definition of a dynamical system cannot be attributed to any one person. Interesting historical remarks on the development of the basic concepts of dynamical systems can be found in the papers by V.V. NEMYTSKII [10] and by G.D. BIRKHOFF [2, vol. 2, p. 710].

The first abstract definition of a dynamical system can be found in the works of A.A. MARKOV [1] and of H. WHITNEY [1, II]. Most concepts were introduced by H. POINCARÉ [1] and by G.D. BIRKHOFF in the framework of the theory of dynamical systems defined by ordinary differential equations.

The theory of dynamical systems received new impetus in the recent years by the publication of the book by V.V. NEMYTSKII and V.V. STEPANOV.

Generalizations to topological transformation groups have appeared since then in the monographs of G.T. WHYBURN, W.H. GOTTSCHALK and G.A. HEDLUND, D. MONTGOMERY and L. ZIPPIN, L. AUSLANDER, L.W. GREEN and F. HAHN.

In this volume we devote ourselves to dynamical systems in a metric space. We do this for simplicity of the proofs. Indeed most of the results that we present can be proved with due care in more general topological spaces.

Various generalizations of the concepts of a dynamical system have been studied in the literature, among them we mention the discrete dynamical systems (also called cascades), i.e., the triple (X, I, π) where I is the set of integers (see for instance W.H. GOTTSCHALK and G.A. HEDLUND [1]), the local dynamical systems $(X, \mathscr{I}_x, \pi)$, which are defined only on intervals $\mathscr{I}_x \subset R$, $0 \in \mathscr{I}_x$ (see for instance T. URA [4] or O. HAJEK [1]).

The definition of local dynamical systems may be considered to have been motivated by ordinary differential equations satisfying conditions of existence and uniqueness of solutions but not necessarily those of global extendability. Example 2.2 shows that local systems defined by ordinary differential equations can be reparametrized to obtain global systems. It is however not known if the same is possible for local systems not defined by ordinary differential equations (see M.I. GRABAR [3]).

A further generalization of dynamical systems is the concept of dynamical systems without uniqueness or semigroups of multivalued mappings: (X, R, f), where $f\colon X \times R \to X$ is multivalued and satisfies certain axioms. These have been investigated by E.A. BARBASHIN, B.M. BUDAK, by M.I. MINKEVITCH, and by G.P. SZEGÖ and G. TRECCANI [1] among others.

Semigroups of multivalued mappings defined only on the positive (or negative) reals (X, R^+, f) have been investigated by I.U. BRONSHTEIN [2] and by E. ROXIN [3, 5, 6]. A complete advanced theory is presented in the recent monograph by N.P. BHATIA and O. HAJEK [1] where the local semi-dynamical systems are investigated. These generalizations of dynamical systems are usually motivated by the need of finding abstract frameworks for the investigation of systems whose behavior does not satisfy the axioms 1.1. For instance ordinary differential equations without global extendability of solutions may define a local dynamical system, ordinary differential equations without uniqueness may define a semigroup of multivalued mappings (R^n, R, f), integral equations (R. K. MILLER and G. SELL [1]) define, in some cases, semi-dynamical systems (X, R^+, π) and functional differential equations are used by N. P. BHATIA and O. HAJEK [1] as example of local semi-dynamical systems.

A natural extension of the theory of dynamical systems and in particular of semigroups of multivalued mappings is the theory of dynamical polysystems, which deals with one-parameter families of dynamical systems (see D. BUSHAW [1, 3]).

A complete presentation of special results for dynamical systems on the plane is due to O. HAJEK. In his recent monograph [1] he was able to extend most results of the Poincaré-Bendixson theory for planar second order differential equations to dynamical systems on the plane.

Lastly, we want to point out that in all the theories mentioned above the mapping π is usually assumed to be continuous, but not continuously differentiable. There exist works (see for instance S. SMALE [2, 3, 5, 6, 7], J. MOSER and D. V. ANOSOV [1]) in which π is assumed to be continuously differentiable. The type of problems solved and the techniques used in these later works are considerably different from the ones used in this book, and close to the works on ergodic problems.

Section 2. 2.1 In order for the differential equation 2.1.1 to define a dynamical system it must satisfy conditions which insure existence, uniqueness and extendability of its solutions. These conditions can be found in most books on differential equations (N. P. BHATIA and G. P. SZEGÖ [1, section 3.1], P. HARTMAN [1, chapter 1—2], E. CODDINGTON and N. LEVINSON [1, chapter 1—2] and V. V. NEMYTSKII and V. V. STEPANOV [1, chapter 1]. Notice that beside the problems of when a differential equation defines a dynamical system, also the more important inverse problem has been discussed, i.e., what are the conditions for a flow to be represented as the system of solutions of differential equations (M. I. GRABAR [1, 2, 3]).

2.3 The differential equation 2.3.1 when μ is a function belonging to a certain set defines a control system, i.e., a particular realization of a dynamical polysystem (D. BUSHAW [1, 2]).

2.4 In addition to the one given here there exists another interesting axiomization of the non-autonomous differential equation 2.4.1 due to G. SELL [5]. He defines first of all a mapping $\pi^*: C \times R \to C$, where C is the space of continuous functions defined by $\pi^*(t, \tau) = f(x, t + \tau)$ and proves that π^* defines a dynamical system on C. Consider next the set $F = \{f(t + \tau): \tau \in R\}$ and the metric space $X = F \times R$ with a suitable metric, where f is a function which ensures uniqueness of solutions of equation 2.4.1. Then the mapping $\pi: X \times R \to X$ defined by $\pi((x, f), \tau) = (\varphi(x, f, \tau), f(t + \tau))$, here $\varphi(x, f, \tau)$ denotes the solution of equation 2.4.1 with $\varphi(x, f, 0) = x$, defines a dynamical system.

2.5 The dynamical system and the space of continuous functions was investigated by M. V. BEBUTOV [3, 4, 5].

Other models of dynamical systems of particular type are used in the so-called symbolic dynamics (see M. MORSE [4]), in the theory of Hamiltonian systems (see G. BIRKHOFF [1], J. MOSER, and C. CONLEY, where additional references can be found), and in the theory of linear dynamical systems (see R. E. KALMAN [1] and F. H. SHOLOHOVICH [2]).

Chapter II

Elementary Concepts

In this chapter we introduce certain fundamental maps $Q: X \to 2^X$ (2^X is the set of all subsets of X) corresponding to a given dynamical system (X, R, π). The basic topological properties such as closedness, and connectedness of the image sets $Q(x)$ are studied. The maps introduced are such that their images have in general certain invariance properties, so that we start with the notion of invariance of a set in section 1. A dynamical system (X, R, π) is assumed given. R^+ and R^- will denote respectively the set of non-negative and non-positive real numbers.

1. Invariant Sets and Trajectories

1.1 Definition. A *set* $M \subset X$ is called *invariant* whenever

1.1.1 $\quad xt \in M \quad$ for all $x \in M$ and $t \in R$.

It is called *positively invariant* whenever 1.1.1 holds with R replaced by R^+, and is called *negatively invariant* if the same holds with R replaced by R^-. Note that 1.1.1 is equivalent to $MR = M$.

The following theorems exhibit some properties of such sets.

1.2 Theorem. Let $\{M_i\}$ be a collection of positively invariant, negatively invariant, or invariant subsets of X. Then their intersection and their union have the same property.

Proof. To fix our ideas let the sets M_i be positively invariant. Let $M' = \cup M_i$, and $M'' = \cap M_i$. For any $x \in M'$, we have $x \in M_i$ for some i. Thus $xt \in M_i$ for all $t \in R^+$ since M_i is positively invariant. Hence indeed $xt \in M'$ for all $t \in R^+$ as $M' \supset M_i$. M' is therefore positively invariant. Now let $x \in M''$. Then $x \in M_i$ for every i and by positive invariance of each M_i, $xt \in M_i$ for each i and each $t \in R^+$. Hence

$xt \in \cap M_i = M''$ for each $t \in R^+$ and M'' is positively invariant. The proof for negative invariance and for invariance are entirely analogous.

1.3 Theorem. Let $M \subset X$ be positively invariant, negatively invariant, or invariant. Then the closure $\overline{M}$ has the same property.

Proof. Consider the case of invariance. Let $x \in \overline{M}$ and $t \in R$. Then there is a sequence $\{x_n\}$ in M such that $x_n \to x$. By invariance of M we have $x_n t \in M$ for each n. Since $x_n t \to xt$ we have $xt \in \overline{M}$. Thus $\overline{M}$ is invariant. The proofs for positive and negative invariance are entirely analogous.

1.4 Theorem. A set $M \subset X$ is positively invariant if and only if the set $X - M$ is negatively invariant. M is invariant if and only if $X - M$ is invariant.

Proof. Let M be positively invariant. If $x \in X - M$ and $t \in R^-$ then we must show that $xt \in X - M$. Suppose not. Then $xt \in M$ and since $-t \in R^+$ we have $xt(-t) = x(t - t) = x\,0 = x \in M$ by positive invariance of M. This contradiction shows that $X - M$ is negatively invariant. Proof of the converse is entirely similar. The proof of the second result follows from the first.

As a final useful result on invariance we have

1.5 Theorem. A set $M \subset X$ is invariant if and only if it is both positively and negatively invariant.

We have left the simple proof as an exercise.

The following important results on invariance are really corollaries of the first three fundamental theorems on invariance established above.

1.6 Theorem. Let $M \subset X$ be invariant. Then so is its boundary ∂M and its interior $\mathscr{I}(M)$. The converse holds whenever M is open or closed.

Proof. If M is invariant then so is $X - M$ by Theorem 1.4. Consequently both $\overline{M}$ and $\overline{X - M}$ are invariant by Theorem 1.3. Hence $\partial M = \overline{M} \cap \overline{X - M}$ is invariant by Theorem 1.2, and $\mathscr{I}(M) = X - \overline{X - M}$ is invariant again by Theorem 1.4. For the converse note that if ∂M is invariant then so is $\mathscr{I}(M)$. For otherwise there is an $x \in \mathscr{I}(M)$ and $t \in R$ with $xt \notin \mathscr{I}(M)$. Since the set (assume for convenience $t > 0$) $x[0, t]$ is connected it follows that $x[0, t] \cap \partial M \neq \emptyset$. Hence there is a

τ, $0 < \tau \leqq t$, with $x\tau \in \partial M$. But then $x = x\tau(-\tau) \in \partial M$ by invariance of ∂M, which is a contradiction to $x \in \mathscr{I}(M)$. We have thus proved that if ∂M is invariant, then so is $\mathscr{I}(M)$. Consequently if M is closed then $M = \partial M \cup \mathscr{I}(M)$ is invariant, and if M is open then $M = \mathscr{I}(M)$ is invariant. The proof that when $\mathscr{I}(M)$ is invariant then so is M for closed or open M is even simpler and left to the reader.

1.7 **Theorem.** If $M \subset X$ is positively or negatively invariant then so is its interior $\mathscr{I}(M)$.

Proof. Assume M is positively invariant. Then $X - M$ and hence $\overline{X - M}$ are negatively invariant. Consequently $\mathscr{I}(M) = X - \overline{X - M}$ is positively invariant.

1.8 *Exercise.* The sets X and $\emptyset$ are positively and negatively invariant.

We now introduce the notion of a trajectory.

1.9 **Definition.** We introduce the maps γ, γ^+, and γ^- from X into 2^X by defining for any $x \in X$,

1.9.1 $\gamma(x) = \{xt: t \in R\}$,

1.9.2 $\gamma^+(x) = \{xt: t \in R^+\}$,

1.9.3 $\gamma^-(x) = \{xt: t \in R^-\}$.

For any $x \in X$, the sets $\gamma(x)$, $\gamma^+(x)$, and $\gamma^-(x)$ are respectively called the *trajectory*, the *positive semi-trajectory*, and the *negative semi-trajectory* through x (or of x). Note that for any $x \in X$, $\gamma(x) = xR$, etc.

1.10 *Exercises.*

1.10.1 For any $x \in X$ the sets $\gamma(x)$, $\gamma^+(x)$, and $\gamma^-(x)$ are, respectively, invariant, positively invariant, and negatively invariant.

1.10.2 For the dynamical system defined by the differential equations 3.3.2 and 3.3.4, prove that the open and the closed unit disc are invariant sets containing more than one trajectory.

For a given map $Q: X \to 2^X$ and $M \subset X$ we let $Q(M) = \cup \{Q(x): x \in M\}$. Thus it is clear that for any $M \subset X$, $\gamma(M)$ is invariant, $\gamma^+(M)$ is positively invariant, and $\gamma^-(M)$ is negatively invariant. This leads to the following characterization of invariance.

1.11 **Proposition.** A set $M \subset X$ is invariant, positively invariant, or negatively invariant if and only if, respectively, $\gamma(M) = M$, $\gamma^+(M) = M$, or $\gamma^-(M) = M$.

Another characterization of invariance is the following.

1.12 Proposition. A set $M \subset X$ is invariant, positively invariant, or negatively invariant if and only if for each $x \in M$, respectively, $\gamma(x) \subset M$, $\gamma^+(x) \subset M$, or $\gamma^-(x) \subset M$.

1.13 *Exercise.* Give detailed proofs of Propositions 1.11 and 1.12.

The sets $\gamma(x)$, $\gamma^+(x)$, $\gamma^-(x)$ as defined above need not in general have elementary topological properties like closedness, compactness, or openness. In the next section we single out two important classes of points whose trajectories are compact.

We close this section by the following useful proposition whose proof is immediate.

1.14 Proposition. A set $M \subset X$ is invariant, positively invariant, or negatively invariant if and only if each of its components has the same property.

2. Critical Points and Periodic Points

2.1 Definition. A point $x \in X$ is said to be a *critical point* (or a rest point, or an equilibrium point) if $x = xt$ for all $t \in R$.

The following theorem contains several characterizations of a critical point.

2.2 Theorem. Let $x \in X$. Then the following are equivalent.

2.2.1 x is critical,

2.2.2 $\{x\} = \gamma(x)$,

2.2.3 $\{x\} = \gamma^+(x)$,

2.2.4 $\{x\} = \gamma^-(x)$,

2.2.5 $\{x\} = x[a, b]$ for some $a < b$.

2.2.6 There is a sequence $\{t_n\}$, $t_n > 0$, $t_n \to 0$ with $x = xt_n$ for each n.

The proofs of the equivalence of 2.2.1 with 2.2.2—2.2.5 are almost trivial once the equivalence of 2.2.1 and 2.2.6 is established. For this we need the following important lemma.

2.3 Lemma. If $x \in X$ and $x = xt$ for some $t \in R$, then $x = x(nt)$ for all integers n.

Proof. If $x = xt$, then indeed $x(-t) = xt(-t) = x(t - t) = x$. Thus we need prove the result for positive integers only. This follows from simple induction. For if $x = x(nt)$ and $x = xt$, then substituting xt for x

in the right hand side of the first equation we get $x = xt(nt) = x(t + nt) = x((n + 1)t)$.

2.4 *Proof of Theorem 2.2.* We will only prove that 2.2.1 is equivalent to 2.2.6. First assume 2.2.1. Then indeed 2.2.6 holds trivially as $x = xt$ for all $t \in R$. Now assume 2.2.6 and let $t \in R$. If $t = kt_n$ for some integers k and n then indeed $x = xt$ by Lemma 2.3. Otherwise for each integer n there is an integer k_n with $k_n t_n < t < (k_n + 1) t_n$ and moreover an integer $m > n$ with

$$k_n t_n < k_m t_m < t < (k_m + 1) t_m < (k_n + 1) t_n.$$

Thus clearly the so constructed sequence $\{k_n t_n\}$ has the property that $k_n t_n \to t$. Now by the continuity axiom

$$x(k_n t_n) \to xt$$

and since $x = x(k_n t_n)$ for each n by Lemma 2.3, we have $x = xt$. Since $t \in R$ was arbitrary, x is critical.

We now prove some theorems about critical points. But first the following lemma.

2.5 **Lemma.** If $x \neq xt$ for some $x \in X$ and $t \in R$ (i.e., x is not critical) then there exist open neighborhoods U of x and V of xt such that $V = Ut$ and $U \cap V = \emptyset$.

Proof. Note that if $W \subset X$ is open, then $Wt = \pi^t(W)$ is open because π^t is a homeomorphism (I, 1.2). Now let W_1 and W_2 be disjoint open sets containing x and xt respectively. For example, if $\varrho(x, xt) = \varepsilon$ (> 0), then take $W_1 = S(x, \varepsilon/2)$ and $W_2 = S(xt, \varepsilon/2)$. Then $W_1 t$ is open and contains xt. Now set $V = W_1 t \cap W_2$ and $U = V(-t)$. Then $V \subset W_2$ and $U \subset W_1$, so that U and V are disjoint. Indeed $x \in U$ and $xt \in V$ so that the open sets U and V are the required neighborhoods of x and xt respectively.

We are now ready to prove some more results about critical points.

2.6 **Theorem.** A point $x \in X$ is critical if and only if every neighborhood of x contains a semi-trajectory.

Proof. The necessity is trivial as $\gamma(x) \Rightarrow \{x\}$ when x is critical so that $\gamma(x)$ is contained in every neighborhood of x. For sufficiency assume that x is not critical. Then there is a $t \in R^+$ with $x \neq xt$ and consequently by Lemma 2.5, there are disjoint open neighborhoods U of x and V of xt with $V = Ut$. Since for each $y \in U$ we have $yt \in V$, U cannot contain a positive semi-trajectory. Similarly U cannot contain a negative semi-trajectory. This contradiction shows that $x = xt$ for every $t \in R$ and so x is critical.

2.7 Theorem. The set of all critical points in X is closed.

Proof. If the set of critical points is not closed then there is a sequence $\{x_n\}$ of critical points with $x_n \to x$ and x is not critical. Thus there is a $t \in R$ with $x \neq xt$ and hence, by Lemma 2.5, there are open sets U containing x and V containing xt with $Ut = V$ and $U \cap V = \emptyset$. Since $x_n \to x$ we have $x_n \in U$ for all sufficiently large n. Then for the t above $x_n t \in V$ and in particular $x_n t \notin U$. But x_n's are critical and therefore $x_n = x_n t \in U$. This contradiction proves the theorem.

As yet another application of Lemma 2.5, we have

2.8 Theorem. Let $x, y \in X$ and $\varrho(yt, x) \to 0$ as $t \to +\infty$ (or as $t \to -\infty$). Then x is critical.

Proof. Let U be a neighborhood of x. Since $\varrho(yt, x) \to 0$ as $t \to +\infty$ there exists $T \geqq 0$ such that $yt \in U$ for all $t \geqq T$. Hence U contains the positive semi-trajectory $\gamma^+(yT)$ and therefore from Theorem 2.6 x is a critical point.

2.9 *Exercise.* Prove that if x is critical and $y \neq x$, then $x \neq yt$ for every $t \in R$. This is sometimes expressed by saying that no motion starting at a point $y \in X$ reaches a critical point $x \neq y$ in a finite time. (Hint: If x is critical, then $\{x\}$ is invariant, and so is $X - \{x\}$.)

We now introduce the periodic points.

2.10 Definition. A *point* $x \in X$ is said to be *periodic* if there is a $T \neq 0$ such that

2.10.1 $\quad xt = x(t + T) \qquad$ for all $t \in R$.

A *number* $T \in R$ for which 2.10.1 holds will be called a *period* of x. If a point $x \in X$ is periodic then both the *motion* π_x and the *trajectory* $\gamma(x)$ are said to be *periodic*.

Note that every $x \in X$ has the period $T = 0$, but it may not be periodic. Further, if $x \in X$ is critical, then every $T \in R$ is a period of x, and indeed x is periodic.

The following characterization of a periodic point is very useful.

2.11 Proposition. $x \in X$ is periodic if and only if there is a $T \neq 0$ with $x = xT$.

Proof. Necessity is obvious. To see sufficiency, take any $t \in R$ and apply the group axiom to obtain:

$$xt = xT(t) = x(t + T).$$

We now study some properties of periodic points.

2.12 Theorem. If $x \in X$ is periodic, but not critical, then there is $T > 0$ such that T is the smallest positive period of x. Further if τ is any other period of x, then $\tau = nT$ for some integer n.

Proof. Consider the set $\{t > 0: t \text{ is a period of } x\} = P$. Note that $P \neq \emptyset$ since if $\tau \neq 0$ is a period of x, then so is $-\tau$, as $x = x\tau$ implies $x(-\tau) = x\tau(-\tau) = x(\tau - \tau) = x\,0 = x$. Since either τ or $-\tau$ is positive, the set P is non-empty. Now set $\inf P = T$. We claim that $T > 0$. Indeed $T \geqq 0$, and if $T = 0$, then there is a sequence $\{t_n\}$ in P with $t_n \to 0$. Since $x = xt_n$ for each n, Theorem 2.2 shows that x is critical, which contradicts our hypothesis. Thus $T > 0$. Note also that T is a period of x. This follows from the fact that there is a sequence $\{t_n\}$ in P with $t_n \to T$ since $T = \inf P$. Now $x = xt_n \to xT$ by the continuity axiom. Hence $x = xT$. Thus T is a period of x. By definition of T it is also the smallest positive period. Finally, let $t \in R$ be any period of x. If $t \neq nT$ for any integer n, then there is an integer n with $nT < t < (n + 1)\,T$. But by Lemma 2.3, nT is a period of x. Thus we have $x = xt = x(nT)$. This gives $xt(-nT) = x(nT)\,(-nT) = x(nT - nT) = x\,0 = x$, i.e., $x = xt(-nT) = x(t - nT)$, showing that $t - nT$ is a period of x. Since $0 < t - nT < T$, we get a contradiction to the fact that T was the smallest positive period of x. This completes the proof.

The above theorem motivates the following definition.

2.13 Definition. If a point x is periodic but not critical, then the smallest positive period of x is called its *fundamental* or *primitive period.*

In contrast to the situation for critical points (Theorem 2.7) it is in general not true that the set of all periodic points is closed. However we can say the following:

2.14 Theorem. Given any $\alpha > 0$, the set of all x such that x is periodic with a positive period $T \leqq \alpha$ is closed.

The proof will be made to depend on the following important lemma.

2.15 Lemma. If $\{x_n\}$ is a sequence of periodic points with positive periodic $T_n \to 0$, and $x_n \to x$, then x is critical.

Proof. For given $t \in R$, there are integers k_n such that

$$k_n T_n \leq t < k_n T_n + T_n.$$

Since $T_n \to 0$, we have $k_n T_n \to t$. Then $x_n = x_n(k_n T_n) \to xt$, and since $x_n \to x$ we have $x = xt$. As $t \in R$ was arbitrary, x is critical.

2.16 *Proof of Theorem 2.14.* Indeed if $\{x_n\}$ is a sequence of periodic points with periods $T_n \leqq \alpha$ and $x_n \to x$, then also $x_n = x_n T_n \to x$. Since

$0 \leqq T_n \leqq \alpha$, either $T_n \to 0$ in which case x is critical by Lemma 2.15 and hence periodic, or there is a subsequence $T_{n_k} \to \tau$, $0 < \tau \leqq \alpha$. But then by the continuity axiom $x_{n_k} T_{n_k} \to x\tau$ and also $x_{n_k} T_{n_k} = x_{n_k} \to x$. Hence $x = x\tau$ and x is periodic with a positive period $\tau \leqq \alpha$. This proves the theorem.

2.17 *Exercise.* If $x \in X$ is periodic, then $\gamma(x)$ is compact. Does the converse hold? Prove also that if x is periodic, then $\gamma^+(x)$ is compact. Does the converse of the last statement hold? (Hint: If x is periodic with period $T > 0$, then $\gamma(x) = \gamma^+(x) = x[0, T]$.)

3. Trajectory Closures and Limit Sets

We have seen above that trajectories of periodic points are compact invariant sets. In general, however, the trajectory of a point $x \in X$ need not even be closed (it is almost never open). Using Theorem 1.3 we see that $\overline{\gamma(x)}$ is invariant for any $x \in X$, in fact it is the smallest closed invariant set containing x. We now introduce two maps $Q\colon X \to 2^X$ such that for each $x \in X$, $Q(x)$ is closed and invariant. These will be the so-called limit sets.

3.1 **Definition.** Define maps Λ^+, Λ^- from X into 2^X by setting for each $x \in X$,

3.1.1 $\Lambda^+(x) = \{y \in X\colon$ there is a sequence $\{t_n\}$ in R with $t_n \to +\infty$ and $xt_n \to y\}$,

3.1.2 $\Lambda^-(x) = \{y \in X\colon$ there is a sequence $\{t_n\}$ in R with $t_n \to -\infty$ and $xt_n \to y\}$.

For any $x \in X$, the set $\Lambda^+(x)$ is called its *positive* (or *omega*) *limit set*, and the set $\Lambda^-(x)$ is called its *negative* (or *alpha*) *limit set.*

3.2 *Exercise.* Show that if $x \in X$ is periodic, then $\Lambda^+(x) = \Lambda^-(x) = \gamma(x)$. (Partial converses of this statement hold. See III, 2.5 and 2.9.)

Before proceeding further we give some examples of limit sets.

3.3 *Examples of Limit Sets.*

3.3.1 Consider the differential system defined in R^2 by the differential equations (polar coordinates)

3.3.2 $$\frac{dr}{dt} = r(1 - r), \qquad \frac{d\theta}{dt} = 1.$$

It can easily be verified that the solutions are unique and all solutions are defined on R. Thus 3.3.2 defines a dynamical system. The trajectories are shown in Fig. 3.3.3. These consist of: (i) a critical point, namely the origin 0, (ii) a periodic trajectory γ coinciding with the unit circle, (iii) spiralling trajectories through each point $P = (r, \theta)$ with $r \neq 0$, $r \neq 1$. For points P with $0 < r < 1$, $\Lambda^+(P)$ is the unit circle and $\Lambda^-(P)$ is the singleton $\{0\}$. For points P with $r > 1$, $\Lambda^+(P)$ is the unit circle and $\Lambda^-(P) = \emptyset$.

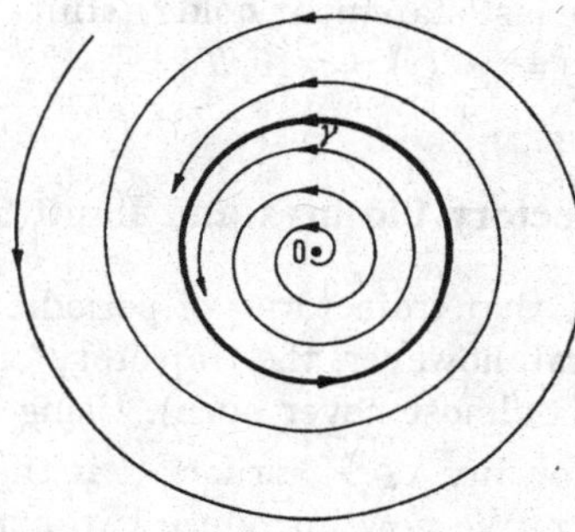

Fig. 3.3.3

3.3.4 Consider another differential system defined in R^2 by (polar coordinates)

$$\frac{dr}{dt} = r(1-r), \quad \frac{d\theta}{dt} = \begin{cases} \sin^2\theta + \dfrac{1}{\log 3} & \text{if } 0 < r \leqq \dfrac{3}{4}, \\[2ex] \sin^2\theta + \dfrac{1}{\log \dfrac{r}{1-r}} & \text{if } \dfrac{3}{4} < r < 1, \\[2ex] \sin^2\theta & \text{if } r = 1, \\[2ex] \sin^2\theta + \dfrac{1}{\log \dfrac{r}{r-1}} & \text{if } r > 1. \end{cases}$$

One may verify that this system also defines a dynamical system on R^2. The trajectories are shown in Fig. 3.3.5. These consist of: (i) three critical points, namely the origin 0 and the points $A = (1, 0)$ and $B = (1, \pi)$ on the unit circle, (ii) a trajectory $\gamma_1 = \{(1, \theta)\colon\ 0 < \theta < \pi\}$ on the unit circle, (iii) a trajectory $\gamma_2 = \{(1, \theta)\colon \pi < \theta < 2\pi\}$ on the unit circle, and (iv) spiralling trajectories through points $P = (r, \theta)$ with $r \neq 0$, $r \neq 1$. For any $P = (r, \theta)$ with $0 < r < 1$, $\Lambda^+(P)$ is the unit circle, and $\Lambda^-(P)$ is $\{0\}$. For any P with $r > 1$, $\Lambda^+(P)$ is the unit circle and $\Lambda^-(P) = \emptyset$. For any $P = (1, \theta)$ with $0 < \theta < \pi$, $\Lambda^+(P) = \{B\}$ and $\Lambda^-(P) = \{A\}$. For any $P = (1, \theta)$ with $\pi < \theta < 2\pi$, $\Lambda^+(P) = \{A\}$ and $\Lambda^-(P) = \{B\}$.

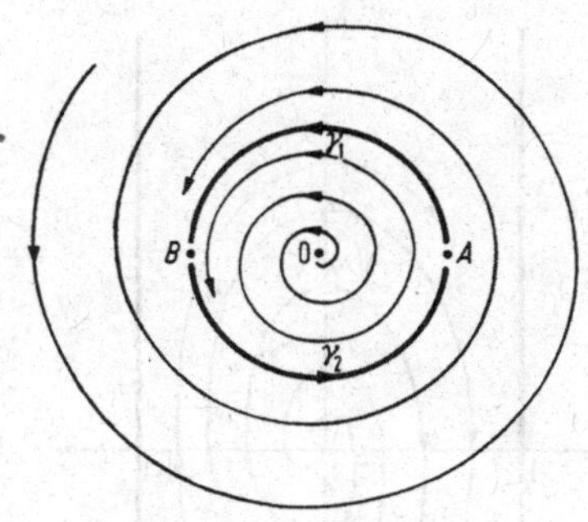

Fig. 3.3.5

3.3.6 Consider in the space R^2 the differential equations

3.3.7 $$\dot{x} = f(x, y), \; \dot{y} = g(x, y),$$

where the functions f and g are given by

3.3.8
$$f(x, y) = \begin{cases} 0 \text{ if } |x| \geqq 1 \\ -\dfrac{y(1-x^2)}{(1+y^2)\,(1-p(x)\,q(y))} \text{ if } |x| < 1, \end{cases}$$
$$g(x, y) = \begin{cases} +1 \text{ if } x \geqq 1 \\ -1 \text{ if } x \leqq -1 \\ x \quad \text{if } |x| < 1. \end{cases}$$

Here $p(x)$, $q(y)$ are any continuously differentiable functions satisfying

3.3.9
$$0 < p(x) < \tfrac{1}{2} \quad \text{if } 0 < x < 1,$$
$$p(x) = 0 \quad \text{if } x \leqq 0,$$
$$p(x) = \tfrac{1}{2} \quad \text{if } x \geqq 1,$$
$$0 < q(y) < \tfrac{1}{2} \quad \text{if } y < 0,$$
$$q(y) = 0 \quad \text{if } y \geqq 0.$$

It is easily verified that such a system describes a dynamical system as shown in Fig. 3.3.10. This dynamical system has the straight lines $x = \pm 1$ as the positive limit set of all points in the strip $-1 < x < +1$, except the origin.

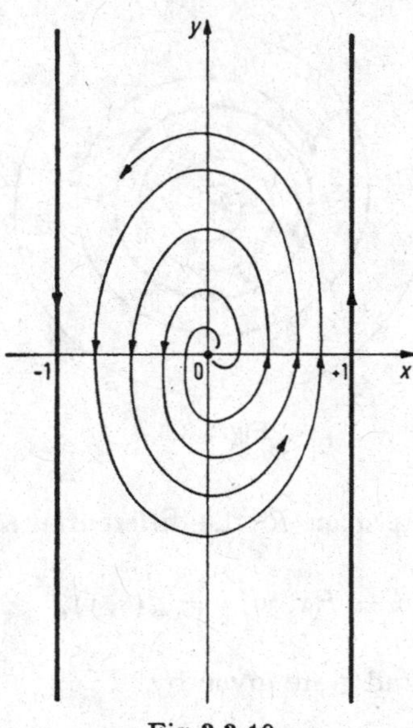

Fig. 3.3.10

We now proceed to establish some important properties of the limit sets and their relationship with the semi-trajectory closures.

3.4 Theorem. For any $x \in X$

3.4.1 $\Lambda^+(x)$ and $\Lambda^-(x)$ are closed invariant sets,

3.4.2 $\overline{\gamma^+(x)} = \gamma^+(x) \cup \Lambda^+(x)$ and $\overline{\gamma^-(x)} = \gamma^-(x) \cup \Lambda^-(x)$.

Proof. Consider the case of $\Lambda^+(x)$. Let $\{y_n\}$ be a sequence in $\Lambda^+(x)$ with $y_n \to y$. We wish to show that $y \in \Lambda^+(x)$. Indeed for each positive integer k, there is a sequence $\{t_n^k\}$ in R with $t_n^k \to +\infty$ and $xt_n^k \to y_k$. We may assume without loss of generality that $\varrho(y_k, xt_n^k) < 1/k$ and $t_n^k \geqq k$ for $n \geqq k$. Consider now the sequence $\{t_n\}$ in R with $t_n = t_n^n$. Then $t_n \to +\infty$ and we claim that $xt_n \to y$. To see this, observe that

$$\varrho(y, xt_n) \leqq \varrho(y, y_n) + \varrho(y_n, xt_n) \leqq \varrho(y, y_n) + 1/n.$$

Since $1/n$ and $\varrho(y, y_n)$ tend to zero we conclude that $\varrho(y, xt_n) \to 0$. Consequently, $xt_n \to y$ and $y \in \Lambda^+(x)$. We have thus proved that $\Lambda^+(x)$ is closed. To see that $\Lambda^+(x)$ is invariant, let $y \in \Lambda^+(x)$ and $t \in R$ be arbitrary. There is a sequence $\{t_n\}$ in R with $t_n \to +\infty$ and $xt_n \to y$. Then by the continuity axiom $xt_n(t) \to yt$. Since $xt_n(t) = x(t_n + t)$ and $\tau_n = (t_n + t) \to +\infty$ we have $yt \in \Lambda^+(x)$ and $\Lambda^+(x)$ is invariant.

We now prove 3.4.2. For this remember that $\overline{\gamma^+(x)} = \overline{xR^+}$. By the definition of $\Lambda^+(x)$ we have $\overline{\gamma^+(x)} \supset \gamma^+(x) \cup \Lambda^+(x)$. To see that $\overline{\gamma^+(x)} \subset \gamma^+(x) \cup \Lambda^+(x)$, let $y \in \overline{\gamma^+(x)}$. Then there is a sequence $\{y_n\}$ in $\gamma^+(x)$ such that $y_n \to y$. Now $y_n = xt_n$ for a $t_n \in R^+$. Either the sequence $\{t_n\}$ has the property that $t_n \to +\infty$, in which case $y \in \Lambda^+(x)$, or there is a subsequence $t_{n_k} \to t \in R^+$ (as R^+ is closed). But then $xt_{n_k} \to xt \in \gamma^+(x)$, and since also $xt_{n_k} \to y$ we have $y = xt \in \gamma^+(x)$. Thus $\overline{\gamma^+(x)} \subset \gamma^+(x) \cup \Lambda^+(x)$. This proves 3.4.2.

3.5 *Exercises.* For any $x \in X$ prove that

3.5.1 $\Lambda^+(x) = \cap \{\overline{\gamma^+(y)}: y \in \gamma^+(x)\} = \cap \{\overline{\gamma^+(xn)}: n \text{ is an integer}\}$.

3.5.2 $\Lambda^+(x) = \Lambda^+(xt)$ for $t \in R$.

We now prove connectedness properties of limit sets. The principal theorem is

3.6 **Theorem.** If the space X is locally compact, then a positive limit set $\Lambda^+(x)$ is connected whenever it is compact. Further, whenever a positive limit set is not compact, then none of its components is compact. (In the following proof and elsewhere in this book we use the notation $S(M, \varepsilon)$, $S[M, \varepsilon]$, and $H(M, \varepsilon)$ for the sets $\{x: \varrho(x, M) < \varepsilon\}$, $\{x: \varrho(x, M) \leq \varepsilon\}$, and $\{x: \varrho(x, M) = \varepsilon\}$, respectively. Here $M \subset X$ and $\varepsilon > 0$.)

Proof. Let $\Lambda^+(x)$ be compact, and let it be not connected. Then $\Lambda^+(x) = P \cup Q$, where P, Q are non-empty, closed, disjoint sets. Since $\Lambda^+(x)$ is compact, so are P and Q. Further, since X is locally compact, there is an $\varepsilon > 0$ such that $S[P, \varepsilon]$, $S[Q, \varepsilon]$ are compact and disjoint. Now let $y \in P$ and $z \in Q$. Then there are sequences $\{t_n\}$, $\{\tau_n\}$, $t_n \to +\infty$, $\tau_n \to +\infty$, such that $xt_n \to y$, and $x\tau_n \to z$.

We may assume without loss of generality, that $xt_n \in S(P, \varepsilon)$, $x\tau_n \in S(Q, \varepsilon)$, and $\tau_n - t_n > 0$ for all n. Since the trajectory segments $x[t_n, \tau_n]$, $n = 1, 2, \ldots$, are compact connected sets, they clearly intersect $H(P, \varepsilon)$ and $H(Q, \varepsilon)$. Thus, in particular, there is a sequence $\{T_n\}$, $t_n < T_n < \tau_n$, such that $xT_n \in H(P, \varepsilon)$ which is compact. We may therefore assume that $xT_n \to \tilde{y}$, and as $T_n \to +\infty$, we have $\tilde{y} \in \Lambda^+(x)$. However, $\tilde{y} \notin P \cup Q$, which is a contradiction. This establishes the first part of the theorem.

To prove the second part of the theorem we need the following topological theorem, which we give without proof.

3.7 **Topological Theorem.** Let S be a Hausdorff continuum (a compact connected Hausdorff space), let U be an open subset of S, and let C be a component of U. Then $U - \overline{U}$ contains a limit point of C.

3.8 *Proof of the Second Part of Theorem 3.6.* Notice that the space X is a locally compact Hausdorff space, and everything that has been said above goes through in such a space. Now X possesses a one-point compactification. So let $\tilde{X} = X \cup \{\omega\}$ be the one-point compactification of X by the ideal point ω. Extend the dynamical system (X, R, π) on X to a dynamical system $(\tilde{X}, R, \tilde{\pi})$ on $\tilde{X}$, where $\tilde{\pi}$ is given by $\tilde{\pi}(x, t) \equiv \pi(x, t)$ for $x \in X$, $t \in R$, and $\tilde{\pi}(\omega, t) = \omega$ for all $t \in R$. If now for $x \in \tilde{X}$, $\tilde{\Lambda}^+(x)$ denotes the positive limit set of x, then clearly $\tilde{\Lambda}^+(x) = \Lambda^+(x) \cup \{\omega\}$, whenever $x \in X$ and $\Lambda^+(x)$ is not compact. However, $\tilde{\Lambda}^+(x)$ is compact, as $\tilde{X}$ is compact, and by the first part of the theorem it is connected. $\tilde{\Lambda}^+(x)$ is therefore a Hausdorff continuum. Further $\Lambda^+(x)$ is an open set in $\tilde{\Lambda}^+(x)$. Now $\tilde{\Lambda}^+(x) - \Lambda^+(x) = \{\omega\}$, and so by Theorem 3.7 every component of $\Lambda^+(x)$ has ω as a limit point, and so is not compact. This proves the theorem completely.

3.9 *Exercises.*

3.9.1 Show that if $\overline{\gamma^+(x)}$ is compact, then $\Lambda^+(x)$ is a non-empty compact and connected set.

3.9.2 Show that if $\overline{\gamma^+(x)}$ is compact, then

$$\varrho(xt, \Lambda^+(x)) \to 0 \quad \text{as} \quad t \to +\infty.$$

3.9.3 Show that in a locally compact space X, $\Lambda^+(x)$ is non-empty and compact if and only if $\overline{\gamma^+(x)}$ is compact.

3.9.4 Show that 3.9.2 may not be true if $\overline{\gamma^+(x)}$ is not compact.

3.9.5 If X is not locally compact, then Theorem 3.6 may not hold. (Hint: Consider the restriction of the dynamical system in example 3.3.4 to the set $R^2 - \gamma_1 \cup \gamma_2$.)

4. The First Prolongation and the Prolongational Limit Set

In this section we introduce yet other maps D^+, D^-, J^+, J^- from X into 2^X. These maps play an important role in the theory of dynamical systems, particularly in the stability theory and the theory of dispersive systems.

4.1 **Definition.** Define maps D^+, D^-, J^+, J^- from X into 2^X by setting for each $x \in X$,

4.1.1 $D^+(x) = \{y \in X$: there is a sequence $\{x_n\}$ in X and a sequence $\{t_n\}$ in R^+ such that $x_n \to x$ and $x_n t_n \to y\}$,

4.1.2 $D^-(x) = \{y \in X$: there is a sequence $\{x_n\}$ in X and a sequence $\{t_n\}$ in R^- such that $x_n \to x$ and $x_n t_n \to y\}$,

4.1.3 $J^+(x) = \{y \in X$: there is a sequence $\{x_n\}$ in X and a sequence $\{t_n\}$ in R^+ such that $x_n \to x$, $t_n \to +\infty$, and $x_n t_n \to y\}$,

4.1.4 $J^-(x) = \{y \in X$: there is a sequence $\{x_n\}$ in X and a sequence $\{t_n\}$ in R^- such that $x_n \to x$, $t_n \to -\infty$, and $x_n t_n \to y\}$.

For any $x \in X$, the set $D^+(x)$ is the *first positive* and the set $D^-(x)$ the *first negative prolongation* of x.

The sets $J^+(x)$ and $J^-(x)$ are called, respectively, the *first positive* and the *first negative prolongational limit set* of x.

It is clear that for any $x \in X$, $D^+(x) \supset \gamma^+(x)$, $D^-(x) \supset \gamma^-(x)$, $J^+(x) \supset \Lambda^+(x)$, and $J^-(x) \supset \Lambda^-(x)$. That these inclusions may be proper is shown by the following simple example.

4.2 *Example.* Consider the planar differential system (cartesian coordinates)

$$\frac{dx_1}{dt} = -x_1, \quad \frac{dx_2}{dt} = x_2.$$

These equations define a dynamical system with the so-called saddle point at the origin. The trajectories are shown in Fig. 4.2.1. For any point $P = (x_1, 0)$, $D^+(P) = \gamma^+(P) \cup \{(x_1, x_2): x_1 = 0\}$, $D^-(P) = \gamma^-(P)$, $J^+(P) = \{(x_1, x_2): x_1 = 0\}$, $J^-(P) = \emptyset$ if $x_1 \neq 0$ and $J^-(P) = \{(x_1, x_2): x_2 = 0\}$ if $x_1 = 0$. For points $P = (x_1, x_2)$ with $x_1 \neq 0$, $x_2 \neq 0$, $D^+(P) = \gamma^+(P)$, $D^-(P) = \gamma^-(P)$, $J^+(P) = J^-(P) = \emptyset$. Clearly, for points $P = (x_1, 0)$, $D^+(P) \neq \overline{\gamma^+(P)}$, and $J^+(P) \neq \Lambda^+(P)$ (what are $\overline{\gamma^+(P)}$ and $\Lambda^+(P)$?).

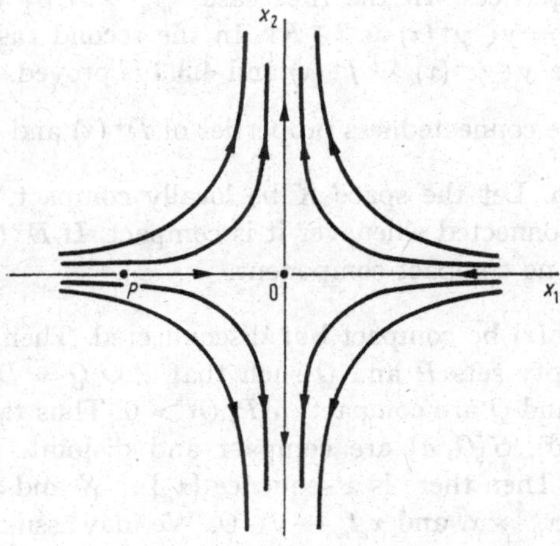

Fig. 4.2.1

We now discuss some basic properties of these sets.

4.3 Theorem. For any $x \in X$,

4.3.1 $D^+(x)$ is closed and positively invariant,

4.3.2 $J^+(x)$ is closed and invariant,

4.3.3 $D^+(x) = \gamma^+(x) \cup J^+(x)$.

Analogous results hold for $D^-(x)$ and $J^-(x)$.

Proof. It is sufficient to prove 4.3.2 and 4.3.3. We first prove 4.3.2. Let $\{y_n\}$ be a sequence in $J^+(x)$ with $y_n \to y$. For each integer k, there are sequences $\{x_n^k\}$ in X and $\{t_n^k\}$ in R^+ with $x_n^k \to x$, $t_n^k \to +\infty$, and $x_n^k t_n^k \to y_k$. We may assume by taking subsequences if necessary that $t_n^k > k$, $\varrho(x_n^k, x) \leqq 1/k$, and $\varrho(x_n^k t_n^k, y_k) \leqq 1/k$ for $n \geqq k$. Now consider the sequences $\{x_n^n\}$, $\{t_n^n\}$. Clearly $x_n^n \to x$, $t_n^n \to +\infty$, and $x_n^n t_n^n \to y$. To see for example that $x_n^n t_n^n \to y$ note that $\varrho(x_n^n t_n^n, y) \leqq \varrho(x_n^n t_n^n, y_n) + \varrho(y_n, y) \leqq 1/n + \varrho(y_n, y)$. Thus $y \in J^+(x)$ and $J^+(x)$ is closed.

To see that $J^+(x)$ is invariant, let $y \in J^+(x)$ and $t \in R$. There is a sequence $\{x_n\}$ in X and a sequence $\{t_n\}$ in R^+ such that $t_n \to +\infty$, $x_n \to x$, and $x_n t_n \to y$. Now consider the sequence $\{t_n + t\}$. Clearly $t_n + t \to +\infty$, and $x_n(t_n + t) = x_n t_n(t) \to yt$. Since $x_n \to x$, we have $yt \in J^+(x)$. As $t \in R$ was arbitrary, $J^+(x)$ is invariant. This completes the proof of 4.3.2.

To see 4.3.3 observe that $D^+(x) \supset \gamma^+(x) \cup J^+(x)$ holds always (see definitions). We will prove $D^+(x) \subset \gamma^+(x) \cup J^+(x)$. Let $y \in D^+(x)$. Then there is a sequence $\{x_n\}$ in X and a sequence $\{t_n\}$ in R^+ with $x_n \to x$, $x_n t_n \to y$. We may assume that either $t_n \to t \in R^+$ or $t_n \to +\infty$, if necessary by taking subsequences. In the first case $x_n t_n \to xt$ by the continuity axiom and so $xt = y \in \gamma^+(x)$ as $t \in R^+$. In the second case $y \in J^+(x)$ by definition. Hence $y \in \gamma^+(x) \cup J^+(x)$ and 4.3.3 is proved.

We now prove connectedness properties of $D^+(x)$ and $J^+(x)$.

4.4 Theorem. Let the space X be locally compact. Then for any $x \in X$, $D^+(x)$ is connected whenever it is compact. If $D^+(x)$ is not compact, then it has no compact component.

Proof. Let $D^+(x)$ be compact but disconnected. Then there are two compact non-empty sets P and Q such that $P \cup Q = D^+(x)$ and $P \cap Q = \emptyset$. Since P and Q are compact, $\varrho(P, Q) > 0$. Thus there is a $\delta > 0$ such that $S[P, \delta]$, $S[Q, \delta]$ are compact and disjoint. Now $x \in P$ or $x \in Q$. Let $x \in P$. Then there is a sequence $\{x_n\}$ in X and a sequence $\{t_n\}$ in R^+ such that $x_n \to x$, and $x_n t_n \to y \in Q$. We may assume $x_n \in S[P, \delta]$ and $x_n t_n \in S[Q, \delta]$. Then the trajectory segments $x_n[0, t_n]$ intersect

$H(P, \delta)$ for every n, and therefore there is a sequence $\{\tau_n\}$, $0 \leqq \tau_n \leqq t_n$, such that $x_n\tau_n \in H(P, \delta)$. Since $H(P, \delta)$ is compact we may assume that $x_n\tau_n \to z \in H(P, \delta)$. Then $z \in D^+(x)$, but $z \notin P \cup Q$ as $z \in H(P, \delta)$. This contradiction shows that $D^+(x)$ is connected.

To prove the second part of the theorem, we consider the one point compactification $\tilde{X} = X \cup \{\omega\}$ of the space X, and for each $x \in X$ define the set $D^*(x)$ by

$$D^*(x) = \{y \in \tilde{X}: \text{there is a sequence } \{x_n\} \text{ in } X \text{ and a sequence } \{t_n\} \text{ in } R^+ \text{ such that } x_n \to x \text{ and } x_nt_n \to y\}.$$

The set $D^*(x)$ is always a compact subset of $\tilde{X}$. Following exactly the same proof as given above for $D^+(x)$ one can show that $D^*(x)$ is connected. Now if $D^+(x)$ is not compact, then clearly

$$D^*(x) = D^+(x) \cup \{\omega\}.$$

Further $D^+(x)$ is an open subset of $D^*(x)$. By the topological Theorem 3.7, each component of $D^+(x)$ has a limit point in $D^*(x) - D^+(x) = \{\omega\}$. Thus no component of $D^+(x)$ is compact. This proves the theorem.

Theorem 4.4 clearly holds for $D^-(x)$. That it also holds for $J^+(x)$ and $J^-(x)$ will be shown next. But first the following lemma.

4.5 Lemma. Let X be locally compact. Then $\Lambda^+(x) \neq \emptyset$ whenever $J^+(x)$ is non-empty and compact.

Proof. If possible let $\Lambda^+(x) = \emptyset$. Then we claim that $\gamma^+(x)$ is closed and disjoint with $J^+(x)$. That $\gamma^+(x)$ is closed follows from $\overline{\gamma^+(x)} = \gamma^+(x) \cup \Lambda^+(x) = \gamma^+(x)$ as $\Lambda^+(x) = \emptyset$. That $\gamma^+(x) \cap J^+(x) = \emptyset$ follows from the fact that if $\gamma^+(x) \cap J^+(x) \neq \emptyset$, then by invariance of $J^+(x)$, $\gamma^+(x) \subset J^+(x)$. Since $J^+(x)$ is compact, we will have $\Lambda^+(x) \neq \emptyset$ and compact (remember that any sequence $\{y_n\}$ in a compact set Q has a convergent subsequence). This again contradicts the assumption $\Lambda^+(x) = \emptyset$. Thus $\gamma^+(x)$ is closed and $\gamma^+(x) \cap J^+(x) = \emptyset$. Since $J^+(x)$ is non-empty and compact we have $\varrho(\gamma^+(x), J^+(x)) > 0$. Thus there is a $\delta > 0$ such that $S[J^+(x), \delta]$ is compact and disjoint with $\gamma^+(x)$. Now choose any $y \in J^+(x)$. There is a sequence $\{x_n\}$ in X and a sequence $\{t_n\}$ in R^+ such that $x_n \to x$, $t_n \to +\infty$, and $x_nt_n \to y$. We may assume that $x \notin S[J^+(x), \delta]$, $x_nt_n \in S[J^+(x), \delta]$ for all n. Then the trajectory segments $x_n[0, t_n]$ intersect $H(J^+(x), \delta)$ and therefore there is a sequence $\{\tau_n\}$, $0 < \tau_n \leqq t_n$, such that $x_n\tau_n \in H(J^+(x), \delta)$. Since $H(J^+(x), \delta)$ is compact, we may assume that $x_n\tau_n \to z \in H(J^+(x), \delta)$. By taking subsequences we may assume that either $\tau_n \to t \in R^+$ or $\tau_n \to +\infty$. If $\tau_n \to t \in R^+$, then by the continuity axiom $x_n\tau_n \to xt = z$, i.e., $z \in \gamma^+(x)$ which contradicts

$\gamma^+(x) \cap S[J^+(x), \delta] = \emptyset$. If $\tau_n \to +\infty$, then $z \in J^+(x)$, but this contradicts $z \in H(J^+(x), \delta)$ as $J^+(x) \cap H(J^+(x), \delta) = \emptyset$. Thus our original assumption $\Lambda^+(x) = \emptyset$ is untenable and the lemma is proved.

As an immediate application we have

4.6 Lemma. Let X be locally compact. Then $J^+(x)$ is non-empty and compact if and only if $D^+(x)$ is compact.

Proof. Let $J^+(x)$ be non-empty and compact. Then $\Lambda^+(x)$ is non-empty and compact. But then $\overline{\gamma^+(x)}$ is compact (prove this). Hence $D^+(x) = \gamma^+(x) \cup J^+(x) = \overline{\gamma^+(x)} \cup J^+(x)$ is compact. The converse is trivial.

We now prove the connectedness properties of $J^+(x)$.

4.7 Theorem. If X is locally compact, then for any $x \in X$ Theorem 4.4 holds for $J^+(x)$.

Proof. Let $J^+(x)$ be compact. If $J^+(x) = \emptyset$ there is nothing to prove. So let $J^+(x) \neq \emptyset$. If $J^+(x)$ is disconnected, then there are non-empty compact sets P, Q such that $J^+(x) = P \cup Q$ and $P \cap Q = \emptyset$. Since $\Lambda^+(x)$ is non-empty and compact, hence connected, we have $\Lambda^+(x) \subset P$ or $\Lambda^+(x) \subset Q$. Let $\Lambda^+(x) \subset P$. Then $\gamma^+(x) \cup P$ is compact and disjoint from Q. That it is compact follows from $\gamma^+(x) \cup P = \overline{\gamma^+(x)} \cup P$ as $\Lambda^+(x) \subset P$ and $\overline{\gamma^+(x)}$ is compact. That it is disjoint from Q follows from the fact that if $Q \cap (\gamma^+(x) \cup P) \neq \emptyset$, then $Q \cap \gamma^+(x) \neq \emptyset$. But Q must be invariant (why?). This will show that $\Lambda^+(x) \subset Q$, a contradiction. Finally, $D^+(x) = (\gamma^+(x) \cup P) \cup Q$ and since Q and $\gamma^+(x) \cup P$ are disjoint compact sets we have a contradiction to the fact that $D^+(x)$ is connected whenever compact. This proves that $J^+(x)$ is connected whenever compact. The proof of the second part of the theorem is exactly similar to proof of the second part of Theorem 4.4 and is left to the reader.

4.8 *Exercises.*

4.8.1 For any $x \in X$, $D^+(x) = \cap \{\overline{S(x, \alpha)\, R^+}: \alpha > 0\}$. Give a similar statement for $J^+(x)$ and $D^-(x)$.

4.8.2 Give examples to show that Theorems 4.4 and 4.7 and Lemmas 4.5 and 4.6 are not in general true for non-locally compact spaces X.

4.8.3 Give a detailed proof of the second part of Theorem 4.7.

4.8.4 Given a dynamical system (X, R, π) on a locally compact space X, let $\tilde{X} = X \cup \{\omega\}$ be the one point compactification of X and define

$\tilde{\pi}: \tilde{X} \times R \to \tilde{X}$ by

$$\tilde{\pi}(x, t) = \begin{cases} \pi(x, t) & \text{if } x \in X \text{ and } t \in R, \\ \omega & \text{if } x = \omega \text{ and } t \in R. \end{cases}$$

Prove rigorously that $(\tilde{X}, R, \tilde{\pi})$ is a dynamical system on $\tilde{X}$.

4.8.5 For any compact $M \subset X$, the set $D^+(M)$ is closed and positively invariant.

We conclude this chapter with a non-trivial property of the maps J^+ and J^- which is generally not possessed by the maps Λ^+ and Λ^-.

4.9 Theorem. Let x, y in X. Then $y \in J^+(x)$ if and only if $x \in J^-(y)$.

Proof. $y \in J^+(x)$ if there is a sequence $\{x_n\}$ in X and $\{t_n\}$ in R such that $x_n \to x$, $t_n \to +\infty$, and $x_n t_n \to y$. Set $x_n t_n = y_n$, and $\tau_n = -t_n$. Then $y_n \to y$, $\tau_n \to -\infty$, and $y_n \tau_n = x_n t_n(\tau_n) = x_n(t_n - t_n) = x_n \to x$. Consequently $x \in J^-(y)$. The converse holds also.

For some applications (section V, 5) we shall need the following concepts of relative prolongation and relative prolongational limit set.

4.10 Definition. Let $M \subset X$ be compact. If $U \subset X$, the (positive) prolongation of M, relative to U is the set

4.10.1 $D^+(M, U) = \bigcup_{x \in M} \{y \in X\colon$ there is a sequence $\{x_n\} \subset U$ and a sequence $\{t_n\} \subset R^+$ such that $x_n \to x$ and $x_n t_n \to y\}$,

while the prolongational limit set of M, relative to U is the set:

4.10.2 $J^+(M, U) = \bigcup_{x \in M} \{y \in X\colon$ there is a sequence $\{x_n\} \subset U$ and a sequence $\{t_n\} \subset R^+$ such that $x_n \to x$ and $t_n \to +\infty$, $x_n t_n \to y\}$.

4.11 *Exercises.* Let $M \subset X$ be compact and $U \subset X$. Prove the following statements.

4.11.1 $D^+(M, U)$ is positively invariant.

4.11.2 $J^+(M, U)$ is invariant.

4.11.3 If $U \subset X$ is closed, then $D^+(M, U)$ and $J^+(M, U)$ are closed.

4.11.4 If $U \subset X$ is positively invariant, then

$$D^+(M, U) = \gamma^+(M \cap U) \cup J^+(M, U).$$

4.11.5 If U is a neighborhood of M, then $D^+(M, U) = D^+(M)$.

Notes and References

Most of the results presented in this chapter are classical.

Section 1. The trajectory $\gamma(x)$ is also called *orbit*. The word orbit designates $\gamma(x)$ mostly when $\overline{\gamma(x)}$ is compact. We do this to keep our terminology close to the classical one. Theorems 1.2—1.7 are to be found in weaker versions in a paper by S. LEFSCHETZ [2].

Section 2. In 2.2.6 one may replace the null sequence $\{t_n\}$ by an arbitrary convergent sequence $\{t_n\}$, $t_n \to t$, $t_n \neq t$.

Theorem 2.6 is a stronger version of a similar theorem found in the book by V. V. NEMYTSKII and N. V. STEPANOV [1]. The problems of existence of critical points of a dynamical system or of a differential equation on manifolds have been discussed by many authors among which we mention L. E. EL'SGOL'C [1], F. HAAS [5], G. S. JONES [4], and G. S. JONES and J. A. YORKE [1]. Some criteria due to N. P. BHATIA, A. LAZER and G. P. SZEGÖ [1] and to N. P. BHATIA and G. P. SZEGÖ are also presented in V,3 of this book in the context of the relations between the properties of sets and of their regions of attraction.

Section 3. The definition of limit sets is due to G. D. BIRKHOFF [1, vol. 1, pp. 654—672]. This concept had been previously used by H. POINCARÉ without a formal definition.

Alternative equivalent definitions (3.5.1) of limit set were used by S. LEFSCHETZ [2] and by T. URA [4]. The second part of Theorem 3.6 is due to N. P. BHATIA [3].

In special phase spaces stronger results than Theorem 3.6 may be proved. For example in R^2 the following result on the structure of limit sets is given by HAJEK [1, p. 184]: every positive limit set consists of critical points and at most a countable number of non-critical trajectories γ_n; each compact subset of R^2 which contains no critical points intersects at most a finite number of γ_n. For further result on the structure of limits sets see SIBIRSKII [1, 2]. (See also V, 6.)

Section 4. The formal definition of prolongation is due to T. URA [2]; he adopts essentially the property 4.8.1 as a definition. This concept has been previously used by H. POINCARÉ [1, vol. 1, p. 44] in a special case (Example 4.2) without a formal definition. The concept of prolongational limit set is due to J. AUSLANDER, N. P. BHATIA and P. SEIBERT [1]. Theorems 4.4 and 4.7 are due to N. P. BHATIA [3]. The notion of relative prolongation is due to T. URA [5]. Most of the concepts defined in this chapter can also be defined in more general situations for instance for the case of local semidynamical systems (see N. P. BHATIA and O. HAJEK [1]) or for the case of dynamical systems without uniqueness (see G. P. SZEGÖ and G. TRECCANI [1]).

The structure of prolongational limit sets on special phase spaces like manifolds or even on R^2 has still not been fully investigated.

N. P. BHATIA [10] has recently introduced a somewhat different concept of relative prolongation than the one in 4.10. This seems more useful for the study of stability and attraction in non-locally compact spaces.

Chapter III

Recursive Concepts

1. Definition of Recursiveness

1.1 **Definition.** A *set* $A \subset X$ will be said to be *positively recursive* with respect to a *set* $B \subset X$ if for each $T \in R$ there is a $t > T$ and an $x \in B$ such that $xt \in A$. Negative recursiveness may be defined by using the inequality $t < T$. We will say that a *set* A is *self positively recursive* whenever it is positively recursive with respect to itself.

The simplest example of a self recursive set is that of a singleton $\{x\}$ where x is a periodic point (prove this!). In this chapter we study those concepts which are connected with the concept of recursiveness defined above. These are the concepts of Poisson stable and non-wandering points, and the concepts of minimal sets and recurrent points. To be sure, one may introduce other concepts by requiring that the entering of points from a set B into the set A happens with some kind of regularity not necessarily as stringent as in the case of a periodic point. This can give rise to various concepts of almost periodicity, but we shall not study this in detail (see GOTTSCHALK and HEDLUND [1] for very general concepts of almost periodicity in topological transformation groups).

2. Poisson Stable and Non-wandering Points

2.1 **Definition.** A *point* $x \in X$ is said to be *positively Poisson stable* if every neighborhood of x is positively recursive with respect to $\{x\}$.

We give some characterizations of positively Poisson stable points.

2.2 **Theorem.** Let $x \in X$. Then the following are equivalent.

2.2.1 x is positively Poisson stable,

2.2.2 given a neighborhood U of x and a $T > 0$, $xt \in U$ for some $t > T$,

2.2.3 $x \in \Lambda^+(x)$,

2.2.4 $\overline{\gamma^+(x)} = \Lambda^+(x)$,

2.2.5 $\gamma(x) \subset \Lambda^+(x)$,

2.2.6 for every $\varepsilon > 0$ there is a $t \geqq 1$ such that $xt \in S(x, \varepsilon)$.

Proof. The only non-trivial statement in the above theorem is the equivalence of 2.2.6 with either of the statements 2.2.1—2.2.5. We shall prove that 2.2.3 is equivalent to 2.2.6. Indeed 2.2.3 implies 2.2.6 trivially. So assume 2.2.6. Then choosing a positive null sequence $\{\varepsilon_n\}$ one sees that there is a sequence $\{t_n\}$, $t_n \geqq 1$, such that $xt_n \in S(x, \varepsilon_n)$ consequently $xt_n \to x$. Now either $\{t_n\}$ contains a convergent subsequence $\{t_{n_k}\}$, $t_{n_k} \to t \geqq 1$, or $t_n \to +\infty$. In the first case $xt_{n_k} \to xt$ by the continuity axiom and so $x = xt$ and x is periodic hence $x \in \Lambda^+(x)$ (see II, 3.2). In the second case $x \in \Lambda^+(x)$ by definition. Hence 2.2.6 implies 2.2.3.

2.3 *Exercises.*

2.3.1 Give complete proof of Theorem 2.2.

2.3.2 If x is positively Poisson stable then so is xt for every $t \in R$,

The following alternative definition of Poisson stability is customary in the literature and is suggested by the above Theorem 2.2.

2.4 **Definition.** A *point* $x \in X$ is *positively* or *negatively Poisson stable* whenever, respectively, $x \in \Lambda^+(x)$ or $x \in \Lambda^-(x)$. It is said to be *Poisson stable* if it is both positively and negatively Poisson stable. If a point $x \in X$ is Poisson stable then both the *motion* π_x and the *trajectory* $\gamma(x)$ are said to be *Poisson stable.*

In view of Theorem 2.2 it is interesting to inquire about the consequences of the condition $\gamma^+(x) = \Lambda^+(x)$. The answer is contained in the following theorem.

2.5 **Theorem.** $\gamma^+(x) = \Lambda^+(x)$ if and only if x is a periodic point.

Proof. Let $\gamma^+(x) = \Lambda^+(x)$. Indeed then $x \in \Lambda^+(x)$ and as $\Lambda^+(x)$ is invariant we see that $\gamma^+(x) = \Lambda^+(x) = \gamma(x)$. Thus $x\tau \in \gamma^+(x)$ for each $\tau < 0$, and, therefore, there is a $\tau' \geqq 0$ such that $x\tau = x\tau'$. Hence by the group axiom $xt = x(t + \tau' - \tau)$ for all $t \in R$, showing that x is periodic with a period $\tau' - \tau$ (> 0). The converse holds trivially and the theorem is proved.

2.6 *Remark.* It is to be noted that if $\gamma^+(x) = \Lambda^+(x)$ then the point x is indeed Poisson stable. It is, therefore, appropriate to inquire whether there exist points which are Poisson stable but are not periodic (i.e., also not a rest point). We give below an example of Poisson stable points which are not periodic.

2.7 *Example.*

2.7.1 Consider a dynamical system defined on a torus by means of the planar differential system

2.7.2 $$\frac{d\varphi}{dt} = f(\varphi, \theta), \quad \frac{d\theta}{dt} = \alpha f(\varphi, \theta),$$

where $f(\varphi, \theta) \equiv f(\varphi + 1, \theta + 1) \equiv f(\varphi + 1, \theta) \equiv f(\varphi, \theta + 1)$, and $f(\varphi, \theta) > 0$ if φ and θ are not both zero (mod 1), $f(0, 0) = 0$. Let $\alpha > 0$ be irrational. It is easily seen that the trajectories of this system on the torus consist of a rest point p corresponding to the point $(0, 0)$. There is exactly one trajectory γ_1 such that $\Lambda^-(\gamma_1) = \{p\}$, and exactly one trajectory γ_2 such that $\Lambda^+(\gamma_2) = \{p\}$. For any other trajectory γ, $\Lambda^+(\gamma) = \Lambda^-(\gamma) =$ the torus. Further $\Lambda^+(\gamma_1) = \Lambda^-(\gamma_2) =$ the torus. In this example points on the trajectory γ_1 are positively Poisson stable, but not negatively Poisson stable. Points on γ_2 are negatively Poisson stable, but not positively Poisson stable. All other points are Poisson stable. Note that no point except the rest point p is periodic.

Fig. 2.7.3 describes this example.

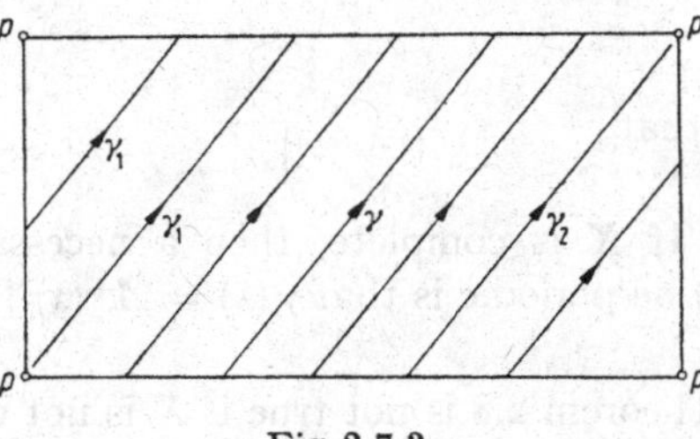

Fig. 2.7.3

2.7.4 If in the above example we have $f(\varphi, \theta) > 0$ for all φ, θ (i.e., also $f(0, 0) > 0$), then every trajectory is dense in the torus and moreover the torus is also the positive and negative limit set of each point. This example describes a compact minimal set which is not a periodic orbit (see section 3 for definition of minimal sets) and indeed each point is positively and negatively Poisson stable.

The following theorem throws some light on positively Poisson stable points x when $\gamma^+(x) \neq \Lambda^+(x)$.

2.8 Theorem. Let X be a complete metric space. Let $x \in X$ be positively Poisson stable, and let it not be a periodic point. Then the set $\Lambda^+(x) - \gamma(x)$ is dense in $\Lambda^+(x)$, i.e., $\overline{\Lambda^+(x) - \gamma(x)} = \Lambda^+(x) = \overline{\gamma(x)}$.

Proof. Since x is positively Poisson stable, we have $\Lambda^+(x) = \overline{\gamma(x)}$. To see that $\overline{\Lambda^+(x) - \gamma(x)} = \Lambda^+(x)$, it is sufficient to show that if $y \in \gamma(x)$ and $\varepsilon > 0$ is arbitrary, then there is a point $z \in \Lambda^+(x) - \gamma(x)$ such that $z \in S(y, \varepsilon)$. To see this notice that since $y \in \Lambda^+(x) \equiv \Lambda^+(y)$, there is a monotone increasing sequence $\{t_n\}$, $t_n \to +\infty$, such that $yt_n \to y$. Choose $\tau_1 > t_1$ such that $y\tau_1 \in S(y, \varepsilon)$. Then $y\tau_1 \notin y[-t_1, t_1]$ (otherwise x will be periodic). Hence $\delta_1 = \varrho(y\tau_1, y[-t_1, t_1]) > 0$. Set $\varepsilon_1 = \min\{\varepsilon/2, \varepsilon - \varrho(y, y\tau_1), \delta_1/2\}$. Then $S(y\tau_1, \varepsilon_1) \subset S(y, \varepsilon)$ and $S(y\tau_1, \varepsilon_1) \cap y[-t_1, t_1] = \emptyset$. Having defined $y\tau_{n-1}$ and ε_{n-1}, choose $\tau_n > t_n$ such that $y\tau_n \in S(y\tau_{n-1}, \varepsilon_{n-1})$ (possible because of positive Poisson stability of x). Then define $\varepsilon_n = \min\{\varepsilon_{n-1}/2, \varepsilon_{n-1} - \varrho(y\tau_{n-1}, y\tau_n), \delta_n/2\}$, where $\delta_n = \varrho(y\tau_n, y[-t_n, t_n])$. Note that $\delta_n > 0$ as the motion is not periodic. Clearly $S(y\tau_n, \varepsilon_n) \subset S(y\tau_{n-1}, \varepsilon_{n-1})$, and $S(y\tau_n, \varepsilon_n) \cap y[-t_n, t_n] = \emptyset$. The sequence $\{y\tau_n\}$ has the property that $\varrho(y\tau_n, y\tau_{n-1}) < \varepsilon_{n-1} \leqq \varepsilon/2^{n-1}$ for $n = 1, 2, \ldots$. $\{y\tau_n\}$ is, therefore, a Cauchy sequence which converges to a point z as the space X is complete. Since $y\tau_n \in \gamma(x)$, and $\tau_n \to +\infty$, we have $z \in \Lambda^+(x)$. Further $\varrho(y, y\tau_n) < \varepsilon$, so that $\varrho(y, z) \leqq \varepsilon$. Notice further that $z \notin \gamma(x)$. For, otherwise, if $z \in \gamma(x) \equiv \gamma(y)$, we will have $z = y\tau$. But there is an n such that $t_n > |\tau|$, so that $z \in y[-t_n, t_n]$. However, $z \in S(y\tau_n, \varepsilon_n)$, and by construction $S(y\tau_n, \varepsilon_n) \cap y[-t_n, t_n] = \emptyset$, i.e., $z \notin y[-t_n, t_n]$. This contradiction proves that $z \notin \gamma(x)$ and the theorem is proved.

It is now clear that

2.9 Theorem. If X is complete, then a necessary and sufficient condition that $\gamma(x)$ be periodic is that $\gamma(x) = \Lambda^+(x)$ $[= \Lambda^-(x)]$.

2.10 *Remark.* Theorem 2.8 is not true if X is not complete. This can be shown for instance by restricting the dynamical system in example 2.7 to a single trajectory say γ_1. Clearly for such a flow $\gamma_1 = \Lambda^+(x)$ for every $x \in \gamma_1$, but no $x \in \gamma_1$ is periodic.

We now introduce the notion of a non-wandering point.

2.11 Definition. A *point* $x \in X$ is said to be *non-wandering* if every neighborhood U of x is self positively recursive.

We give some characterizations of non-wandering points.

2.12 Theorem. For any $x \in X$, the following are equivalent.

2.12.1 x is non-wandering,

2.12.2 $x \in J^+(x)$,

2.12.3 every neighborhood of x is self negatively recursive,

2.12.4 $x \in J^-(x)$.

Proof. Assume 2.12.1. Consider a null sequence $\{\varepsilon_n\}$, $0 < \varepsilon_n$, $\varepsilon_n \to 0$, and a sequence $\{t_n\}$ in R with $t_n \to +\infty$. Since each $S(x, \varepsilon_n)$ is self positively recursive, we have an $x_n \in S(x, \varepsilon_n)$ and a $\tau_n > t_n$ with $x_n\tau_n \in S(x, \varepsilon_n)$. Since $\varepsilon_n \to 0$ we have $x_n \to x$ and $x_n\tau_n \to x$ and since $\tau_n \to +\infty$ we conclude $x \in J^+(x)$. Thus 2.12.2 holds. Now assume 2.12.2. Then there is a sequence $\{x_n\}$ in X and a sequence $\{t_n\}$ in R with $x_n \to x$, $t_n \to +\infty$, and $x_n t_n \to x$. Now for any neighborhood U and $T > 0$ there is an N such that $t_n > T$ for $n \geqq N$ and $x_n \in U$ and $x_n t_n \in U$ for $n \geqq N$. Thus U is self positively recursive. Consequently x is non-wandering and 2.12.1 holds. Equivalence of 2.12.3 and 2.12.4 is proved in the same way. To complete the proof, it is sufficient to show that 2.12.2 and 2.12.4 are equivalent. But this is trivial. A more general result was proved in II, 4.9.

We shall now prove a theorem which shows the connection between Poisson stable points and non-wandering points. First the following elementary result.

2.13 Theorem. Let $x \in X$. Every $y \in \Lambda^+(x)$ is non-wandering.

Proof. We have to show that if $y \in \Lambda^+(x)$ for some $x \in X$, then $y \in J^+(y)$. Indeed there is a sequence $t_n \to +\infty$ with $xt_n \to y$. Since $t_n \to +\infty$, we may assume, if necessary by taking a subsequence, that $t_{n+1} - t_n \geqq n$ for each n. Then setting $\tau_n = t_{n+1} - t_n$ and $xt_n = x_n$ we have $x_n \to y$, $x_n(t_{n+1} - t_n) = xt_{n+1} \to y$, and $\tau_n = t_{n+1} - t_n \to +\infty$. Thus $y \in J^+(y)$ and y is non-wandering.

A slightly deeper result is the following.

2.14 Theorem. Let $P \subset X$ be such that every $x \in P$ is either positively or negatively Poisson stable. Then every $x \in \overline{P}$ is non-wandering.

Proof. Let $\{x_n\}$ in P and $x_n \to x$. We must prove that $x \in J^+(x)$. Indeed for each n we have either $x_n \in \Lambda^+(x_n)$ or $x_n \in \Lambda^-(x_n)$. Thus by taking a subsequence, we may assume that either $x_n \in \Lambda^+(x_n)$ for all n or $x_n \in \Lambda^-(x_n)$ for all n. Assume $x_n \in \Lambda^+(x_n)$ for all n. For each n there is a $t_n > n$ with $\varrho(x_n, x_n t_n) < 1/n$. Then clearly $\varrho(x, x_n t_n) \leqq \varrho(x, x_n) +$

$\varrho(x_n, x_n t_n) \leqq \varrho(x, x_n) + 1/n$. This shows that $x_n t_n \to x$ and consequently $x \in J^+(x)$. In the second case similar considerations show that $x \in J^-(x)$. Thus by Theorem 2.12 every $x \in \overline{P}$ is non-wandering.

We now give a partial converse of Theorem 2.14.

2.15 Theorem. Let X be complete. Let every $x \in X$ be non-wandering. Then the set of Poisson stable points P is dense in X.

Proof. Let $x \in X$ and $\varepsilon > 0$. We will find a point $y \in S(x, \varepsilon)$ such that $y \in \Lambda^+(y)$ and $y \in \Lambda^-(y)$. Let $S(x, \varepsilon) = U$. Since U is self positively recursive, there is a $t_1 > 1$ such that $U \cap Ut_1 \neq \emptyset$. Indeed $U \cap Ut_1$ is open. So choose an x_1 in $U \cap Ut_1$ and positive $\varepsilon_1 < 1/2$ such that $S(x_1, \varepsilon_1) \subset U \cap Ut_1$. Set $U_1 = S(x_1, \varepsilon_1)$. Since x_1 is non-wandering, U_1 is self negatively recursive, and there is a $t_2 < -2$ such that $U_1 \cap U_1 t_2 \neq \emptyset$. Choose x_2 in $U_1 \cap U_1 t_2$ and a positive $\varepsilon_2 < 1/2^2$ such that $S(x_2, \varepsilon_2) \subset U_1 \cap U_1 t_2$. Set $U_2 = S(x_2, \varepsilon_2)$. Since x_2 is non-wandering, U_2 is self positively recursive and there is a $t_3 > 3$ such that $U_2 \cap U_2 t_3 \neq \emptyset$. Choose x_3 in $U_2 \cap U_2 t_3$ and positive $\varepsilon_3 < 1/2^3$ such that $S(x_3, \varepsilon_3) \subset U_2 \cap U_2 t_3$. Set $U_3 = S(x_3, \varepsilon_3)$. Proceeding in this fashion we obtain a sequence of open sets $\{U_n\}$, with the property that $U_n \supset U_{n+1}$ and $\cap \{U_n : n = 1, 2, 3, \ldots\}$ is a singleton $\{y\}$ with $y \in U$ (prove these facts which are consequences of construction and completeness of X). We claim that y is Poisson stable. To see this notice that by construction, for any integer $n \geqq 2$, $U_n(-t_n) \subset U_{n-1}$. In particular, since $y \in U_n$ for every n, $y(-t_n) \in U_{n-1}$. Clearly, then the sequences $\{y(-t_{2n})\}$ and $\{y(-t_{2n+1})\}$ both converge to y. Since $-t_{2n} \to +\infty$ and $-t_{2n+1} \to -\infty$ we have $y \in \Lambda^+(y)$ and $y \in \Lambda^-(y)$.

2.16 *Remark.* The above theorem remains true in locally compact spaces. The proof is the same as above except that one starts with a relatively compact neighborhood of the point x. We also remark that the closure of the set of Poisson stable points is called the *set of central motions.*

3. Minimal Sets and Recurrent Points

In this section we study certain subsets of the set of Poisson stable points. These sets have the property that each point in them returns to its arbitrary neighborhood with a certain regularity. The points need not however be periodic. We start with the definition of a minimal set.

3.1 Definition. A *set* $M \subset X$ is called *minimal,* if it is non-empty, closed, and invariant, and no proper subset of M has these properties.

A useful characterization of minimal sets is given by

3.2 **Theorem.** A non-empty set $M \subset X$ is minimal if and only if $\overline{\gamma(x)} = M$ for every $x \in M$.

Proof. Let M be minimal. Then for each $x \in M$, $\overline{\gamma(x)} \subset M$ as M is closed and invariant. Since $\overline{\gamma(x)}$ is a closed invariant set we must have $\overline{\gamma(x)} = M$, for otherwise $\overline{\gamma(x)}$ will be a non-empty proper subset of M, a contradiction to minimality of M. Now assume that $\overline{\gamma(x)} = M$ for every $x \in M$. If M is not minimal then there is non-empty closed invariant set $P \subset M$ with $P \neq M$. Choose $x \in P$. Then $\overline{\gamma(x)} \subset P$, hence $\overline{\gamma(x)} \neq M$ but $x \in M$. This contradiction shows that M is minimal.

Minimal sets possess the following important property.

3.3 **Theorem.** If $M \subset X$ is minimal and $\mathscr{I}(M) \neq \emptyset$, then $M = \mathscr{I}(M)$.

Proof. Let $x \in \mathscr{I}(M)$, and $y \in M$. Since $x \in \overline{\gamma(y)}$, there is a $t \in R$ with $yt \in \mathscr{I}(M)$. Then $y \in \mathscr{I}(M)(-t)$. The later is however an open subset of M, so that y is an interior point of M. Thus every point in M is an interior point of M. This proves the theorem.

Compact minimal sets may be characterized in various ways. We have

3.4 **Theorem.** Let $M \subset X$ be non-empty and compact. Then the following are equivalent.

3.4.1 M is minimal,

3.4.2 $\overline{\gamma(x)} = M$ for every $x \in M$,

3.4.3 $\overline{\gamma^+(x)} = M$ for every $x \in M$,

3.4.4 $\overline{\gamma^-(x)} = M$ for every $x \in M$,

3.4.5 $\Lambda^+(x) = M$ for every $x \in M$,

3.4.6 $\Lambda^-(x) = M$ for every $x \in M$.

The proof is immediate from elementary considerations and is left to the reader.

A rest point and a periodic trajectory are examples of compact minimal sets. By the above theorem, every point in a compact minimal set is Poisson stable. Example 2.7 indicated that the closure of a Poisson stable trajectory need not be a minimal set (in that example the closure of every Poisson stable trajectory except the rest point is the whole torus, which is not minimal as it contains a rest point). G. D. BIRKHOFF discovered an intrinsic property of motions in a compact minimal set, which is usually called the property of recurrence. We now study this concept.

3.5 Definition. For any $x \in X$, the *motion* π_x is said to be *recurrent* if for each $\varepsilon > 0$ there exists a $T = T(\varepsilon) > 0$, such that

$$\gamma(x) \subset S(x[t - T, t + T], \varepsilon)$$

for all $t \in R$.

3.6 *Remark.* It is clear that if a motion π_x is recurrent then every motion π_y with $y \in \gamma(x)$ is also recurrent. Thus we shall also speak of the *trajectory* $\gamma(x)$ being *recurrent.* Moreover, a *point* $x \in X$ may be called *recurrent* whenever the motion π_x is recurrent.

3.7 *Exercises.*

3.7.1 Show that every recurrent motion is Poisson stable.

3.7.2 In example 2.7 every point is positively Poisson stable, but the only recurrent point is the rest point.

That the concept of recurrence is basic in the theory of compact minimal sets is seen from the following theorem of Birkhoff.

3.8 Theorem. Every trajectory in a compact minimal set is recurrent. Thus every compact minimal set is the closure of a recurrent trajectory.

Proof. Let M be a compact minimal set. Suppose that there is an $x \in M$ such that the motion π_x is not recurrent. Then there is an $\varepsilon > 0$ and sequences $\{T_n\}$, $\{t_n\}$, $\{\tau_n\}$, with $T_n > 0$, $T_n \to +\infty$, and

$$x\tau_n \notin S(x[t_n - T_n, t_n + T_n], \varepsilon), \qquad n = 1, 2, \ldots .$$

This shows that

$$\varrho(x\tau_n, x(t_n + t)) \geqq \varepsilon \text{ whenever } |t| \leqq T_n, \ n = 1, 2, \ldots .$$

The sequences $\{xt_n\}$, $\{x\tau_n\}$ are contained in the compact set M and may without loss of generality be assumed to be convergent. So let $xt_n \to y$ and $x\tau_n \to z$. Then $y, z \in \overline{\gamma(x)} = M$. Consider now the motion π_y. For any $t \in R$, the inequality

$$\varrho(yt, z) \geqq \varrho(x(t_n + t), x\tau_n) - \varrho(yt, x(t_n + t)) - \varrho(x\tau_n, z)$$

yields

$$\varrho(yt, z) \geqq \varepsilon$$

in view of the previous inequality. Thus $z \notin \overline{\gamma(y)}$, i.e., $z \notin M$ as $M = \overline{\gamma(y)}$. This contradiction proves the theorem.

3.9 Theorem. If a trajectory $\gamma(x)$ is recurrent and $\overline{\gamma(x)}$ is compact then $\overline{\gamma(x)}$ is also minimal.

Proof. Set $\overline{\gamma(x)} = M$. Let if possible M be not minimal. Then there exists a non-empty compact invariant subset N of M, $N \neq M$. Clearly $x \notin N$ (otherwise $\gamma(x) \subset N$ which is impossible). Now let $\varrho(x, N) = \varepsilon\ (> 0)$. As π_x is recurrent, there is a $T > 0$ such that

3.9.1 $\quad \gamma(x) \subset S(x[t - T, t + T], \varepsilon/3) \quad$ for all $t \in R$.

Now choose any $y \in N$. Since $y \in M = \overline{\gamma(x)}$, and $y \notin \gamma(x)$, we have $y \in \Lambda^+(x)$ or $y \in \Lambda^-(x)$. Let $y \in \Lambda^+(x)$. Then there is a sequence $\{t_n\}$, $t_n \to +\infty$, such that $xt_n \to y$. By the continuity axiom, there is a $\delta > 0$ such that $\varrho(yt, zt) < \varepsilon/3$ whenever $\varrho(y, z) < \delta$ and $|t| \leqq T$. This shows that there is an n with

$$\varrho(yt, x(t_n + t)) < \varepsilon/3 \quad \text{for} \quad |t| \leqq T.$$

From this it follows that

$$\varrho(x, x(t_n + t)) \geqq \varrho(x, yt) - \varrho(yt, x(t_n + t)) \geqq \varepsilon - \varepsilon/3 = 2\varepsilon/3$$
$$\text{for} \quad |t| \leqq T.$$

This however contradicts 3.9.1. The theorem is proved.

3.10 Corollary. If the space X is complete, then the closure $\overline{\gamma(x)}$ of any recurrent trajectory $\gamma(x)$ is a compact minimal set.

The proof follows from the observation that the conditions imply compactness of $\overline{\gamma(x)}$, so that the result follows from Theorem 3.9. The details are left to the reader as an exercise.

Another way of defining a recurrent motion is provided via the concept of a *relatively dense set* of numbers.

3.11 Definition. A *set* D of real numbers is called *relatively dense* if there is a $T > 0$ such that

$$D \cap (t - T, t + T) \neq \emptyset \quad \text{for all } t \in R.$$

3.12 Theorem. For an $x \in X$ let $\overline{\gamma(x)}$ be compact. Then the motion π_x is recurrent if and only if for each $\varepsilon > 0$ the set

$$K_\varepsilon = \{t \colon \varrho(x, xt) < \varepsilon\}$$

is relatively dense.

Proof. Let for each $\varepsilon > 0$ the set K_ε be relatively dense. For any $\varepsilon > 0$ there is by definition a $T_\varepsilon = T > 0$ such that

$$K_\varepsilon \cap (t - T, t + T) \neq \emptyset \qquad \text{for all } t \in R.$$

As $\overline{\gamma(x)}$ is compact, to show that the motion π_x is recurrent we need show only that $\overline{\gamma(x)}$ is minimal. Let $\overline{\gamma(x)}$ be not minimal. Then there is minimal subset M of $\overline{\gamma(x)}$, $M \neq \overline{\gamma(x)}$ (see Theorem 4.4). Clearly $x \notin M$ (otherwise $\overline{\gamma(x)} \subset M$ which will imply $\overline{\gamma(x)} = M$). Set $\varrho(x, M) = 3\varepsilon\ (> 0)$. Choose any $y \in M$. Then there is a $\delta > 0$ such that $\varrho(yt, zt) < \varepsilon$ whenever $\varrho(y, z) < \delta$ and $|t| \leqq T = T_\varepsilon$. As $y \in M \subset \overline{\gamma(x)}$, and $y \notin \gamma(x)$, we conclude that $y \in \Lambda^+(x)$ or $y \in \Lambda^-(x)$. Let $y \in \Lambda^+(x)$. Then there is a sequence $\{t_n\}$, $t_n \to +\infty$ and $xt_n \to y$. Thus for all sufficiently large n we have $\varrho(yt, x(t_n + t)) < \varepsilon$ for $|t| \leqq T = T_\varepsilon$. But then for $t \in [t_n - T, t_n + T]$ we have

$$\varrho(x, xt) \geqq \varrho(x, y(t - t_n)) - \varrho(xt, y(t - t_n)) \geqq 3\varepsilon - \varepsilon = 2\varepsilon.$$

This shows that

$$K_\varepsilon \cap [t_n - T_\varepsilon, t_n + T_\varepsilon] = \emptyset,$$

which is a contradiction. This shows that $\overline{\gamma(x)}$ is minimal and hence the motion π_x is recurrent. The converse holds trivially. The theorem is proved.

The above discussion was centered on compact minimal sets. Not very much is known about the properties of non-compact minimal sets. It can be shown that all minimal sets in R^2 consist of single trajectories with empty limit sets (Bhatia and Szegö [1], Theorem 1.3.6). However, compact minimal sets contain in general more than one trajectory.

3.13 Theorem. There exist non-compact minimal sets which contain more than one trajectory.

Proof. Consider example 2.7 of the dynamical system on a torus T. Consider the dynamical system obtained by restricting the given dynamical system to the complement of the rest point in this example. The resulting space X is not compact, but for each $x \in X$, $\overline{\gamma(x)} = X$, so that X is minimal. This proves the theorem.

3.14 *Remark*. In the example in the proof above the motions π_x are not recurrent. This shows that Theorem 3.8 is not necessarily true for non-compact minimal sets.

4. Lagrange Stability and Existence of Minimal Sets

Given a dynamical system (X, R, π) it is in general not known if X contains minimal sets. The purpose of this section is to give a necessary and sufficient condition for the existence of compact minimal sets.

4.1 Definition. For any $x \in X$, the motion π_x is said to be *positively Lagrange stable* if $\overline{\gamma^+(x)}$ is compact. Further, if $\overline{\gamma^-(x)}$ is compact, then the motion π_x is called *negatively Lagrange stable.* It is said to be *Lagrange stable* if $\overline{\gamma(x)}$ is compact.

4.2 *Remark.* If $X = R^n$, then the above statements are equivalent to the sets $\gamma^+(x)$, $\gamma^-(x)$, $\gamma(x)$ being bounded, respectively.

4.3 *Exercises.*

4.3.1 If X is locally compact, then a motion π_x is positively Lagrange stable if and only if $\Lambda^+(x)$ is a non-empty compact set.

4.3.2 If a motion π_x is positively Lagrange stable, then $\Lambda^+(x)$ is compact and connected.

4.3.3 If a motion π_x is positively Lagrange stable, then $\varrho(xt, \Lambda^+(x)) \to 0$ as $t \to +\infty$.

The notion of Lagrange stability plays an important role in the study of minimal sets as Theorem 3.9 and 3.12 would indicate. That this notion is also critical in the existence of compact minimal sets is demonstrated by the following theorems.

4.4 Theorem. Every non-empty compact invariant set contains a compact minimal set.

Proof. The proof follows by considering the family of all non-empty closed invariant subsets of the given set. This family is partially ordered by set inclusion and since the original set is compact, Zorn's lemma shows that there is a non-empty compact invariant subset which contains no proper subsets which are closed and invariant. Such a set is then compact and minimal.

Finally we have

4.5 Theorem. The space X contains a compact minimal set if and only if there is an $x \in X$ such that either $\overline{\gamma^+(x)}$ or $\overline{\gamma^-(x)}$ is compact.

Proof. Let $\overline{\gamma^+(x)}$ be compact. Then $\Lambda^+(x)$ is a non-empty compact invariant subset of $\overline{\gamma^+(x)}$ and hence of X. The existence of a compact minimal set now follows from Theorem 4.4. Similarly if $\overline{\gamma^-(x)}$ is compact, then $\Lambda^-(x)$ is a non-empty compact invariant set and the existence of a compact minimal set follows again from Theorem 4.4. The converse is trivial, for every trajectory in a compact minimal set is Lagrange stable.

Notes and References

The concepts of Poisson stability, non-wandering points, minimal sets, recurrent points, and Lagrange stability are all classical. The theorems given here are also generally well known (see for example G. D. Birkhoff [1], and Nemytskii and Stepanov [1]). The use of prolongational limit sets to describe non-wandering points is due to J. Auslander [3], who also introduced the notion of generalized recurrence via higher order prolongational limit sets. Auslander's work is discussed in VII, 3.

G. D. Birkhoff did a considerable amount of work on the notion of central motions (Remark 2.16) which he introduced. That work is of great interest in ergodic theory and for this we refer the reader to Nemytskii and Stepanov [1].

Theorems 2.5 and 2.13 do not seem to be in the classical works.

For further developments in the direction of this chapter we mention H. Chu [1, 2], Dowker [1], Ellis [5, 8, 10, 11, 12], England and Kent [1], Garcia and Hedlund [1], Gottschalk [11], Hilmy [1–7], and Sibirskii [1, 2].

Sections 3, 4. The concepts of non-wandering points and recurrent points are due to G. D. Birkhoff [1] who also proved the important relationship between compact minimal sets and recurrent points (Theorems 3.8, 3.9, Corollary 3.10, Theorem 4.4). The important result (Theorem 2.8) on Poisson stable points is from Nemytskii and Stepanov [1].

It is worthwhile to note that the only recurrent motions in R^2 are the periodic ones. Consequently, all compact minimal sets in R^2 are the trajectories of periodic points. In fact O. Hajek [1, p. 183] shows that all positively Poisson stable points in R^2 are periodic. We recall also that the only non-compact minimal sets in R^2 consist of a single trajectory with empty positive and negative limit sets (Bhatia and Szegö [1], Theorem 1.3.6). Thus Theorem 3.3 implies that all minimal sets in R^2 have empty interiors. Theorem 3.3 is attributed by Nemytskii and Stepanov [1] to G. S. Tumarkin. This theorem poses an important problem, viz., which phase spaces (for example which manifolds) can be minimal.

For further study of minimal sets in topological transformation groups we refer to W. H. Gottschalk [11] as a starting point.

A special case of recurrence, namely almost periodicity, is deferred to V, 6, because it is intimately connected with stability of motion. Note also that the concepts of this chapter can be generalized to non-metric topological spaces, whereas almost periodicity requires a uniformity on the space.

Chapter IV

Dispersive Concepts

This chapter is devoted to dynamical systems (X, R, π) which are notably marked by the absence of Lagrange stable motions or Poisson stable points or non-wandering points and in general the absence of recursiveness. Such concepts (for the lack of a better term) may be called dispersive concepts. The first section contains those concepts which are definable in terms of the machinery introduced in Chapter II. The second section studies the theory of parallelizable dynamical systems.

1. Unstable and Dispersive Dynamical Systems

1.1 **Definition.** Let $x \in X$. Then

1.1.1 the motion π_x is said to be *positively Lagrange unstable* whenever $\overline{\gamma^+(x)}$ is not compact. It is called *negatively Lagrange unstable* if $\overline{\gamma^-(x)}$ is not compact. Finally, it is called *Lagrange unstable* if it is both positively and negatively Lagrange unstable.

1.1.2 the point x is called *positively Poisson unstable* whenever $x \notin \Lambda^+(x)$, *negatively Poisson unstable* whenever $x \notin \Lambda^-(x)$, and *Poisson unstable* whenever it is both positively and negatively Poisson unstable.

1.1.3 the point x is called *wandering* whenever $x \notin J^+(x)$.

We are now in a position to formally define some of the concepts of dispersiveness of a dynamical system (X, R, π).

1.2 **Definition.** The dynamical system (X, R, π) is said to be

1.2.1 *Lagrange unstable* if for each $x \in X$ the motion π_x is Lagrange unstable,

1.2.2 *Poisson unstable* if each $x \in X$ is Poisson unstable,

1.2.3 *completely unstable* if every $x \in X$ is wandering,

1.2.4 *dispersive* if for every pair of points $x, y \in X$ there exist neighborhoods U_x of x and U_y of y such that U_x is not positively recursive with respect to U_y.

1.3 *Remark.* In 1.1.3 it is unnecessary to define concepts such as positively wandering or negatively wandering with the obvious requirement that $x \notin J^+(x)$ or $x \notin J^-(x)$. This is so because $x \notin J^+(x)$ if and only if $x \notin J^-(x)$. Thus we would not get any new concepts.

1.4 *Exercises.*

1.4.1 Call a dynamical system (X, R, π) *positively Lagrange unstable* whenever for each $x \in X$, $\gamma^+(x)$ is non-compact. Prove that (X, R, π) is Lagrange unstable if and only if it is positively Lagrange unstable.

1.4.2 Call a dynamical system (X, R, π) *positively Poisson unstable* whenever for each $x \in X$, $x \notin \Lambda^+(x)$ holds. Show that there exists a positively Poisson unstable dynamical system which is not Poisson unstable.

1.4.3 Give examples, of systems for which one of the conditions 1.2.1 through 1.2.4 holds but none of the subsequent ones hold.

1.4.4 Show that each of the conditions 1.2.1 through 1.2.4 implies the preceding one.

1.4.5 Prove that a point $x \in X$ is a wandering point if and only if there is a neighborhood U of x and a $T > 0$ such that $U \cap Ut = \emptyset$ for all t, $|t| \geqq T$.

To emphasize the fact that none of the concepts 1.2.1—1.2.4 implies the following one we give some examples.

1.5 *Examples.*

1.5.1 Consider a dynamical system in a euclidean (x_1, x_2)-plane, whose phase portait is as in Fig. 1.5.2. The unit circle contains a rest point p and a trajectory γ such that for each point $q \in \gamma$ we have $\Lambda^+(q) = \Lambda^-(q) = \{p\}$. All trajectories in the interior of the unit circle $(= \{p\} \cup \gamma)$ have the same property as γ. All trajectories in the exterior of the unit circle spiral to the unit circle as $t \to +\infty$, so that for each point q in the exterior of the unit circle we have $\Lambda^+(q) = \{p\} \cup \gamma$, and $\Lambda^-(q) = \emptyset$. Notice that if we consider the dynamical system obtained from this one by deleting the rest point p (the dynamical system is thus defined on $R^2 - \{p\}$) then this system is Lagrange unstable and Poisson unstable, but it is not completely unstable because for each $q \in \gamma$ we have $J^+(q) = \gamma$, i.e., $q \in J^+(q)$.

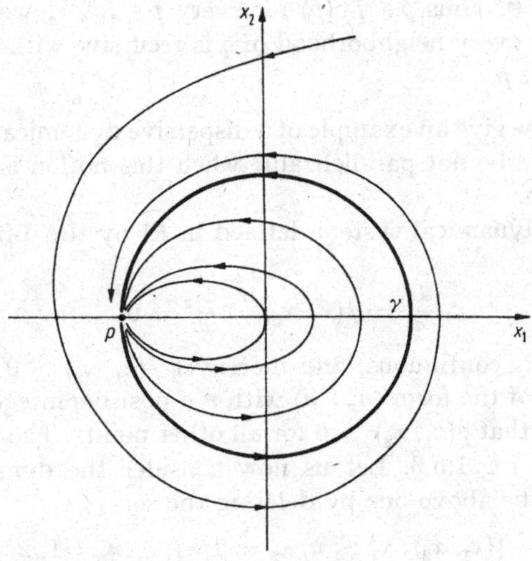

Fig. 1.5.2

1.5.3 Consider a planar dynamical system defined by the differential equations (cartesian coordinates)

$$\dot{x}_1 = \sin x_2, \ \dot{x}_2 = \cos^2 x_2.$$

The dynamical system contains, in particular, trajectories γ_k given by

$$\gamma_k = \{(x_1, x_2) : x_2 = k\pi + \pi/2\}, \ \ k = 0, \pm 1, \pm 2, \ldots .$$

These are lines parallel to the x_1-axis. Between any two consecutive γ_k's the trajectories are given by $\gamma = \{(x_1, x_2) : x_1 + c = \sec x_2\}$, where c is some constant depending on the trajectory. The phase portrait between the lines $x_2 = -\pi/2$ and $x_2 = +\pi/2$ is shown in Fig. 1.5.4. This system

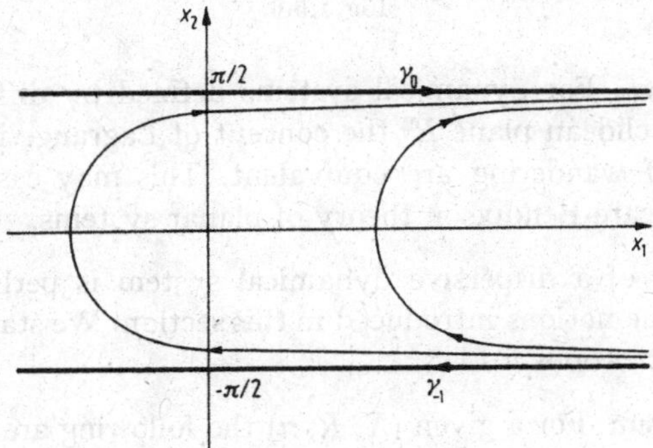

Fig. 1.5.4

is completely unstable but not dispersive. This follows by noticing, for example, that for each point $p \in \gamma_{-1}$, $J^+(p) = \gamma_0$, and for all other points $p \notin \gamma_k$, $J^+(p) = \emptyset$. Thus $p \notin J^+(p)$ for every $p \in R^2$. However, if $p \in \gamma_{-1}$ and $q \in \gamma_0$, then every neighborhood of q is recursive with respect to any neighborhood of p.

1.5.5 We now give an example of a dispersive dynamical system which will be shown to be not parallelizable when this notion is introduced in the next section.

Consider a dynamical system defined in R^2 by the differential equations

$$\frac{dx_1}{dt} = f(x_1, x_2), \qquad \frac{dx_2}{dt} = 0,$$

where $f(x_1, x_2)$ is continuous, and moreover $f(x_1, x_2) = 0$ whenever the point (x_1, x_2) is of the form $(n, 1/n)$ with n a positive integer. For simplicity we assume that $f(x_1, x_2) > 0$ for all other points. The phase portrait is as shown in Fig. 1.5.6. Let us now consider the dynamical system obtained from the above one by deleting the sets

$$I_n = \{(x_1, x_2) : x_1 \leqq n, x_2 = 1/n\}, \qquad n = 1, 2, 3, \ldots$$

from the plane R^2. This system is dispersive.

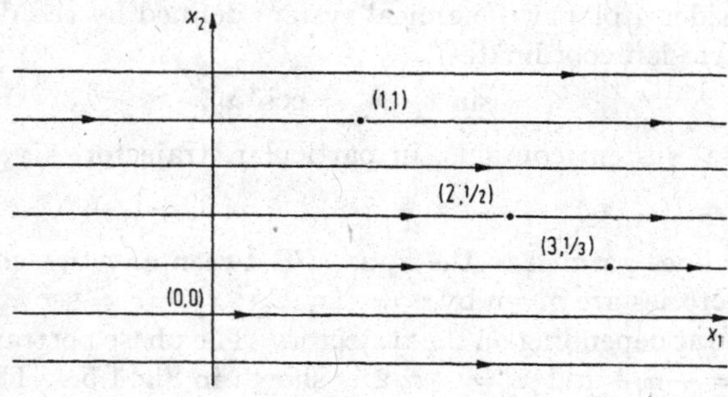

Fig. 1.5.6

1.6 *Remark.* For dynamical systems defined by differential equations in the euclidean plane R^2 the concept of Lagrange instability and the concept of wandering are equivalent. This may easily be proved using the Poincaré-Bendixşon theory of planar systems.

The notion of a dispersive dynamical system is perhaps the most important of the notions introduced in this section. We start with several of its characterizations.

1.7 **Theorem.** For a given (X, R, π) the following are equivalent.

1.7.1 (X, R, π) is dispersive.

1.7.2 For any two points x, y in X there are neighborhoods U_x of x and U_y of y and a constant $T > 0$ such that $U_x \cap U_y t = \emptyset$ for all t, $|t| \geqq T$.

1.7.3 For any two points x, y in X, $y \notin J^+(x)$.

The proof easily follows from the definitions and is left to the reader.

We now develop a criterion for dispersive flows.

1.8 Theorem. A dynamical system (X, R, π) is dispersive if and only if for each $x \in X$, $J^+(x) = \emptyset$.

Proof. Let (X, R, π) be dispersive. Let if possible $x \in X$ and $J^+(x) \neq \emptyset$. Then if $y \in J^+(x)$, there are sequences $\{x_n\}$, $\{t_n\}$, $x_n \to x$, $t_n \to +\infty$, and $x_n t_n \to y$. This shows that for any neighborhoods U_x, U_y of x and y respectively $U_x t_n \cap U_y \neq \emptyset$ as the element $x_n t_n = y_n$ is contained in this intersection. Since $t_n \to +\infty$, this contradicts the definition of a dispersive flow as U_y is positively recursive relative to U_x. Hence $J^+(x) = \emptyset$ for each $x \in X$. Conversely, let $J^+(x) = \emptyset$ for each $x \in X$. We claim that for $\{x, y\} \subset X$ there are neighborhoods U_x of x and U_y of y and a $T \geqq 0$ such that $U_x t \cap U_y = \emptyset$ for all $t \geqq T$. For if not, then there will be sequences $\{x_n\}$, $\{y_n\}$, $\{t_n\}$, $x_n \to x$, $y_n = x_n t_n$, $y_n \to y$, and $t_n \to +\infty$, so that $y \in J^+(x)$. This is absurd as $J^+(x) = \emptyset$. Note that U_y is not positively recursive relative to U_x. This proves the theorem.

1.9 *Remark.* Using the above theorem the dynamical system described in Example 1.5.5 is clearly seen to be dispersive.

We now give another criterion for dispersive flows, which is sometimes more useful than the one given above.

1.10 Theorem. The dynamical system (X, R, π) is dispersive if and only if for each $x \in X$, $D^+(x) = \gamma^+(x)$ and there are no rest points or periodic trajectories.

Proof. If (X, R, π) is dispersive, then $J^+(x) = \emptyset$ for each $x \in X$. Consequently $D^+(x) = \gamma^+(x) \cup J^+(x) \equiv \gamma^+(x)$ for each $x \in X$, and there are no rest points or periodic orbits. For if x is a rest point or $\gamma(x)$ is periodic then $\gamma(x) \equiv \Lambda^+(x) \subset J^+(x)$. Conversely, if $D^+(x) = \gamma^+(x)$ and there are no rest points or periodic orbits, then $J^+(x) = \emptyset$. For indeed $D^+(x) \equiv \gamma^+(x) \cup J^+(x) = \gamma^+(x)$ implies that $J^+(x) \subset \gamma^+(x)$. $J^+(x)$ being closed and invariant, we conclude that if $J^+(x)$ is not empty, then $\gamma(x) \subset J^+(x) \subset \gamma^+(x)$, i.e. $\gamma(x) = \gamma^+(x)$. This shows that if $\tau < 0$ is arbitrary, then there is a $\tau' \geqq 0$ such that $x\tau = x\tau'$, i.e. $x = x(\tau' - \tau)$. Since

$\tau' - \tau > 0$, the last equality shows that the trajectory $\gamma(x)$ is closed and has a period $\tau' - \tau$. Since we assumed that there are no rest points or periodic orbits, we have arrived at a contradiction. Thus $J^+(x) = \emptyset$ for each $x \in X$, and the dynamical system is dispersive. The theorem is proved.

We will now introduce parallelizable dynamical systems.

2. Parallelizable Dynamical Systems

2.1 **Definition.** A dynamical system (X, R, π) is called *parallelizable* if there exists a set $S \subset X$ and a homeomorphism $h\colon X \to S \times R$ such that $SR = X$ and $h(xt) = (x, t)$ for every $x \in S$ and $t \in R$.

For the study of parallelizable dynamical systems we need to develop a theory of sections. Thus we start with

2.2 **Definition.** A set $S \subset X$ is called a *section* of (X, R, π) if for each $x \in X$ there is a unique $\tau(x)$ such that $x\tau(x) \in S$.

Not every dynamical system has a section. Indeed any (X, R, π) has a section if and only if it has no rest points or periodic orbits.

The function $\tau(x)$ will be basic in what follows. In general $\tau(x)$ is not continuous, but the existence of a section S with continuous $\tau(x)$ implies certain properties of the dynamical system which we sum up in the following lemma.

2.3 **Lemma.** If S is a section of the dynamical system (X, R, π) with $\tau(x)$ continuous on X then

2.3.1 S is closed in X,

2.3.2 S is connected, arcwise connected, simply connected if and only if X is respectively connected, arcwise connected, simply connected.

2.3.3 If $K \subset S$ is closed in S, then Kt is closed in X for every $t \in R$.

2.3.4 If $K \subset S$ is open in S, then KI, where I is any open interval in R, is open in X.

Proof of 2.3.1. If $\{x_n\}$ in S, and $x_n \to x \in X$, then $\tau(x_n) \to \tau(x)$ by continuity. Since $\tau(x_n) = 0$ for each n, we get $\tau(x) = 0$. Thus $x\tau(x) = x\,0 = x \in S$ by definition of $\tau(x)$ and S. That is, S is, closed in X.

Proof of 2.3.2. We shall prove only the first part. The interested reader can supply the proofs of the remaining parts. Let S be not connected. Then there are disjoint closed sets S_1, S_2 such that $S_1 \cup S_2 = S$. As

$X = SR$, we have $X = S_1R \cup S_2R$. Note that S_1R and S_2R are disjoint. We prove that they are closed. Consider S_1R, and let $\{x_n\}$ in S_1R, $x_n \to x$. Then $\tau(x_n) \to \tau(x)$, and by the continuity axiom $x_n\tau(x_n) \to x\tau(x)$. Since $\{x_n\tau(x_n)\}$ is in S_1 and S_1 is closed we conclude that $x\tau(x) \in S_1$. Then $x = x\tau(x)\,(-\tau(x)) \in x\tau(x)\,R \subset S_1R$. Thus S_1R is closed. Similarly we can prove that S_2R is closed. Thus X being the union of two disjoint non-empty closed sets is not connected. We conclude that if X is connected, so must be S. The converse follows similarly.

Proof of 2.3.3. The proof follows by observing that if K is closed in S, then it is closed in X.

Proof of 2.3.4. The simple proof is left to the reader.

The following theorem now gives a criterion for parallelizable dynamical systems.

2.4 Theorem. A dynamical system (X, R, π) is parallelizable if and only if it has a section S with $\tau(x)$ continuous on X.

Proof. Sufficiency. Indeed $SR = X$. Define $h\colon X \to S \times R$ by $h(x) = (x\tau(x), -\tau(x))$. Then h is $1-1$ and continuous by the continuity of $\tau(x)$ and the continuity axiom. The inverse $h^{-1}\colon S \times R \to X$ is given by $h^{-1}(x, t) = xt$ and is clearly continuous. This h is thus a homeomorphism of X onto $S \times R$, i.e. (X, R, π) is parallelizable. To see necessity, we note that if the dynamical system is parallelizable, then the set S in its definition is a section of X. Since for any $x \in X$, $x = yt$ for some $y \in S$ and $t \in R$, we set $\tau(x) = -t$. Then continuity of $\tau(x)$ follows from that of h. The proof is completed.

2.5 *Remark.* The above theorem shows that the dynamical system of Example 1.5.5 is not parallelizable. Notice however that the phase space in this example is not locally compact.

The following is the most important theorem in this section.

2.6 Theorem. A dynamical system (X, R, π) on a locally compact separable metric space X is parallelizable if and only if it is dispersive.

The proof of this theorem depends on properties of certain sections which we now describe.

2.7 Definition. An open set U in X will be called a *tube*, if there exists a $\tau > 0$ and a subset $S \subset U$ such that

2.7.1 $SI_\tau \subset U$, and

2.7.2 for each $x \in U$ there is a unique $\tau(x)$, $|\tau(x)| < \tau$, such that $x\tau(x) \in S$. Here $I_\tau \equiv (-\tau, \tau)$.

It is clear that if 2.7.1 and 2.7.2 hold, then $U = SI_\tau$. U is also called a *τ-tube* with section S, and S a $(\tau - U)$-*section* of the tube U.

If $I_\tau = R$, then U is an ∞-tube, and S an $(\infty - U)$-section. In this last case indeed $U = SR$. Note also that the function $\tau(x)$ which maps U into I_τ is $1-1$ along each trajectory in U.

2.8 **Lemma.** Let U be a τ-tube with section S. If $K \subset S$ is compact, then the function $\tau(x)$ is continuous on KI_s for any s, $0 < s < \tau$.

Proof. To show is: if $\{x_n\}$ in KI_s and $x_n \to x \in KI_s$, then $\tau(x_n) \to \tau(x)$. Note that the sequence $\{x_n\tau(x_n)\}$ is in K, and we may assume that it is convergent as K is compact. Further $\{\tau(x_n)\}$ is in I_s and hence bounded so that we may also assume that $\{\tau(x_n)\}$ is convergent. Thus let $x_n\tau(x_n) \to x^* \in K$, and $\tau(x_n) \to \tau^* \in \bar{I}_s$. Since $x_n \to x$, we have $x^* = x\tau^*$. Since $|\tau^*| \leq s < \tau$, we have $\tau^* = \tau(x)$ by uniqueness. The lemma is proved.

The next theorem shows that if $x \in X$ is not a rest point, then there is a tube containing x.

2.9 **Theorem.** If $x \in X$ is not a rest point, then there exists a tube containing x.

Proof. Since x is not a rest point, there is a $T_0 > 0$ such that $\varrho(x, xT_0) > 0$. Consider the function

$$\psi(y, t) = \int_t^{t+T_0} \varrho(x, y\tau)\, d\tau.$$

It follows that

$$\psi(y, t_1 + t_2) = \int_{t_1+t_2}^{t_1+t_2+T_0} \varrho(x, y\tau)\, d\tau = \int_{t_2}^{t_2+T_0} \varrho(x, y(\tau + t_1))\, d\tau$$

$$= \int_{t_2}^{t_2+T_0} \varrho(x, yt_1(\tau))\, d\tau = \psi(yt_1, t_2).$$

Further the function $\psi(y, t)$ is continuous in (y, t) and has the continuous partial derivative

$$\psi_t(y, t) = \varrho(x, y(t + T_0)) - \varrho(x, yt).$$

Since

$$\psi_t(x, 0) = \varrho(x, xT_0) > 0,$$

there is an $\varepsilon > 0$ such that $\psi_t(y, 0) > 0$ for $y \in S(x, \varepsilon)$. Define $\tau_0 > 0$ such that $x[-3\tau_0, 3\tau_0] \subset S(x, \varepsilon)$. Then, in particular, $\psi(x, \tau_0) > \psi(x, 0) > \psi(x, -\tau_0)$. Now choose $\zeta > 0$ such that

$$(S[x\tau_0, \zeta] \cup S[x(-\tau_0), \zeta]) \subset S(x, \varepsilon),$$

and such that for $y \in S(x\tau_0, \zeta)$ we have $\psi(y, 0) > \psi(x, 0)$, and for $y \in S(x(-\tau_0), \zeta)$ we have $\psi(y, 0) < \psi(x, 0)$. Finally determine $\delta > 0$ such that

$$S[x, \delta]\,\tau_0 \subset S(x\tau_0, \zeta), \qquad S[x, \delta]\,(-\tau_0) \subset S(x(-\tau_0), \zeta),$$

and

$$S[x, \delta]\,[-3\tau_0, 3\tau_0] \subset S(x, \varepsilon).$$

We will show that if $y \in S[x, \delta]$, then there is exactly one $\tau(y)$, $|\tau(y)| < \tau_0$ such that $\psi(y, \tau(y)) = \psi(x, 0)$. This follows from the fact that $\psi(y, t) = \psi(yt, 0)$ is a strictly increasing function of t, and $\psi(y, \tau_0) > \psi(x, 0) > \psi(y, -\tau_0)$.

Consider now the open set $U = S(x, \delta)\, I_{\tau_0}$, and set $S = \{y \in U: \psi(y, 0) = \psi(x, 0)\}$. We claim that S is a $(2\tau_0 - U)$-section. For this we need prove that if $y \in U$, then there is a unique $\tau(y)$, $|\tau(y)| < 2\tau_0$ such that $y\tau(y) \in S$. Indeed for any $y \in U$, there is a t' $|t'| < \tau_0$ such that $y' = yt' \in S(x, \delta)$, and for $y' \in S(x, \delta)$ there is a t'', $|t''| < \tau_0$, such that $y't'' \in S$. Thus $y(t' + t'') = y\tau(y) \in S$, where $\tau(y) = t' + t''$, and $|\tau(y)| \leqq |t'| + |t''| < 2\tau_0$. Now let if possible there be two numbers, $\tau'(y)$, $\tau''(y)$, $|\tau'(y)| < 2\tau_0$, $|\tau''(y)| < 2\tau_0$, such that $y\tau'(y) \in S$ and $y\tau''(y) \in S$, and let $y' = yt' \in S(x, \delta)$, where $|t'| \leqq \tau_0$. Then $\psi(y', \tau'(y) - t') = \psi(y, \tau'(y)) = \psi(y\tau'(y), 0)$, and $\psi(y', \tau''(y) - t') = \psi(y, \tau''(y)) = \psi(y\tau''(y), 0)$, so that $\psi(y', \tau'(y) - t') = \psi(y', \tau''(y) - t') = \psi(x, 0)$. Now $|\tau'(y) - t'| \leqq 3\tau_0$, and $|\tau''(y) - t'| \leqq 3\tau_0$, and $\psi_t(y', t) > 0$ for $|t| \leqq 3\tau_0$, i.e., $\psi(y', t)$ is strictly increasing for $|t| \leqq 3\tau_0$. Hence $\tau'(y) - t' = \tau''(y) - t'$, or $\tau'(y) = \tau''(y)$. The theorem is proved.

2.10 *Remark.* If X is locally compact, then we can restrict $\delta > 0$ in the above proof to ensure that $S[x, \delta]$ is compact. Thus the $(2\tau_0 - U)$-section S constructed in the above proof will also be locally compact. By Lemma 2.8 we may further assume the function $\tau(x)$ corresponding to the section S to be continuous on U.

In fact the following more general theorem can now be proved.

2.11 **Theorem.** Let $x \in X$ be not a rest point. Let $\tau > 0$ be given, restricted only by $\tau < T/4$ if the motion π_x is periodic with least period T. Then there exists a tube U containing x with a $(\tau - U)$-section S. Further, if X is locally compact, then the function $\tau(x)$ corresponding to the section S can be assumed continuous on U.

The proof of this theorem is left to the reader.

For wandering points $x \in X$ one can prove:

2.12 Theorem. If $x \in X$ is a wandering point, i.e., $x \notin J^+(x)$, and moreover X is locally compact, then there exists a tube U containing x, with an $(\infty - U)$-section S, and with $\tau(x)$ continuous on U.

Proof. Indeed there is a tube W containing x, with a $(\tau - W)$-section S, and $\tau(x)$ continuous on W. Since x is wandering, we claim that there is a $\delta > 0$ such that $S(x, \delta) \cap S = S^*$ is an $(\infty - U)$-section of the open set $U = S^*R$, which is an ∞-tube containing x. To see this notice that there is a $\delta > 0$, such that every trajectory $\gamma(y)$ with $y \in S^*$, intersects S^* only at the point y. For otherwise, there will be a sequence $\{y_n\}$ in S, $y_n \to x$, and a sequence $\{t_n\}$ in R, $t_n \to +\infty$ (or $t_n \to -\infty$), such that $y_n t_n \to x$, i.e., either $x \in J^+(x)$, or $x \in J^-(x)$, both of which are ruled out by the assumption that x is wandering. We have thus shown that there is a $\delta > 0$ such that $S^* = S(x, \delta) \cap S$ is an $(\infty - U)$-section of $U = S^*R$. Furthermore U is open, and continuity of $\tau(x)$ on U follows from its continuity on $W \cap U$, and continuity of the phase map π. This we leave to the reader to verify. The theorem is proved.

For further development we need the following definition.

2.13 Definition. Given an open ∞-tube U with a section S and $\tau(x)$ continuous on U, and given sets N, K, $N \subset K \subset S$, where N is open in S and K is compact, we shall call KR the *compactly based tube* over K. Then indeed $\tau(x)$ restricted to KR is continuous on KR.

2.14 *Remark.* A compactly based tube need not be closed in X. As an example, one may consider a dynamical system defined in the euclidean plane R^2, as shown in Fig. 2.14.1. The x_2-axis consists entirely of rest points, all other trajectories are parallel to the x_1-axis, with each having a rest point on the x_2-axis as the only point in its positive limit set, whereas the negative limit sets are empty. Here, for example the set $\{(x_1, x_2) : 0 \leqq x_2 \leqq 1, x_1 > 0\}$ is a compactly based tube, which is not closed in X.

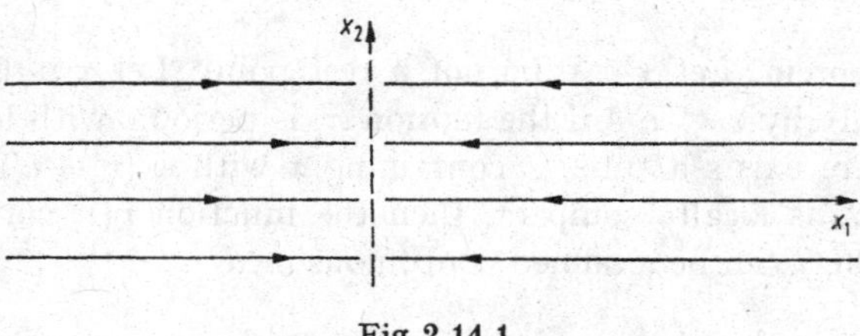

Fig. 2.14.1

We can now prove the following theorem.

2.15 Theorem. If X is locally compact and separable, and if every $x \in X$ is a wandering point, then there exists a countable covering $\{K_n R\}$ of X, by compactly based tubes $K_n R$ each with $\tau_n(x)$ continuous on $K_n R$.

Proof. The proof is immediate, when we notice that by using Theorem 2.11, one can find a compactly based tube containing a wandering point of X. The rest follows by the assumption of separability of X.

We gave an example above to show that a compactly based tube need not be closed in X. One may wonder if for a wandering dynamical system (X, R, π), a compactly based tube is not closed. Here is a counter-example.

2.16 *Example.* Consider again Example 1.5.3. Any compactly based tube containing a point $p \in \gamma_{-1}$ is not closed, because its closure will contain γ_0 which is not in such a tube. This is an example of a wandering dynamical system which is not dispersive. In the case that (X, R, π) is dispersive one obtains the following:

2.17 Lemma. A compactly based ∞-tube U with section K of a dispersive dynamical system (X, R, π) is closed in X.

Proof. $U = KR$ and if $\{x_n\}$ is a sequence in KR, then there are sequences $\{y_n\}$ in K and $\{\tau_n\}$ in R such that $x_n = y_n \tau_n$. We may assume that $y_n \to y \in K$ as K is compact. If now $x_n \to x$, we claim that the sequence $\{\tau_n\}$ is bounded, so that $x \in yR \subset KR$. For otherwise if $\{\tau_n\}$ contains an unbounded subsequence $\{\tau_{n_k}\}$, say $\tau_{n_k} \to +\infty$, then clearly $x \in J^+(y)$, which is absurd, as $J^+(y) = \emptyset$ for each $y \in X$ by Theorem 1.8. The lemma is proved.

We now prove the last lemma required to prove Theorem 2.6.

2.18 Lemma. Let U_1, U_2 be two compactly based tubes of a dispersive dynamical system with sections K_1, K_2 and continuous functions $\tau_1(x)$ and $\tau_2(x)$ respectively. If $U_1 \cap U_2 \neq \emptyset$ then $U = U_1 \cup U_2$ is a compactly based tube with a section $K \supset K_1$ and a continuous function $\tau(x)$. Moreover, if the time distance between K_1 and K_2 along orbits in $U_1 \cap U_2$ is less than τ (> 0), the time distance between K and K_2 along orbits in U is also less than τ.

Proof. U_1 and U_2 are invariant and closed. Therefore $U_1 \cap U_2$ is invariant and closed. Further, $K_2 \cap U_1$ is compact and non-empty, as shown in Fig. 2.18.1. Set $S_2 = K_2 \cap U_1$ and $S_1 = K_1 \cap U_1$. Any orbit in $U_1 \cap U_2$ intersects K_2 and hence S_2 in exactly one point, and also intersects K_1 and hence S_1 in exactly one point. Thus for any $x \in U_1 \cap U_2$ we have $\tau_1(x) = \tau_2(x) + \tau_1(x\tau_2(x))$. This is so because $x\tau_1(x) = x\tau_2(x)(\tau_1(x\tau_2(x))) = x(\tau_2(x) + \tau_1(x\tau_2(x)))$, and there are no rest points or periodic orbits. The function $\tau_1(x)$ is continuous on S_2 (which is compact), and by Tietze's theorem it can be extended to a continuous function $\tau(x)$ defined on K_2, where $\tau(x) \equiv \tau_1(x)$ for $x \in S_2$. Further if $|\tau_1(x)| < \tau$ for $x \in S_2$, we can have $|\tau(x)| < \tau$ on K_2. Notice now that $\{x\tau(x): x \in S_2\} = S_1$, and $\tau(x)$ being continuous $\{x\tau(x): x \in K_2\}$ is compact as K_2 is compact. We set now $K = K_1 \cup \{x\tau(x): x \in K_2\}$, and define $\tau^*(x)$ on $KR = K_1R \cup K_2R$ as follows. $\tau^*(x) = \tau_1(x)$ for $x \in K_1R$, and $\tau^*(x) = \tau_2(x) + \tau(x\tau_2(x))$ if $x \in K_2R$. $\tau^*(x)$ is continuous on KR and we need only verify that if $x \in U_1 \cap U_2$, then $\tau_1(x) = \tau_2(x) + \tau(x\tau_2(x))$, which has already been proved. The lemma is proved.

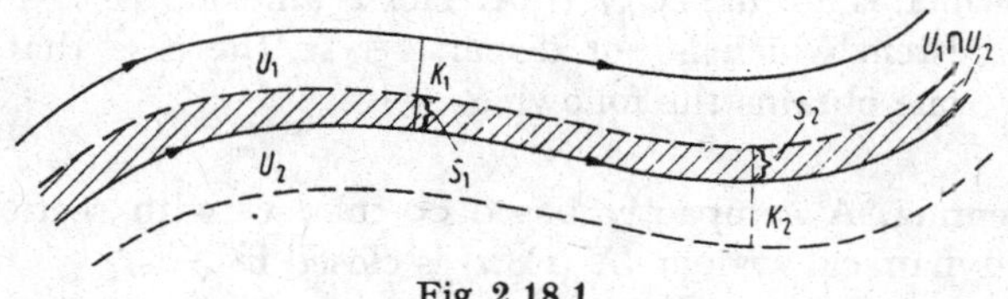

Fig. 2.18.1

2.19 *Proof of Theorem 2.6.* Only the sufficiency part needs proof. It is sufficient to prove that X has a section S with $\tau(x)$ continuous on X. By Theorem 2.15 there is a countable covering $\{U_n\}$ of X by compactly based tubes U_n with sections K_n and continuous functions $\tau_n(x)$. We replace this covering by a like covering $\{U^n\}$ of compactly based tubes which we construct as follows. Set $K_1 = K^1$, and $U_1 = U^1$. Beginning with U^1 and U_2 we use Lemma 2.18 to enlarge K^1 to a compact set K^2, thus obtaining the compactly based tube $U^2 = U^1 \cup U_2$ with $\tau^2(x)$ continuous on U^2. This leaves K^1 unaltered. Having found U^n, we take it together with U_{n+1} and construct similarly U^{n+1} with $K^{n+1} \supset K^n$, and $\tau^{n+1}(x)$ continuous on U^{n+1}. Now set $S = \cup K^n$, then $X = SR$, and the function $\tau(x)$ defined by $\tau(x) = \tau^n(x)$ for $x \in U^n$ is continuous on X, with the property that $x\tau(x) \in S$. Moreover $\tau(x)$ is unique for each $x \in X$. Thus X has a section S with continuous $\tau(x)$ defined on X. The system (X, R, π) is thus parallelizable and the theorem is proved.

2.20 *Exercise.* Show that a parallelizable dynamical system is dispersive.

Notes and References

The concepts in this chapter are all classical.

Section 1. The important concept of dispersiveness may be attributed to WHITNEY [1, 2, 3], but it also appears in NEMYTSKII [10]. All definitions (except 1.2.4) appear in NEMYTSKII and STEPANOV [1]. The classical definition of a wandering point is contained in Exercise 1.4.5. Theorems 1.7, 1.8, 1.10 were proved by BHATIA [3]. Theorem 1.10 was first conjectured by J. AUSLANDER [3].

Section 2. The concept of a parallelizable system is due to WHITNEY [1, 2, 3]. The crucial Theorem 2.6 was obtained by NEMYTSKII and STEPANOV [1] in a different form using the concept of complete instability and Nemytskii's notion of an improper saddle point (or a saddle point at infinity). The present form of the theorem is found in NEMYTSKII [10] who showed that dispersiveness was equivalent to the conjunction of complete instability and absence of improper saddle points. The development in this section follows closely that of ANTOSIEWICZ and DUGUNDJI [1]. Theorems 2.9 and 2.11 are due to BEBUTOV (see NEMYTSKII and STEPANOV [1]).

For further developments on the theory of sections see T. URA [6].

We would like to emphasize that in the discussion of dispersive concepts these distinct points of view together with distinct techniques have been advanced: (i) the absence of recursive motions, (ii) the absence of improper saddle points, and (iii) the theory of sections leading to parallel systems. All these are intimately connected with the theory of separatrices which is still in its infancy (see MARKUS [11]).

Chapter V

Stability Theory

This chapter is devoted to the study of various notions of stability and attraction and the characterization of some of them in terms of the so-called Liapunov functions.

Sections 1, 2, 3 expose the theory for compact sets in locally compact phase spaces. This theory seems to be fairly complete.

Section 4 is devoted to a study of stability and attraction properties of closed sets. Here we need not to restrict to locally compact spaces. However, it seemed to us quite unnecessary to try to document all the stability and attraction phenomena that can occur in this case, so that we have restricted ourselves to only those which seemed to be of most interest.

Section 5 exposes a concept of relative stability that may be useful in several situations. The concepts of stability studied in sections 1—5 may be termed as orbital stability concepts in contrast to the concept of stability of motion and its effect on recurrent motions as studied in section 6.

1. Stability and Attraction for Compact Sets

Throughout this section, the phase space X will be assumed to be locally compact. Some of the results do not require this assumption. Such results will be indicated in remarks or exercises. M will denote a non-empty compact subset of X.

1.1 Definitions. With a given M we associate the sets

1.1.1 $A_\omega(M) = \{x \in X: \Lambda^+(x) \cap M \neq \emptyset\}$,

1.1.2 $A(M) = \{x \in X: \Lambda^+(x) \neq \emptyset \text{ and } \Lambda^+(x) \subset M\}$,

1.1.3 $A_u(M) = \{x \in X: J^+(x) \neq \emptyset \text{ and } J^+(x) \subset M\}$.

The sets $A_\omega(M)$, $A(M)$, $A_u(M)$ are respectively called the *region of weak attraction, attraction, and uniform attraction of the set* M. Moreover, any *point* x in $A_\omega(M)$, $A(M)$, or $A_u(M)$ may respectively be said to be *weakly attracted, attracted, or uniformly attracted to* M.

The motivation for the above definition is the following proposition.

1.2 **Proposition.** Given M, a point x is

1.2.1 weakly attracted to M if and only if there is a sequence $\{t_n\}$ in R with $t_n \to +\infty$ and $\varrho(xt_n, M) \to 0$,

1.2.2 attracted to M if and only if $\varrho(xt, M) \to 0$ as $t \to +\infty$,

1.2.3 uniformly attracted to M if and only if for every neighborhood V of M there is a neighborhood U of x and a $T > 0$ with $Ut \subset V$ for $t \geqq T$.

Proof. 1.2.1 and 1.2.2 easily follow from the definitions. We prove 1.2.3. First assume that for an $x \in X$, $J^+(x) \neq \emptyset$ and $J^+(x) \subset M$. Then $\Lambda^+(x) \neq \emptyset$ and $\Lambda^+(x) \subset M$ (II, 4.5). Assume now that there is a neighborhood $V = S(M, \alpha)$ of M $(\alpha > 0)$ such that for every neighborhood U of x and $T > 0$ there is a $t > T$ with $Ut \not\subset V$. By 1.2.2 there is a $t_0 \geqq 0$ such that $xt \in V$ for $t > t_0$. This shows that there exist sequences $\{t_n\}$, $\{\tau_n\}$ in R, $t_n < \tau_n$, $t_n \to +\infty$ and a sequence $\{x_n\}$ in X, $x_n \to x$, such that $x_n t_n \in S(M, \alpha)$, but $x_n \tau_n \in H(M, \alpha)$. Since $H(M, \alpha)$ may be assumed compact (the space is locally compact) we conclude that there is a sequence $\{x_n\}$, $x_n \to x$, and a sequence $\{\tau_n\}$, $\tau_n \to +\infty$, with $x_n \tau_n \to y \in H(M, \alpha)$. Then $y \in J^+(x)$ but $y \notin M$. This is a contradiction. Now assume the converse requirement that for every neighborhood V of M there is a neighborhood U of x and a $T > 0$ with $Ut \subset V$ for $t \geqq T$. Then indeed $\overline{Ut} \subset \overline{V}$. This shows indeed that $xt \in \overline{V}$ for $t \geqq T$, and since we may take $\overline{V}$ to be compact, we have $\Lambda^+(x) \neq \emptyset$. Consequently $J^+(x) \neq \emptyset$. Clearly $J^+(x) \subset \overline{U[t, +\infty)}$ for every neighborhood U of x and every $t \in R^+$. Hence $J^+(x) \subset \overline{V}$ for every neighborhood V of M. Thus $J^+(x) \subset \cap \{\overline{V}\colon V \text{ is a neighborhood of } M\} = M$. This proves the proposition.

The following theorem expresses an elementary property of the sets $A_\omega(M)$, $A(M)$, and $A_u(M)$.

1.3 **Theorem.** For any given M,

1.3.1 $A_\omega(M) \supset A(M) \supset A_u(M)$,

1.3.2 the sets $A_\omega(M)$, $A(M)$, and $A_u(M)$

are invariant.

Proof. 1.3.1 is quite apparent from Definition 1.1, and 1.3.2 is easily seen from the following lemma.

1.4 Lemma. Let X be any metric space (not necessarily locally compact) and $x \in X$. Then

1.4.1 $\Lambda^+(x) = \Lambda^+(x)\, t = \Lambda^+(xt)$ for every $t \in R$,

1.4.2 $J^+(x) = J^+(x)\, t = J^+(xt)$ for every $t \in R$.

Proof. The first equalities in 1.4.1 and 1.4.2 follow from invariance of the sets $\Lambda^+(x)$ and $J^+(x)$. To see the second equalities, consider, for example, the case $\Lambda^+(x)\, t = \Lambda^+(xt)$. Let $z \in \Lambda^+(x)\, t$. Then there is a $y \in \Lambda^+(x)$ with $z = yt$ and a sequence $\{t_n\}$, $t_n \to +\infty$, with $xt_n \to y$. Then by the continuity axiom $xt_n(t) \to yt$. But $xt_n(t) = xt(t_n)$ and since $t_n \to +\infty$, we must have $yt \in \Lambda^+(xt)$. Thus $\Lambda^+(x)\, t \subset \Lambda^+(xt)$. The argument is clearly reversible, so that also $\Lambda^+(xt) \subset \Lambda^+(x)\, t$. This proves 1.4.1. The proof of the second equality in 1.4.2 is entirely analogous.

We now introduce the fundamental definitions of attraction and stability.

1.5 Definition. A given *set* M is said to be

1.5.1 a *weak attractor* if $A_\omega(M)$ is a neighborhood of M,

1.5.2 an *attractor* if $A(M)$ is a neighborhood of M,

1.5.3 a *uniform attractor* if $A_u(M)$ is a neighborhood of M,

1.5.4 *stable* if every neighborhood U of M contains a positively invariant neighborhood V of M,

1.5.5 *asymptotically stable* if it is stable and is an attractor,

1.5.6 *unstable,* if it is not stable.

1.6 *Remark.* A weak attractor will be called a global weak attractor whenever $A_\omega(M) = X$. Similarly for attractors, uniform attractors, or asymptotically stable sets, the adjective global is used to indicate that the corresponding region of attraction is the whole space.

We will presently show that the concepts introduced have distinct meaning. For this we consider some examples.

1.7 *Examples.*

1.7.1 Consider Example II, 3.3.1. Take any singleton $\{x\}$ with x in the unit circle. Then $\{x\}$ is a weak attractor but it is none of the other kind of objects introduced in 1.5.

1.7.2 Consider Example II, 3.3.4. The set $M = \{A, B\}$ is again a weak attractor but none of the other kinds of sets in 1.5.

1.7.3 In the above example let M be the set consisting of the points A and B and the trajectory γ_1. Then again M is a weak attractor but none of the rest.

1.7.4 Consider a planar dynamical system defined by the differential equations (in polar coordinates)

1.7.5 $$\dot{r} = r(1 - r), \qquad \dot{\theta} = \sin^2 (\theta/2).$$

The trajectories are shown in Fig. 1.7.6. These consist of two rest points $p_1 = (0, 0)$ and $p_2 = (1, 0)$, a trajectory γ on the unit circle with $\{p_2\}$ as the positive and the negative limit set of all points on the unit circle. All points p, $p \neq p_1$, have $\Lambda^+(p) = \{p_2\}$. Thus $\{p_2\}$ is an attractor. But it is neither stable nor a uniform attractor. Note that for any $p = (\alpha, 0)$, $\alpha > 0$, $J^+(p)$ is the unit circle.

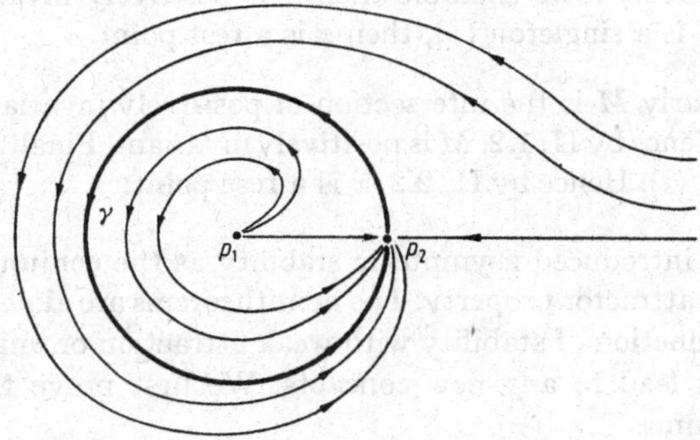

Fig. 1.7.6

1.7.7 In Examples 1.7.1, 1.7.2, and 1.7.4, if we take M to be the unit circle, then M is clearly asymptotically stable and also is a uniform attractor. As an example of a uniform attractor which is not stable we consider in the present examples any set consisting of the unit circle and sny po int p different from the origin, but not on the unit circle.

1.7.8 Fig. 1.7.9 describes an example of a global attractor in the plane which is not asymptotically stable. The point 0 is the only positive limit point of each point in the plane.

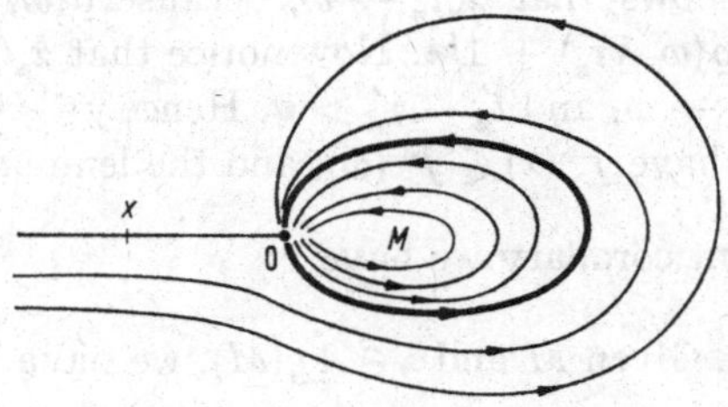

Fig. 1.7.9

We are now ready to prove some theorems about the concepts introduced.

1.8 Theorem. If M is a weak attractor, attractor, or uniform attractor, then the corresponding set $A_\omega(M)$, $A(M)$, or $A_u(M)$ is open (indeed an open neighborhood of M).

Proof. Let N denote any one of the sets $A_\omega(M)$, $A(M)$, or $A_u(M)$. Then N is an invariant neighborhood of M. Consequently ∂N is invariant and disjoint from M and is indeed closed. Note now that for each $x \in N$, $\Lambda^+(x) \cap M \neq \emptyset$, whereas for each $x \in \partial N$, $\Lambda^+(x) \subset \partial N$. Since $\partial N \cap M = \emptyset$ we conclude that $N \cap \partial N = \emptyset$. Thus N is open.

1.9 Theorem. If M is stable then it is positively invariant. Consequently, if M is a singleton $\{x\}$, then x is a rest point.

Proof. Clearly M is the intersection of positively invariant neighborhoods of M. Hence by II, 1.2, M is positively invariant. Finally, if $M = \{x\}$, then $\{x\} = \gamma^+(x)$. Hence by II, 2.2, x is a rest point.

In 1.5 we introduced asymptotic stability as the conjunction of stability and the attractor property. The next theorems are aimed at showing that the conjunction of stability with weak attraction or uniform attraction does not lead to any new concepts. We first prove the following important lemma.

1.10 Lemma. Let X be an arbitrary metric space. Let $x \in X$ and $\omega \in \Lambda^+(x)$. Then $J^+(x) \subset J^+(\omega)$.

Proof. Given $\omega \in \Lambda^+(x)$, and any $y \in J^+(x)$, there exist sequences $\{\tau_n'\}$, $\tau_n' \to +\infty$, $x\tau_n' \to \omega$, and $\{t_n'\}$ and $\{x_n\}$, $x_n \to x$, $t_n' \to +\infty$, $x_n t_n' \to y$. We can assume, if necessary by choosing subsequences, that $t_n' - \tau_n' > n$ for each n. Consider for each fixed k, $k = 1, 2, \ldots$, the sequence $\{x_n \tau_k'\}$. By the continuity axiom $x_n \tau_k' \to x\tau_k'$, $k = 1, 2, \ldots$. We may, therefore, assume without loss of generality that for each fixed k, $\varrho(x_n\tau_k', x\tau_k') \leqq 1/k$ for $n \geqq k$. This shows that $x_n \tau_n' \to \omega$, because $\varrho(\omega, x_n\tau_n') \leqq \varrho(\omega, x\tau_n') + \varrho(x\tau_n', x_n\tau_n') \leqq \varrho(\omega, x\tau_n') + 1/n$. Now notice that $x_n t_n' = x_n\tau_n'(t_n' - \tau_n')$, and $x_n t_n' \to y$, $x_n \tau_n' \to \omega$, and $t_n' - \tau_n' > n$. Hence $y \in J^+(\omega)$. As $y \in J^+(x)$ was arbitrary, we have $J^+(x) \subset J^+(\omega)$, and the lemma is proved.

As an important corollary we have

1.11 Corollary. Given M and $x \in A_\omega(M)$, we have

$$J^+(x) \subset J^+(M) \subset D^+(M).$$

Proof. Indeed for $x \in A_\omega(M)$, $\Lambda^+(x) \cap M \neq \emptyset$. So take any $\omega \in \Lambda^+(x) \cap M$ and apply the above lemma.

We now give an important characterization of stability of a set M.

1.12 Theorem. A set M is stable if and only if

$$D^+(M) = M.$$

Proof. Let $D^+(M) = M$, and suppose if possible that M is not stable. Then there is an $\varepsilon > 0$, a sequence $\{x_n\}$, and a sequence $\{t_n\}$, with $t_n \geqq 0$, $\varrho(x_n, M) \to 0$, and $\varrho(x_n t_n, M) \geqq \varepsilon$. We may assume without loss of generality that $\varepsilon > 0$ has been chosen so small that $S[M, \varepsilon]$ and hence $H(M, \varepsilon)$ is compact (this is possible as X is locally compact). Further, we may assume that $x_n \to x \in M$. We can now choose a sequence $\{\tau_n\}$, $0 \leqq \tau_n \leqq t_n$, such that $x_n \tau_n \in H(M, \varepsilon)$, $n = 1, 2, \ldots$. Since $H(M, \varepsilon)$ is compact, we may assume that $x_n \tau_n \to y \in H(M, \varepsilon)$. Then clearly $y \in D^+(x) \subset D^+(M)$, but $y \notin M$. This contradiction shows that M is stable. Now assume that M is stable. Given any neighborhood V of M there is a positively invariant neighborhood U of M with $U \subset V$. Since for any $x \in M$, $D^+(x) \subset \overline{WR^+}$ for any neighborhood W of x, we get $D^+(x) \subset \overline{U}$ since U is positively invariant. Thus $D^+(M) \subset \overline{V}$ for any neighborhood V of M. Hence $D^+(M) \subset \cap \{\overline{V}\colon V$ is a neighborhood of $M\} = M$ as M is compact. Since $M \subset D^+(M)$ holds always, we have $D^+(M) = M$. This proves the theorem.

In view of Theorem 1.12 we can give the following definition.

1.13 Definition. If a given *set* M is unstable, the non-empty set $D^+(M) - M$ will be called the *region of instability* of M. If $D^+(M)$ is not compact, then M is said to be *globally unstable*.

1.14 Theorem. If M is stable and is a weak attractor, then M is an attractor and consequently asymptotically stable.

Proof. Indeed $A_\omega(M)$ is a neighborhood of M. We need show that $A_\omega(M) \subset A(M)$. Let $x \in A_\omega(M)$. Then $\Lambda^+(x) \neq \emptyset$ and $\Lambda^+(x) \subset J^+(x) \subset D^+(M)$ by Corollary 1.11. Since M is stable we have $D^+(M) = M$ by Theorem 1.12, and consequently $\Lambda^+(x) \subset M$. Thus M is an attractor. This proves the theorem.

1.15 Theorem. Let M be positively invariant and a uniform attractor, then M is stable. Consequently M is asymptotically stable.

Proof. Note that $M \subset A_u(M)$. Thus, in particular, $D^+(M) = MR^+ \cup J^+(M) = M \cup M = M$, as $M = MR^+$ by positive invariance of M, and $J^+(M) \subset M$, as $M \subset A_u(M)$. Hence by Theorem 1.12, M is stable.

1.16 Theorem. If M is asymptotically stable then M is a uniform attractor.

Proof. We need show that $A(M) \subset A_u(M)$. Let $x \in A(M)$. Then $\Lambda^+(x) \neq \emptyset$ and $\Lambda^+(x) \subset M$. Choose any $\omega \in \Lambda^+(x)$ and apply Lemma 1.10 to get $J^+(x) \subset J^+(\omega) \subset J^+(M) \subset D^+(M)$. But $D^+(M) = M$ by stability of M. Hence $J^+(x) \subset M$ for $x \in A(M)$. This proves the theorem.

The various implications that have been proved above are shown in the following diagram.

1.17 Diagram.

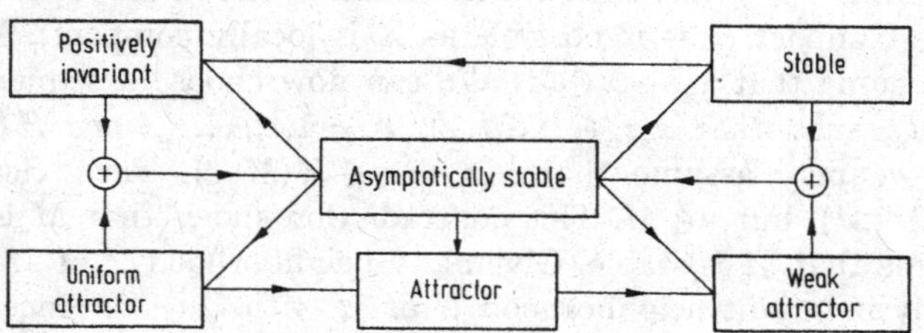

1.18 *Exercises.*

1.18.1 If M is stable, then $J^-(X-M) \cap M = \emptyset$.

1.18.2 If X is locally compact, then a compact set M is stable if and only if $J^-(X-M) \cap M = \emptyset$.

1.18.3 In an arbitrary metric space X let M be a compact subset of X. Define $A_\omega(M)$, $A(M)$, $A_u(M)$ as before. Show that they are invariant sets. Further if M is a weak attractor or attractor then the respective sets $A_\omega(M)$, $A(M)$ are open. However, if M is a uniform attractor, i.e. $A_u(M)$ is a neighborhood of M, then $A_u(M)$ need not be open.

1.18.4 Give examples to show that Theorems 1.12 and 1.15 do not hold in arbitrary metric spaces.

1.18.5 Show that Theorem 1.14 holds in arbitrary metric spaces.

1.18.6 Show that if M is compact, then $D^+(M)$ is closed.

The next few theorems are devoted to the problem of determining whether the components of a stable or asymptotically stable set inherit the same properties. It is very simple to show that in general the properties of weak attraction, attraction, and uniform attraction are not inherited by the components.

1.19 Theorem. A set M is stable if and only if every component of M is stable.

Proof. Note that if M is compact, then every component of M is compact. Further if M is positively invariant, so is every one of its components. Now let $M = \cup \{M_i : i \in \mathscr{I}\}$ where $\mathscr{I}$ is an index set, and M_i are components of M. Let each M_i be stable, i.e., $D^+(M_i) = M_i$. Then $D^+(M) = \cup D^+(M_i) = \cup M_i = M$ and M is stable. To see the converse, let $D^+(M) = M$, i.e., M is stable. Let M_i be a component of M. Then $D^+(M_i)$ is a compact connected set, and $D^+(M_i) \subset M$. Since M_i is a component of M we have $D^+(M_i) \subset M_i$. Clearly then $D^+(M_i) = M_i$ as $M_i \subset D^+(M_i)$ holds always. Thus M_i is stable. This proves the theorem.

1.20 *Remark.* Theorem 1.19 holds even if X is not locally compact (N. P. Bhatia [9]).

1.21 **Theorem.** Let M be asymptotically stable, and let M^* be a component of M. Then M^* is asymptotically stable if and only if it is an isolated component.

Proof. Let M^* be isolated from the other components of M. Then there is a closed neighborhood U of M^* with $U \subset A(M)$ and $U \cap (M - M^*) = \emptyset$. Since M^* is stable by Theorem 1.19, there exists a positively invariant neighborhood V of M^* with $V \subset U$. Now for every $x \in V$, $\Lambda^+(x) \neq \emptyset$ as $x \in A(M)$, and $\Lambda^+(x) \subset \overline{V} \subset U$. But $\Lambda^+(x) \subset M$ and since $U \cap (M - M^*) = \emptyset$ we must have $\Lambda^+(x) \subset M^*$. Thus M^* is an attractor since $V \subset A(M^*)$ and so $A(M^*)$ is a neighborhood of M^*. Conversely, let M^* be asymptotically stable. We claim that $A(M^*) \cap (M - M^*) = \emptyset$, so that M^* is isolated. To see this, note that if M_0 is a component of M other than M^*, then M_0 is positively invariant and compact. Moreover, $M^* \cap M_0 = \emptyset$. Thus if $A(M^*) \cap M_0 \neq \emptyset$, then for an $x \in A(M^*) \cap M_0$, $\Lambda^+(x) \neq \emptyset$ and $\Lambda^+(x) \subset M^*$ as $x \in A(M^*)$ on one hand, and $\Lambda^+(x) \subset M_0$ as $x \in M_0$ and M_0 is closed and positively invariant. This contradicts $M^* \cap M_0 = \emptyset$. The theorem is proved.

As a corollary to the above theorem we can obtain the following very important theorem.

1.22 **Theorem.** Let X be locally compact and locally connected. Let M be asymptotically stable. Then M has a finite number of components, each of which is asymptotically stable.

Proof. $A(M)$ is open (and invariant). Since X is locally connected each component of $A(M)$ is open. The components of $A(M)$ are therefore an open cover for the compact set M. Hence a finite number of components of $A(M)$ cover M. It is now easily seen that each component of $A(M)$ contains precisely one component of M.

1.23 *Exercise.* Give a detailed proof of Theorem 1.22.

1.24 *Remark.* Theorem 1.22 is not true without local connectedness of X. Consider the set X consisting of the points $0, 1, 1/2, \ldots, 1/n, \ldots$ on the real line. X is a compact metric space with the usual euclidean distance between real numbers. Consider any dynamical system on X. Then one sees easily that each point of X is a rest point. The compact set X is asymptotically stable and each point of X is a component of x. However, the point 0 is not asymptotically stable. However, local compactness is not needed as shown by N. P. BHATIA [9].

As a final result in this section we prove an important property of weak attractors.

1.25 **Theorem.** Let M be a compact weak attractor. Then $D^+(M)$ is a compact asymptotically stable set, with $A(D^+(M)) \equiv A_\omega(M)$. Moreover, $D^+(M)$ is the smallest asymptotically stable set containing M.

For the proof we shall need the following lemma.

1.26 **Lemma.** Let M be a compact weak attractor and let $\alpha > 0$. Then there is $T > 0$ such that

$$D^+(M) \subset S[M, \alpha]\,[0, T] \equiv \pi(S[M, \alpha], [0, T]),$$

where π is the map defining the dynamical system.

Proof. Choose ε, $0 < \varepsilon \leqq \alpha$, such that $S[M, \varepsilon]$ is a compact subset of $A_\omega(M)$. For $x \in H(M, \varepsilon)$, define $\tau_x = \inf\{t > 0\colon\ xt \in S(M, \varepsilon)\}$; since $x \in A_\omega(M)$, τ_x is defined. Set $T = \sup\{\tau_x\colon\ x \in H(M, \varepsilon)\}$. We claim that $T < +\infty$. If this is not the case, there is a sequence $\{x_n\}$ in $H(M, \varepsilon)$ for which $\tau_{x_n} \to +\infty$. We may assume that $x_n \to x \in H(M, \varepsilon)$. Let $\tau > 0$ such that $x\tau \in S(M, \varepsilon)$. For sufficiently large n, we then have $\tau_{x_n} < \tau$, which contradicts $\tau_{x_n} \to +\infty$. Now let $y \in D^+(M) - S[M, \varepsilon]$. Then there are sequences $\{x_n\}$, $\{t_n\}$ with $x_n \to x \in M$, and $t_n \geqq 0$ such that $x_n t_n \to y$. Then for all sufficiently large n there is a τ_n, $0 < \tau_n < t_n$, such that $x_n \tau_n \in H(M, \varepsilon)$, and $xt \notin S[M, \varepsilon]$ for $\tau_n < t \leqq t_n$. By the first part of this proof $0 < t_n - \tau_n < T$. Then $x_n t_n = x_n \tau_n (t_n - \tau_n) \in S[M, \varepsilon]\,[0, T]$. Therefore, $y \in S[M, \varepsilon]\,[0, T]$, since this set is closed. The lemma is proved.

1.27 *Proof of Theorem 1.25.* Notice that if $S[M, \varepsilon]$ is compact, then $S[M, \varepsilon]\,[0, T]$ is compact for any $T > 0$. Thus $D^+(M)$, being a closed subset of the compact set $S[M, \varepsilon]\,[0, T]$ (by the above lemma), is compact. Further, as $\varepsilon > 0$ is chosen, such that $S[M, \varepsilon] \subset A_\omega(M)$ we have $D^+(M) \subset S[M, \varepsilon]\,[0, T] \subset A_\omega(M)$. Thus $A_\omega(M)$ is an open invariant set containing $D^+(M)$, and is, therefore, a neighborhood of $D^+(M)$. By Corollary 1.11 $x \in A_\omega(M)$ implies $\Lambda^+(x) \neq \emptyset$, and $\Lambda^+(x) \subset D^+(M)$. Therefore, $D^+(M)$ is an attractor. Notice that $A_\omega(M) = A(D^+(M))$, for if there is an $x \in A(D^+(M))$, then there is a $t > 0$ such that $xt \in A_\omega(M)$ (this being a neighborhood of $D^+(M)$), and since $\Lambda^+(x) \equiv \Lambda^+(xt)$, we have $x \in A_\omega(M)$. To show that $D^+(M)$ is stable, let $x \in D^+(M)$. Since $x \in A_\omega(M)$, we can choose an $\omega \in \Lambda^+(x) \cap M$. Then $J^+(x) \subset D^+(\omega) \subset D^+(M)$ by Lemma 1.10. Thus

$$D^+(x) = \gamma^+(x) \cup J^+(x) \subset D^+(M) \cup D^+(M) = D^+(M),$$

for $D^+(M)$ is positively invariant. This shows that $D^+(D^+(M)) = D^+(M)$, i.e., $D^+(M)$ is stable. We have thus proved that $D^+(M)$ is asymptotically stable. Finally, let M^* be any compact set such that $M \subset M^* \subset D^+(M)$. Then $D^+(M) \subset D^+(M^*) \subset D^+(D^+(M)) = D^+(M)$, and so $D^+(M^*) = D^+(M)$. If M^* is stable, then $M^* = D^+(M^*) = D^+(M)$. Thus $D^+(M)$ is the smallest stable (also asymptotically stable) set containing M. The theorem is proved.

1.28 *Exercises.*

1.28.1 Show that if a singleton $\{x\}$ is a weak attractor, then $D^+(x)$ is invariant. Show also that for every $y \in D^+(x)$, $x \in \Lambda^-(y)$ holds.

1.28.2 If M is invariant and a weak attractor then for every $x \in D^+(M)$, $\Lambda^-(x) \cap M \neq \emptyset$. Give an example to show that if M is not invariant, then the result may not hold.

1.28.3 In Examples 1.7 calculate $D^+(M)$ of the weak attractors and attractors and verify that it is an asymptotically stable compact set.

1.29 *Remark.* All the concepts introduced in this section were introduced using the so-called positive concepts from Chapter II, i.e., using positive semi-trajectories, positive limit sets, and positive prolongations. Thus let us for the moment identify these concepts by using the adjective "positive" before the terms introduced already, i.e., positively stable, positive weak attractor, etc. instead of stable, or weak attractor. It is clear that by using the dual concepts of a negative semi-trajectory, negative limit set, negative prolongation, and negative prolongational limit set, one can define in an obvious fashion the concepts negatively stable, negative weak attractor, negatively asymptotically stable, etc. In the future, whenever we need such a concept we will explicitly write it with the adjective "negative".

1.30 *Exercise.*

1.30.1 Let M be minimal and asymptotically stable. Then for every $x \in A(M)$, $\overline{\gamma^+(x)}$ is asymptotically stable with $A(\overline{\gamma^+(x)}) = A(M)$.

1.30.2 Show that in a locally compact space X, a compact set M is a uniform attractor if and only if it is a weak attractor and for every compact set U in $A_\omega(M)$ and every neighborhood V of M there is a $T > 0$ with $Ut \subset V$ for $t \geqq T$.

2. Liapunov Functions: Characterization of Asymptotic Stability

Throughout this section the space X is assumed locally compact unless otherwise specified.

The basic feature of the stability theory à la Liapunov is that one seeks to characterize stability and asymptotic stability of a given set in terms of a non-negative scalar function defined on a neighborhood of the given set and decreasing along its trajectories. It is in general not possible to characterize stability and the various attractor properties by means of continuous functions. However, in the case of asymptotic stability one can give very strong theorems. These we describe in this section.

We need the following lemma.

2.1 Lemma. Let the phase space X be arbitrary and $K \subset X$. Let φ be any continuous real-valued function defined on K such that $\varphi(xt) \leqq \varphi(x)$ whenever $x[0,t] \subset K$, $t \geqq 0$. Then if for some x, $\overline{\gamma^+(x)} \subset K$, we have $\varphi(y) = \varphi(z)$ for every $y, z \in \Lambda^+(x)$.

Proof. Assume $\varphi(y) < \varphi(z)$. There are indeed sequences $\{t_n\}$ and $\{\tau_n\}$ in R such that $t_n \to +\infty$, $\tau_n \to +\infty$, and $xt_n \to y$, $x\tau_n \to z$. We may assume by taking a subsequence that $\tau_n > t_n$ for each n. Then clearly $\varphi(xt_n) \geqq \varphi(x\tau_n)$ as $x\tau_n = xt_n(\tau_n - t_n)$, $\tau_n - t_n > 0$, and $x_n t_n[0, \tau_n - t_n] \subset K$. Thus proceeding to the limit we have by continuity of φ, $\varphi(y) \geqq \varphi(z)$. This contradicts the original assumption and the lemma is proved.

The best known result on asymptotic stability is

2.2 Theorem. A compact set $M \subset X$ is asymptotically stable if and only if there exists a continuous real-valued function Φ defined on a neighborhood N of M such that

2.2.1 $\Phi(x) = 0$ if $x \in M$ and $\Phi(x) > 0$ if $x \notin M$;

2.2.2 $\Phi(xt) < \Phi(x)$ for $x \notin M$, $t > 0$ and $x[0,t] \subset N$.

Proof. Assume that a function Φ as required is given. Choose $\alpha > 0$ such that $S[M,\alpha] \subset N$ and is compact. Let $m = \min\{\Phi(x) : x \in H(M,\alpha)\}$. Then by 2.2.1 and continuity of Φ, $m > 0$. Set $K = \{x \in S[M,\alpha] : \Phi(x) \leqq m\}$. Then K is indeed compact and because of 2.2.2 K is positively invariant. This establishes that M is stable as K is a positively invariant neighborhood of M. To see that M is an attractor, choose any compact positively invariant neighborhood K of M with $K \subset N$. Then for any $x \in K$, $\emptyset \neq \Lambda^+(x) \subset K$, and Lemma 2.1 shows that Φ is constant on $\Lambda^+(x)$. But this shows using 2.2.2 that $\Lambda^+(x) \subset M$. Thus M is an attractor and consequently asymptotically stable. Now let M be asymptotically stable and $A(M)$ its region of attraction. For each $x \in A(M)$ define

$$\varphi(x) = \sup\{\varrho(xt, M) \colon t \geqq 0\}.$$

Indeed $\varphi(x)$ is defined for each $x \in A(M)$, because if $\varrho(x, M) = \alpha$, then there is a $T > 0$ with $x[T, +\infty) \subset S(M, \alpha)$. Thus

$$\varphi(x) \equiv \sup\{\varrho(xt, M) \colon 0 \leqq t \leqq T\}.$$

As $\varrho(xt, M)$ is a continuous function of t, $\varphi(x)$ is defined. This $\varphi(x)$ has the properties: $\varphi(x) = 0$ for $x \in M$, $\varphi(x) > 0$ for $x \notin M$, and $\varphi(xt) \leqq \varphi(x)$ for $t \geqq 0$. This is clear when we remember that M is stable and hence positively invariant and that $A(M)$ is invariant. Thus if $\varphi(x)$ is defined for any $x \in A(M)$, it is defined for all xt with $t \in R^+$. We further claim that this $\varphi(x)$ is continuous in $A(M)$. Indeed stability of M implies continuity of $\varphi(x)$ on M. For $x \notin M$ we can prove the continuity of $\varphi(x)$ as follows. For $x \notin M$, set $\varrho(x, M) = \alpha$ (> 0) and choose ε, $0 < \varepsilon < \alpha/4$, such that $S[x, \varepsilon]$ is a compact subset of $A(M)$; this is possible as X is locally compact and $A(M)$ is open. Since M is a uniform attractor, there is a $T > 0$ such that $S[x, \varepsilon]\, t \subset S(M, \alpha/4)$ for all $t \geqq T$ (see 1.30.2). Thus for $y \in S[x, \varepsilon]$ we have

$$\begin{aligned} \varphi(x) - \varphi(y) &= \sup\{\varrho(xt, M) \colon t \geqq 0\} - \sup\{\varrho(yt, M) \colon t \geqq 0\} \\ &= \sup\{\varrho(xt, M) \colon 0 \leqq t \leqq T\} - \sup\{\varrho(yt, M) \colon 0 \leqq t \leqq T\}. \end{aligned}$$

Therefore

$$\begin{aligned} |\varphi(x) - \varphi(y)| &\leqq \sup\{|\varrho(xt, M) - \varrho(yt, M)| \colon 0 \leqq t \leqq T\} \\ &\leqq \sup\{\varrho(xt, yt) \colon 0 \leqq t \leqq T\}. \end{aligned}$$

The continuity axiom implies that the right hand side of the above inequality tends to zero as $y \to x$, for T is fixed for $y \in S[x, \varepsilon]$. The function $\varphi(x)$ is therefore continuous in $A(M)$. However the above function may not be strictly decreasing along parts of trajectories in $A(M)$ which are not in M and so may not satisfy 2.2.2. Such a function can be obtained by setting

$$\Phi(x) = \int_0^\infty \varphi(x\tau) \exp(-\tau)\, d\tau.$$

That $\Phi(x)$ is continuous and satisfies 2.2.1 in $A(M)$ is clear. To see that $\Phi(x)$ satisfies 2.2.2, let $x \notin M$ and $t > 0$. Then indeed $\Phi(xt) \leqq \Phi(x)$

holds, because $\varphi(xt) \leqq \varphi(x)$ holds. To rule out $\Phi(xt) = \Phi(x)$, observe that in this case we must have $\varphi(x(t+\tau)) \equiv \varphi(x\tau)$ for all $\tau \geqq 0$. Thus in particular, letting $\tau = 0, t, 2t, \ldots$ we get $\varphi(x) = \varphi(x(nt))$, $n = 1, 2, 3, \ldots$. But asymptotic stability of M implies that for $x \in A(M)$, $\varrho(xt, M) \to 0$ as $t \to \infty$. Thus $\varphi(x(nt)) \to 0$ as $n \to \infty$, as $\varphi(x)$ is continuous. This shows that $\varphi(x) = 0$. But as $x \notin M$, we must have $\varphi(x) > 0$, a contradiction. We have thus proved that $\Phi(xt) < \Phi(x)$ for $x \notin M$ and $t > 0$. The theorem is proved.

2.3 *Remark.* Theorem 2.2 says nothing about the size of the region of attraction of M. Thus if a function $\Phi(x)$ as in Theorem 2.2 is known to exist in a neighborhood N of M, we need not have either $N \subset A(M)$ or $A(M) \subset N$. We will give an example to elucidate this point. In particular this means that the above theorem *cannot* immediately be stated as a theorem on global asymptotic stability.

2.4 *Example.* Consider a dynamical system defined in the real euclidean plane by the differential equations

2.4.1 $$\dot{x} = f(x, y), \qquad \dot{y} = g(x, y),$$

where

2.4.2 $$g(x, y) = -y \text{ for all } (x, y), \text{ and } f(x, y) = \begin{cases} x \text{ if } x^2y^2 \geqq 1 \\ 2x^3y^2 - x \text{ if } x^2y^2 < 1. \end{cases}$$

These equations are integrable by elementary means and the phase portrait is as in Fig. 2.4.3.

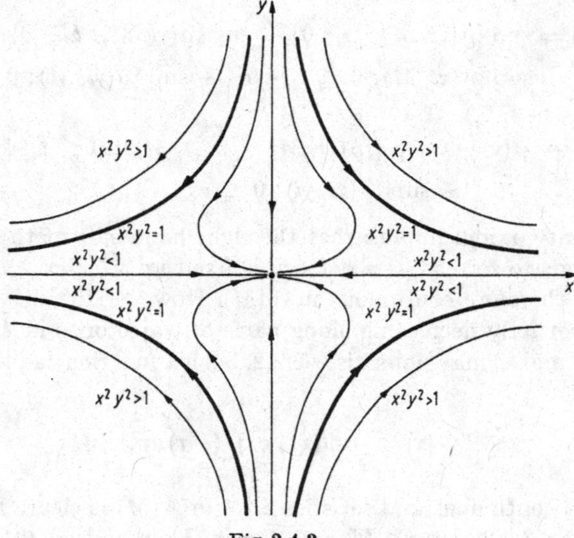

Fig. 2.4.3

The origin (0, 0) is asymptotically stable, with the set $\{(x, y): x^2y^2 < 1\}$ as its region of attraction. Consider now the function

2.4.4 $$\Phi(x,y) = y^2 + \frac{x^2}{1+x^2}.$$

This function satisfies conditions of Theorem 2.2 in the whole euclidean plane. To see this one may find the derivative $\dot{\Phi}(x, y) = \frac{\partial \Phi}{\partial x} f(x, y) + \frac{\partial \Phi}{\partial y} g(x, y)$ of Φ for the given system as is standard practice for differential equations. It can easily be verified that $\dot{\Phi}(x, y) < 0$ for all $(x, y) \neq (0, 0)$, which implies that this Φ satisfies conditions of Theorem 2.2 in every neighborhood of the origin. But not every neighborhood of the origin is contained in the region of attraction, nor does it contain the region of attraction.

Since Theorem 2.2 guarantees asymptotic stability of M, it is clear that $A(M) \cap N$ is a neighborhood of M. How far the extent of this neighborhood can be determined by a function $\Phi(x)$, as in Theorem 2.2, is the subject of the following proposition.

2.5 Theorem. Let $\Phi(x)$ be a function defined in an open neighborhood N of M and having properties as in Theorem 2.2. Then if $K \subset N$ is any compact positively invariant set, we have $K \subset A(M)$. Further for all sufficiently small $\alpha > 0$ the set $K_\alpha = \overline{\{x \in N: \Phi(x) \leqq \alpha\}}$ has a compact positively invariant subset P_α such that $P_\alpha \cap \overline{K_\alpha - P_\alpha} = \emptyset$, P_α is a neighborhood of M, and $P_\alpha \subset A(M)$. In particular if K_α is compact, then $K_\alpha \subset A(M)$.

Proof. The first part follows from Lemma 2.1. To prove the second part, let $\varepsilon > 0$ be chosen such that $S[M, \varepsilon]$ is a compact subset of $\mathscr{I}(N)$. Set $m(\varepsilon) = \min\{\Phi(x): x \in H(M, \varepsilon)\}$. Clearly $m(\varepsilon) > 0$. Now let $0 < \alpha < m(\varepsilon)$, and set $P_\alpha = K_\alpha \cap S[M, \varepsilon]$. We claim that P_α is a set predicted in the lemma. Clearly $P_\alpha \neq \emptyset$ as $P_\alpha \supset M$. P_α is compact, for if $\{x_n\}$ is a sequence in P_α, then we may assume that it is convergent, because $\{x_n\}$ is in the compact set $S[M, \varepsilon]$. If now $x_n \to x$, then clearly $x \in S[M, \varepsilon]$ on one hand, and $\Phi(x) \leqq \alpha$, because $\Phi(x_n) \leqq \alpha$ for each n, showing that $x \in K_\alpha$. It follows that $x \in (K_\alpha \cap S[M, \varepsilon]) = P_\alpha$. That P_α is positively invariant may be seen by first observing that if $x \in P_\alpha$, and $xR^+ \not\subset S(M, \varepsilon)$, then there is a $t > 0$ such that $xt \in H(M, \varepsilon)$ and $x[0, t] \subset S[M, \varepsilon] \subset N$. By 2.2.2 we get $\Phi(xt) \leqq \Phi(x) \leqq \alpha$, and $xt \in H(M, \varepsilon)$ implies $\Phi(xt) \geqq m(\varepsilon) > \alpha$, a contradiction. Hence $xR^+ \subset S(M, \varepsilon) \subset N$. It follows that $\Phi(xt) \leqq \Phi(x) \leqq \alpha$ for each $t \geqq 0$, which shows that $xR^+ \subset K_\alpha$. Thus $xR^+ \subset K_\alpha \cap S(M, \varepsilon) \subset P_\alpha$. That is, P_α is positively invariant. By the first part of the theorem we have $P_\alpha \subset A(M)$. That P_α is a neighborhood of M follows from the continuity of Φ and the fact that $\Phi(x) = 0$ for $x \in M$. Finally to see that $P_\alpha \cap \overline{K_\alpha - P_\alpha} = \emptyset$, observe that $\bar{K}_\alpha \cap H(M, \varepsilon) = \emptyset$. The last part follows by observing that if K_α is compact, then it is positively invariant.

2.6 *Remark.* For any given $\varepsilon, \alpha > 0$ chosen as above, the set P_α defined in the above proof is the largest, compact positively invariant set in K_α. For if $S \subset K_\alpha$ were a larger set, then $S - P_\alpha \subset \mathscr{C}(S[M, \varepsilon])$, so that $\overline{S - P_\alpha} \cap P_\alpha = \emptyset$. It follows that $S - P_\alpha$ is compact and positively invariant. This is impossible unless $S - P_\alpha = \emptyset$. For, if $S - P_\alpha$ were a non-empty subset of N, by the above theorem, $S - P_\alpha \subset A(M)$. Therefore, $(S - P_\alpha) R^+ \cap P_\alpha \neq \emptyset$ as P_α is a neighborhood of M. But $(S - P_\alpha) R^+ = S - P_\alpha$, and $(S - P_\alpha) \cap P_\alpha = \emptyset$.

2.7 **Theorem.** Let $M \subset X$ be a compact asymptotically stable set. Then its region of attraction $A(M)$ contains a countable dense subset. In particular, if $A(M) = X$, then X is separable.

Proof. Take any $\Phi(x)$ defined in a neighborhood N of M and satisfying conditions of Theorem 2.2. Choose $\varepsilon, \alpha > 0$ as in the proof of Theorem 2.5 and construct P_α as before. As P_α is a compact positively invariant neighborhood of M, and $P_\alpha \subset A(M)$,

$$A(M) = P_\alpha R = P_\alpha R^+ \cup P_\alpha R^- = P_\alpha \cup P_\alpha R^- = P_\alpha R^-.$$

Thus $A(M) = \cup \{P_\alpha[-n, 0]: n = 1, 2, \ldots\} \equiv P_\alpha R^-$.

Since each $P_\alpha[-n, 0]$ is compact and thus contains a countable dense subset, the result follows.

The above proposition motivated the introduction of the following definition.

2.8 **Definition.** A *continuous scalar function* $\Phi(x)$ defined on a set $N \subset X$ will be said to be *uniformly unbounded* on N if given any $\alpha > 0$ there is a compact set $K \subset N$, $K \neq N$, such that $\Phi(x) \geqq \alpha$ for $x \notin K$.

2.9 **Theorem.** Let $M \subset X$ be a compact asymptotically stable set and let M be invariant. Then there exists a continuous uniformly unbounded function $\Phi(x)$ defined on $A(M)$ such that

2.9.1 $\Phi(x) = 0$ for $x \in M$, and $\Phi(x) > 0$ for $x \notin M$,

2.9.2 $\Phi(xt) = e^{-t}\Phi(x)$ for all $x \in A(M)$ and $t \in R$.

Proof. Consider any function $\varphi(x)$ defined in any neighborhood N of M and satisfying conditions of Theorem 2.2. Choose $\varepsilon, \alpha > 0$ as in Theorem 2.5, and consider the set P_α. Note that $\partial P_\alpha = \{x \in S[M, \varepsilon] : \varphi(x) = \alpha\}$. We claim that for every point $x \in A(M) - M$, there is a unique $\tau(x) \in R$ such that $x\tau(x) \in \partial P_\alpha$. The uniqueness of $\tau(x)$ follows from the fact that if $x\tau(x) \in \partial P_\alpha$, then $\varphi(x(\tau(x) + t)) < \varphi(x\tau(x)) = \alpha$ for $t > 0$. As P_α is positively invariant and $\varphi(y) = \alpha$ for $y \in \partial P_\alpha$ we conclude that $x(\tau(x) + t)$ is in the interior of P_α for all $t > 0$. This establishes uniqueness of $\tau(x)$ for all x for which $\tau(x)$ is defined. That $\tau(x)$ is defined for all $x \in A(M) - P_\alpha$ follows from the fact that P_α is a neighborhood of M, so that any trajectory $\gamma(x) \subset A(M)$ with $x \notin P_\alpha$, must intersect ∂P_α. For $x \in \partial P_\alpha$ indeed $\tau(x)$ is defined and $\tau(x) = 0$. For $x \in \mathscr{I}(P_\alpha) - M$, if there is no $\tau(x)$ such that $x\tau(x) \in \partial P_\alpha$, then $\gamma(x) \subset \mathscr{I}(P_\alpha) - M \subset P_\alpha$. P_α being compact $\Lambda^-(x) \neq \emptyset$, $\Lambda^-(x) \subset P_\alpha$, but $\Lambda^-(x) \cap M = \emptyset$ (otherwise M will be unstable). Now $\Lambda^-(x)$ is compact and invariant, so that if $y \in \Lambda^-(x)$ we have $\Lambda^+(y) \neq \emptyset$, and $\Lambda^+(y) \subset \Lambda^-(x)$. Then on one hand we have $\Lambda^+(y) \cap M = \emptyset$ and on the other hand $\Lambda^+(y) \subset M$ as $y \in A(M)$. This contradicts $\Lambda^+(y) \neq \emptyset$. Thus $\gamma(x) \cap \partial P_\alpha \neq \emptyset$, and $\tau(x)$ is defined for each $x \in A(M) - M$. Note that $A(M) - M$ is invariant, as M and $A(M)$ are both invariant. For $x \in A(M) - M$ and $\tau \in R$ observe now that

$$\tau(xt) \equiv \tau(x) - t.$$

This follows from the fact that any trajectory $\gamma(x)$ in $A(M) - M$ intersects ∂P_α at exactly one point. Thus $xt(\tau(xt)) = x(\tau(x))$, i.e., by the homeomorphism axiom

2.9.3 $\qquad x(t + \tau(xt)) = x(\tau(x))$.

As $\gamma(x)$ can neither be periodic nor a rest point, we have $t + \tau(xt) = \tau(x)$. This shows further that $\tau(xt)$ is a continuous function of t and $\tau(xt) \to \pm\infty$ as $t \to \mp\infty$. We now claim that $\tau(x)$ is continuous on $A(M) - M$. For any $x \in A(M) - M$, and $\varepsilon > 0$, the point $y \equiv x(\tau(x) + \varepsilon) \in \mathscr{I}(P_\alpha)$. There is therefore a neighborhood N_y of y such that $N_y \subset P_\alpha$. Then $N^+ = N_y(-\tau(x) - \varepsilon)$ is a neighborhood of x and note that for each $\omega \in N^+$, $\tau(\omega) \leqq \tau(x) + \varepsilon$. Again $z \equiv x(\tau(x) - \varepsilon) \in A(M) - P_\alpha$, the last set being open. Thus there is a neighborhood N_z of z such that $N_z \subset (A(M) - P_\alpha)$. Then $N^- = N_z(-\tau(x) + \varepsilon)$ is a neighborhood of x and note that for each $\omega \in N^-$ we have $\tau(\omega) \geqq \tau(x) - \varepsilon$. Thus if ω is in the neighborhood $N_\varepsilon = N^+ \cap N^-$ of x, we have

$$\tau(x) - \varepsilon \leqq \tau(\omega) \leqq \tau(x) + \varepsilon.$$

This proves continuity of $\tau(x)$ in $A(M) - M$. We now show that $\tau(x) \to -\infty$ as $x \to M$, $x \in A(M) - M$. If this were not true, then there will be a $T > 0$ and a sequence $\{x_n\}$ in $A(M) - M$, such that $x_n \to x \in M$ and v$-T \leqq \tau(x_n) \leqq 0$. Since $\{\tau(x_n)\}$ is a bounded sequence it contains a convergent subsequence. We may therefore assume that $\tau(x_n) \to \tau$, where

$-T \leqq \tau \leqq 0$. Then by the continuity axiom $x_n \tau(x_n) \to x\tau$. As M is invariant $x\tau \in M$, on the other hand $x_n \tau(x_n) \in \partial P_\alpha$ which is compact. Therefore $x\tau \in \partial P_\alpha$. But $\partial P_\alpha \cap M = \emptyset$, a contradiction. We now define the function $\Phi(x)$ on $A(M)$ as follows

$$\Phi(x) = 0 \quad \text{for} \quad x \in M,$$

and

$$\Phi(x) = e^{\tau(x)} \quad \text{for} \quad x \in A(M) - M.$$

The above observations show that this function is continuous on $A(M)$. It is clearly positive for $x \notin M$, and

2.9.4 $\Phi(xt) = e^{\tau(xt)} = e^{\tau(x)-t} = \Phi(x)\, e^{-t}$.

Lastly to see that this $\Phi(x)$ is uniformly unbounded, recall that $A(M) = \bigcup \{P_\alpha[-n, 0]: n = 1, 2, 3, \ldots\}$. Each $P_\alpha[-n, 0]$ is compact and positively invariant. Observe that if $x \notin P_\alpha[-n, 0]$, then $\tau(x) > n$, so that $\Phi(x) > e^n$. This proves the theorem completely.

If M is not invariant, then we have:

2.10 Theorem. If $M \subset X$ is any compact asymptotically stable set, there exists a continuous, uniformly unbounded function $\Phi(x)$ on $A(M)$ such that

2.10.1 $\Phi(x) = 0$ for $x \in M$ and $\Phi(x) > 0$ for $x \notin M$,

2.10.2 $\Phi(xt) < \Phi(x)$ for $x \notin M$ and $t > 0$.

Proof. Consider any function $\varphi(x)$ defined in a neighborhood N of M and satisfying conditions of Theorem 2.2. Choose ε, $\alpha > 0$ as before and consider P_α. For each $x \in A(M) - P_\alpha$ define $\tau(x)$ as before. This $\tau(x)$ is continuous and $\tau(x) \to 0$ as $x \to P_\alpha$, $x \notin P_\alpha$. Now define

$$\Phi(x) \equiv \varphi(x) \quad \text{for} \quad x \in P_\alpha,$$

and

$$\Phi(x) = \alpha e^{\tau(x)} \quad \text{for} \quad x \in A(M) - P_\alpha.$$

This $\Phi(x)$ has the desired properties as may easily be verified.

2.11 Theorem. Let $M \subset X$ be compact and let there exist a continuous uniformly unbounded function $\Phi(x)$ defined on an open neighborhood N of M such that

2.11.1 $\Phi(x) = 0$ for $x \in M$ and $\Phi(x) > 0$ for $x \notin M$,

2.11.2 $\Phi(xt) < \Phi(x)$ for $x \notin M$, $t > 0$ and $x[0, t] \subset N$.

Then M is asymptotically stable and $N \subset A(M)$. If, in addition, any condition guaranteeing the invariance of N holds, then $N = A(M)$.

Proof. The proof follows from the observation that if $\Phi(x)$ is uniformly unbounded on N, then for any $\alpha > 0$, the set $K_\alpha = \{x: \Phi(x) \leqq \alpha\}$ is compact. Then by Theorem 2.5, $K_\alpha \subset A(M)$. Now $N = \cup \{K_n: n = 1, 2, \ldots\}$, and since each $K_n \subset A(M)$ we have $N \subset A(M)$. Lastly if N is invariant we must have $N = A(M)$ as N is a neighborhood of M. The remaining details of the proof will be the same as in the proof of sufficiency of Theorem 2.2. These we leave to the reader.

For global asymptotic stability we can state the following two theorems as corollaries of the above results.

2.12 Theorem. A compact invariant set $M \subset X$ is globally asymptotically stable if and only if there exists a continuous uniformly unbounded function $\Phi(x)$ defined on X such that

2.12.1 $\Phi(x) = 0$ for $x \in M$, $\Phi(x) > 0$ for $x \notin M$,

2.12.2 $\Phi(xt) = e^{-t}\Phi(x)$.

Proof. The sufficiency follows from Theorem 2.11 as X is an invariant neighborhood of M. The necessity follows from Theorem 2.9.

2.13 Theorem. A compact set $M \subset X$ is globally asymptotically stable if and only if there exists a continuous uniformly unbounded function $\Phi(x)$ defined on X such that

2.13.1 $\Phi(x) = 0$ for $x \in M$, $\Phi(x) > 0$ for $x \notin M$,

2.13.2 $\Phi(xt) < \Phi(x)$ for $x \notin M$ and $t > 0$.

Proof. The sufficiency follows from Theorem 2.11, the necessity from Theorem 2.10.

2.14 *Remark*. In dynamical systems defined in locally compact metric spaces, one may define ultimate boundedness of the dynamical system by the property that there is a compact set $K \subset X$ with $\Lambda^+(x) \neq \emptyset$, and $\Lambda^+(x) \subset K$ for each $x \in X$, i.e., whenever there exists a compact global attractor in X. It is shown in Theorem 1.25 that if $K \subset X$ is a compact weak attractor, then $D^+(K)$ (the first positive prolongation of K) is a compact positively invariant set which is asymptotically stable and has the same region of attraction as K. One can show now that the largest invariant set in $D^+(K)$ is compact and asymptotically stable with the same region of attraction as of K. These observations will allow one to write theorems on ultimate boundedness which are similar to those on global asymptotic stability. We leave these to the reader.

As a final result on asymptotic stability we wish to prove the following theorem whose proof illustrates yet another technique of proving the existence of special kinds of Liapunov functions.

2.15 Theorem. If $M \subset X$ is a compact asymptotically stable set with the region of attraction $A(M)$, then for any $\beta > 0$ there exists a continuous uniformly unbounded function $\Phi(x)$ on $A(M)$ having the following properties.

2.15.1 $\Phi(x) = 0$ for $x \in M$ and $\Phi(x) > 0$ for $x \notin M$,

2.15.2 $\Phi(xt) \leqq e^{-\beta t}\Phi(x)$ for $x \in A(M)$ and $t \geqq 0$.

Proof. Consider any compact positively invariant neighborhood P of M, $P \subset A(M)$. Then there is an integer n_0 such that $S[M, 1/n] \subset P$ for $n \geqq n_0$. For each $n \geqq n_0$ we construct a function φ_n: $P \to R^+$ as follows

2.15.3 $\varphi_n(x) = \sup\{\varrho(xt, S[M, 1/n])\, e^{\beta t} \colon t \geqq 0\}$.

We claim that this function is well defined. In fact, for each $n \geqq n_0$, there is a $T_n > 0$ such that $xt \in S[M, 1/n]$ for all $x \in P$ and $t \geqq T_n$. For otherwise for some n^* there will be a sequence x_n in P and a sequence $t_n \to +\infty$, such that $x_n t_n \in P - S[M, 1/n^*]$. We may assume that $x_n \to x \in P$, and $x_n t_n \to y \in P - S(M, 1/n^*)$. Then $y \in J^+(x)$, $x \in A(M)$, but $y \notin M$. This contradicts uniform attraction of M (Theorem 1.16). Thus in fact for each $n \geqq n_0$

2.15.4 $\varphi_n(x) = \sup\{\varrho(xt, S[M, 1/n])\, e^{\beta t} \colon 0 \leqq t \leqq T_n\}$,

and $\varphi_n(x)$ is defined. One now easily sees that $\varphi_n(x)$ is continuous and moreover, satisfies the inequality

2.15.5 $\varphi_n(xt) \leqq e^{-\beta t}\varphi_n(x)$ for $x \in P$, $t \geqq 0$.

To see this note that

$$\begin{aligned}\varphi_n(xt) &= \sup\{\varrho(x(t+\tau), S[M, 1/n])\, e^{\beta\tau} \colon 0 \leqq \tau\} \\ &= \sup\{\varrho(x\tau, S[M, 1/n])\, e^{\beta(\tau - t)} \colon t \leqq \tau\} \\ &= e^{-\beta t}\sup\{\varrho(x\tau, S[M, 1/n])\, e^{\beta\tau} \colon \tau \geqq t\} \\ &\leqq e^{-\beta t}\varphi_n(x), \text{ if } t \geqq 0.\end{aligned}$$

Finally let $\max\{\varphi_n(x) \colon x \in P\} = M_n$. Define φ: $P \to R^+$ by

2.15.6 $\varphi(x) = \sum_{n=n_0}^{+\infty} \frac{1}{2^n}\frac{\varphi_n(x)}{M_n}$.

This $\varphi(x)$ is continuous on P, and satisfies $\varphi(xt) \leqq \varphi(x)\, e^{-\beta t}$ for $x \in P$ and $t \geqq 0$. Moreover, $\varphi(x) = 0$ if $x \in M$, and $\varphi(x) > 0$ if $x \in P - M$.

Now choose any $\alpha > 0$ sufficiently small such that

$$P_\alpha = \{x \in P \colon \varphi(x) \leqq \alpha\}$$

is in the interior of P. Now define $\Phi\colon A(M) \to R^+$ by

2.15.7 $$\Phi(x) = \begin{cases} \varphi(x) \text{ if } x \in P_\alpha, \\ \alpha e^{\beta\tau(x)} \text{ if } x \notin P_\alpha, \end{cases}$$

where $\tau(x)$ is the unique number such that $x\tau(x) \in \partial P_\alpha$. This $\Phi(x)$ has all the properties required in the theorem as is easily verified.

We shall give next, theorems which allow the identification of $\partial A(M)$ in the case of an asymptotically stable compact, connected set $M \subset X$.

2.16 Theorem. Consider a dynamical system (X, R, π) on a locally compact metric space X. Let $M \subset X$ be a compact, connected invariant set. Then a necessary condition for M to be asymptotically stable and the open invariant set $A(M)$ its region of attraction is the existence of two real-valued functions $\varphi(x)$ and $\theta(x)$ which have the following properties:

2.16.1 $\varphi(x)$ is defined and continuous in $A(M)$.

2.16.2 $\theta(x)$ is defined and continuous in X.

2.16.3 $-1 < \varphi(x) < 0$ if $x \in A(M) - M$.

2.16.4 $\theta(x) > 0$ if $x \in X - M$.

2.16.5 $\varphi(x) = \theta(x) = 0$ if $x \in M$.

2.16.6 $\varphi(x)$ is strictly increasing as $\varrho(x, M)$ increases, i.e. given any $\gamma_1 > 0$, there exists $\gamma_2 > 0$ such that $\varphi(x) < -\gamma_2$ for $\varrho(x, M) > \gamma_1$.

2.16.7 For each sequence $\{x^n\} \subset A(M)$ such that either $x^n \to x \in \partial A(M)$ or $\varrho(x^n, M) \to +\infty$, $\varphi(x^n) \to -1$.

2.16.8 $$\left(\frac{d\varphi(xt)}{dt}\right)_{t=0} = \lim_{t\to 0} \frac{\varphi(xt) - \varphi(x)}{t} = \theta(x)\,(1 + \varphi(x)).$$

Proof. Let $\delta_1 > 0$ be such that $S[M, \delta_1] \subset A(M)$. Assume

2.16.9 $\lambda(t) = \operatorname{Sup}\{\varrho(xt, M) : x \in S(M, \delta_1) - M\}$,

for each $t \geqq 0$ such $\lambda(t)$ exists and is positive. Since M is asymptotically stable and $S[M, \delta_1] \subset A(M)$, $\lambda(t) \leqq \varepsilon_1$ for $0 \leqq t < +\infty$, and $\lim\limits_{t\to+\infty} \lambda(t) = 0$. Then for each integer k one can find $t_k > 0$ such that

2.16.10 $\lambda(t) < \varepsilon_1/2^k$, for all t with $t \geq t_k$.

In addition we can find an increasing subsequence $\{t_k\}$ such that $t_k \to +\infty$.

Assume next

2.16.11 $l_k(t) = [2\varepsilon_1 t_{k+1} - \varepsilon_1(t_k + t)]/2^k(t_{k+1} - t_k)$

for $t_k \leqq t \leqq t^{k+1}$ and $k \geqq 1$.

The function $l_k(t)$ is increasing in the interval of definition. In addition, since

2.16.12 $l_k(t) \geqq [2\varepsilon_1 t_{k+1} - \varepsilon_1(t_k + t_{k+1})]/2^k(t_{k+1} - t_k) = \frac{\varepsilon_1}{2^{k-1}} > \lambda(t)$,

in the interval of definition, we have

2.16.13 $\lambda(t) < l_k(t)$ for $t_k \leqq t \leqq t_{k+1}$ and $k \geqq 1$.

For $k = 1$ we can assume that the function $l_1(t)$ is defined on the interval $(-\infty, t_2)$. Since $l_1(t) > \varepsilon_1 \geqq \lambda(t)$, the function

2.16.14 $L(t) = \begin{cases} l_k(t) \text{ for } t_k \leqq t \leqq t_{k+1},\ k \geqq 1 \\ l_1(t) \text{ for } -\infty < t \leqq t_2 \end{cases}$

for each $t > 0$ satisfies the inequality

2.16.15 $\lambda(t) < L(t)$.

Such function $L(t)$ is defined and continuous on the whole real line and in addition it is strictly decreasing from $+\infty$ to 0. From the equation 2.16.9 and the inequality 2.16.15 it then follows that

2.16.16 $\varrho(xt, M) < L(t)$ for $t \in R^+$ and $x \in S(M, \delta_1)$.

Let $L^{-1}(s)$ be the inverse function of $L(t)$: from the properties of $L(t)$ it follows that $L^{-1}(s)$ exists, is continuous and strictly decreasing on the interval $0 < s < +\infty$. Consider the function

2.16.17 $\xi(s) = s\,e^{-L^{-1}(s)}, \quad 0 < s < +\infty$,

which is positive, continuous, strictly increasing and such that $\xi(s) \to 0$ as $s \to 0$. We shall next assume $\xi(0) = 0$ and therefore define $\xi(s)$ on the whole interval $0 \leqq s < +\infty$.

Consider the function

2.16.18 $\theta(x) = \xi(\varrho(x, M)) = \varrho(x, M)\, e^{-L^{-1}(\varrho(x,M))}$.

This function satisfies all conditions of the theorem since it is defined and continuous for all $x \in X$, it is positive for all $x \in X - M$, it vanishes for $x \in M$ and it is strictly increasing as $\varrho(x, M)$ increases. Consider next the function

2.16.19 $\varphi(x) = \exp\left[-\int_0^{+\infty} \xi(\varrho(xt, M))\, dt\right] - 1$,

we shall prove that, if for each $x \in A(M)$, $\varphi(x)$ exists, we have

2.16.20 $1 + \varphi(xt) = [1 + \varphi(x)] \exp\left[\int_0^t \xi(\varrho(x\tau, M))\, d\tau\right]$.

From 2.16.19 it follows

2.16.21 $1 + \varphi(x) = \exp\left[-\int_0^t \xi(\varrho(x\tau, M))\, d\tau\right] \exp\left[-\int_t^{+\infty} \xi(\varrho(x\tau, M))\, d\tau\right]$.

Hence, if we let $\tau = t + s$, we have

$$2.16.22 \qquad (1+\varphi(x)) \exp\left[\int_0^t \xi(\varrho(x\tau, M))\, d\tau\right]$$

$$= \exp\left[-\int_0^{+\infty} \xi(\varrho(x(t+s), M))\, ds\right]$$

$$= \exp\left[-\int_0^{+\infty} \xi(\varrho((xt)\, s, M))\, ds\right] = 1 + \varphi(xt)$$

where the equality $x(t+s) = (xt)\, s$ has been used.

In order for 2.16.20 to be true it is then enough that for each $x \in A(M)$, $\varphi(x)$ exists, i.e. that the integral $\int_0^{+\infty} \xi(\varrho(xt, M))\, dt$ exists for each $x \in A(M)$. For each $x \in A(M)$ it is possible to find $\tau(x) > 0$, such that, for all $t \geqq \tau(x)$, we have $x\tau(x) \in S(M, \delta_1)$. It follows that

$$2.16.23 \qquad \int_0^{\infty} \xi(\varrho(xt, M))\, dt = \int_0^{\tau(x)} \xi(\varrho(xt, M))\, dt + \int_{\tau(x)}^{+\infty} \xi(\varrho(xt, M))\, dt.$$

If we change the integration variable in the second integral from t into $s + \tau(x)$ and notice that, as usual, $x(s + \tau(x)) = (x\tau(x))\, s$, we have that

$$2.16.24 \qquad \int_{\tau(x)}^{+\infty} \xi(\varrho(x, t, M))\, dt = \int_0^{+\infty} \xi(\varrho(x(s+\tau(x)), M))\, ds$$

$$= \int_0^{+\infty} \xi(\varrho((x\tau(x))\, s, M))\, ds.$$

Now $x\tau(x) \in S(M, \delta_1)$ and hence from 2.16.16 we have

$$2.16.25 \quad L(s) > \varrho((x\tau(x))\, s, M) \qquad \text{for} \qquad s \in R^+.$$

Now, since L^{-1} is strictly decreasing, from the inequality 2.16.25 it follows that

$$2.16.26 \quad s = L^{-1}(L(s)) < L^{-1}\{\varrho((x\tau(x))\, s, M)\},$$

from which we have that

$$2.16.27 \quad \varrho((x\tau(x))\, s, M) e^{-s} > \varrho((x\tau(x))\, s, M) \exp\left[-L^{-1}\{\varrho((x\tau(x))\, s, M)\}\right]$$

$$= \xi\{\varrho((x\tau(x))\, s, M)\}.$$

Thus, since $L(t) < \delta$ for $t > 0$, because of the inequalities 2.16.16 and 2.16.27, it follows that the integral 2.16.24 is less than δ and therefore converges. It follows that for each $x \in A(M)$, $\int_0^{+\infty} \xi(\varrho(xt, M))\, dt$ exists, hence for each $x \in A(M)$, $\varphi(x)$ exists and the equality is proved.

We want to show next that $\varphi(x)$ is continuous for each $x \in A(M)$. For that it is enough to prove that the function

2.16.28 $$\varphi^*(x) = \log(1 + \varphi(x)) = -\int_0^{+\infty} \xi(\varrho(xt, M))\, dt$$

is continuous for each $x \in A(M)$, i.e. that for each $x \in A(M)$ and $\varepsilon > 0$ there exists $\delta_1 > 0$ such that, for $\varrho(y, x) < \delta_1$

2.16.29 $$|\varphi^*(y) - \varphi^*(x)| = \left| \int_0^{\infty} \{\xi(\varrho(yt, M)) - \xi(\varrho(xt, M))\}\, dt \right| < \varepsilon.$$

For $\varrho(y, x)$ sufficiently small, the points x and y belong to $A(M)$ and therefore there exists $T > 0$ such that

2.16.30 $$\int_T^{+\infty} \xi(\varrho(yt, M))\, dt < \frac{\varepsilon}{3},$$

2.16.31 $$\int_T^{+\infty} \xi(\varrho(yt, M))\, dt < \frac{\varepsilon}{2};$$

these inequalities follow from the convergence of the integral 2.16.23. Now, since the map π of the dynamical system is continuous, given $\delta_2 > 0$, we can find $\delta_3 > 0$ such that

2.16.32 $\varrho(xt, yt) < \delta_2$ for $\varrho(x, y) < \delta_3$ and $0 \leqq t \leqq T$.

Since $\xi(\varrho(x, M)) = \theta(x)$, from the continuity of $\theta(x)$ it follows that for sufficiently small $\delta_2 > 0$, we have

2.16.33 $|\xi(\varrho(xt, M)) - \xi(\varrho(yt, M))| < \varepsilon/2T, \quad 0 \leqq t \leqq T.$

Thus, for $\varrho(x, y) < \delta_3$, we have

2.16.34 $|\varphi_1^*(x) - \varphi_1^*(y)| < \varepsilon,$

and the properties 2.16.1—2.16.5 of the theorem have therefore been proved.

We shall now prove that also the property 2.16.6 is satisfied. For that take $\gamma_1 > 0$. Let $T > 0$ and $\alpha > 0$ be such that $\varrho(xt, M) > \alpha$ for $0 \leqq t \leqq T$ and $\varrho(x, M) > \gamma_1$.

Then there exists $\beta > 0$ such that $\xi(\varrho(xt, M)) \geqq \beta$ for $0 \leqq t \leqq T$, hence

2.16.35 $$\varphi(x) < \exp\left[-\int_0^T \beta\, dt\right] - 1 < -\gamma_2$$

which concludes the proof of the property 2.16.6.

It remains to be proved that also the property 2.16.7 is true. For that let $x \in A(M)$ and consider a sequence $\{x^n\} \subset A(M)$ with $x^n \to x$. Let $\delta_1 > 0$ be such that $H(M, \delta_1) \subset A(M)$ and $\{\tau^n\} \subset R^+$ such that $x^n t \notin S(M, \delta_1)$ for $0 \leqq t \leqq \tau^n$, $x^n \tau^n \in H(M, \delta_1)$. Clearly such a sequence $\{\tau^n\}$

exists and, in addition $\tau^n \to +\infty$, if not $x \in A(M) \cap \partial A(M)$, which is absurd.

Let $x^n \tau^n \to x^* \in H(M, \delta_1)$. From the continuity of $\varphi(x)$ and the property 2.16.3, it then follows that $-1 < \varphi(x^*) < 0$. From the property 2.16.4 it then follows that there exists a real number $\alpha > 0$, such that $\xi(\varrho(x^n t, M)) \geqq \alpha$ for $0 \leqq t \leqq \tau^n$. From the equality 2.16.20 it then follows that

$$2.16.36 \quad (1 + \varphi(x^n \tau^n)) \exp\left[-\int_0^{\tau n} \xi(\varrho(x^n t, M))\, dt\right] = (1 + \varphi(x^n)).$$

Now the limit of the right-hand side of the 2.16.36 exists and it is equal to zero, hence $\varphi(x^n) \to -1$. The same reasoning applies if $\{x^n\} \subset A(M)$ and $\|x^n\| \to +\infty$, which concludes the proof of the property 2.16.7 and of the theorem.

2.17 *Remarks.*

2.17.1 The importance of Theorem 2.15 is only in connection with asymptotically stable sets M which are not necessarily invariant. For invariant sets $M, \Phi(x)$ may be more easily obtained by setting $\Phi(x) = e^{\beta\tau(x)}$ for all $x \in A(M) - M$ (see proof of Theorem 2.9).

2.17.2 Theorems analogous to those proved in this section can also be derived for negatively asymptotically stable sets. A negatively asymptotically stable set is sometimes called completely unstable. But we shall avoid this term in general.

2.17.3 The definitions introduced in section 1 can be reformulated to include closed sets having a compact boundary. Then all the theorems in this and the previous section hold for such sets also.

2.17.4 Also Corollary 4.24 is useful in the identification of the region of attraction of compact asymptotically stable sets.

3. Topological Properties of Regions of Attraction

In this section we shall present some additional properties of weak attractors, their regions of attraction, and the level lines of certain Liapunov functions for asymptotically stable sets. Again the space X will be locally compact.

We start by recalling the result in Theorem 2.5 which may be stated as

3.1 **Theorem.** Let M be a compact asymptotically stable subset of X and Φ a function defined on a neighborhood N of M and satisfying conditions of Theorem 2.2. Let $\varepsilon > 0$ be such that $S[M, \varepsilon]$ is a compact subset of N. Let α, $0 < \alpha < m(\varepsilon)$, where

3.1.1 $m(\varepsilon) = \min\{\Phi(x) : x \in H(M, \varepsilon)\}$.

Then the set

3.1.2 $P_\alpha = K_\alpha \cap S[M, \varepsilon]$,

where

3.1.3 $K_\alpha = \{x \in N : \Phi(x) \leqq \alpha\}$,

is a compact positively invariant subset of $A(M)$.

The following lemma expresses an important topological property of the sets P_α defined above.

3.2 Lemma. The set P_α defined in the above theorem is a retract of $A(M)$. Consequently P_α is a retract of every set N, $P_\alpha \subset N \subset A(M)$.

Proof. This is so because we can define a map $h: A(M) \to P_\alpha$ by $h(x) = x$ if $x \in P_\alpha$, and $h(x) = x\tau(x)$ if $x \notin P_\alpha$, where $\tau(x)$ is defined as in the proof of Theorem 2.9. Because of the continuity of $\tau(x)$ and of the phase map π, and the fact that $\tau(x) = 0$ for $x \in \partial P_\alpha$, it follows that h is a continuous map of $A(M)$ into P_α which is an identity on P_α. Thus by definition P_α is a retract of $A(M)$, and hence also a retract of every subset of $A(M)$ which contains P_α.

From the definition of stability it is clear that if a singleton $\{x\}$ is stable, then x is a critical or rest point (Theorem 1.9). In particular, if $\{x\}$ is asymptotically stable, then x is a rest point. Our next theorem concerns an important topological property of the region of attraction of a rest point in a euclidean n-space R^n. The proof depends on the following lemma in the topology of n-spaces.

3.3 Lemma. Let $\{U_n\}$ be a monotone sequence of open n-cells in R^n, i.e., $U_n \subset U_{n+1}$, $n = 1, 2, \ldots$. Then $\cup\{U_n, n = 1, 2, \ldots\}$ is an open n-cell.

3.4 Theorem. If a rest point $p \in R^n$ is asymptotically stable, then $A(p)$ is homeomorphic to R^n.

Proof. Since $A(p)$ is a neighborhood of p, there is an $\varepsilon > 0$ such that the closed ball $S[p, \varepsilon] \subset A(p)$. For each $t \in R$ (the transition π^t being a homeomorphism of R^n onto R^n), the image $S(p, \varepsilon)\, t$ of the open ball $S(p, \varepsilon)$ by π^t is an open n-cell. We claim now that for any given t_1 there exists a t_2, $t_2 < t_1$ such that $S(p, \varepsilon)\, t_1 \subset S(p, \varepsilon)\, t_2$. This is so because $\overline{S(p, \varepsilon)\, t_1}$ being a subset of the compact set $S[P, \varepsilon]\, t_1$ is itself compact.

Further, $S[p, \varepsilon]\, t_1 \subset A(p)$, and $A(p)$ is open. Since p is uniformly attracting there exists a $T > 0$ such that $S[p, \varepsilon]\,(t_1 + t) \subset S(p, \varepsilon)\, t_1$ for $t \geqq T$. In particular, $S[p, \varepsilon]\,(t_1 + T) \subset S(p, \varepsilon)\, t_1 \subset S[p, \varepsilon]\, t_1$. Hence $S[p, \varepsilon]\, t_1 \subset S(p, \varepsilon)\,(t_1 - T) \subset S[p, \varepsilon]\,(t_1 - T)$. Setting $t_2 = t_1 - T$, we get $S[p, \varepsilon]\, t_1 \subset S(p, \varepsilon)\, t_2$. The above analysis shows that we can choose a sequence $\{t_n\}$, $t_n \to -\infty$, such that $\{S(p, \varepsilon)\, t_n\}$ is a monotone sequence of open n-cells. By Lemma 3.3 $\cup \{S(p, \varepsilon)\, t_n;\ n = 1, 2, \ldots\}$ is an open n-cell. But this last union is $A(p)$, so that $A(p)$ is an open n-cell and hence homeomorphic to R^n.

3.5 Corollary. If p is an asymptotically stable rest point in R^n, then $A(p) - \{p\}$ is homeomorphic to $R^n - \{0\}$, where 0 is the origin in R^n.

We can now present the following result.

3.6 Theorem. Let $M \subset R^n$ be a compact invariant globally asymptotically stable set in R^n. Then $R^n - M = \mathcal{C}(M)$ is homeomorphic to $R^n - \{0\}$.

For the proof of this theorem we need to establish some further results which are important in themselves.

3.7 Lemma. Let $M \subset X$ be a compact positively invariant set. Let $\{\gamma_n\}$ be a sequence of periodic trajectories with periods T_n, such that $\gamma_n \subset M$, and $T_n \to 0$. Then M contains a rest point.

Proof. Consider any sequence of points $\{x_n\}$, with $x_n \in \gamma_n$, $n = 1, 2, \ldots$. We may assume without loss of generality that $x_n \to x \in M$, as M is compact. We will demonstrate that x is a rest point. For suppose that this is not the case. Then there is a $\tau > 0$, such that $x \neq x\tau$. Let $\varrho(x, x\tau) = \alpha\ (> 0)$. The spheres $S(x, \alpha/4)$, and $S(x\tau, \alpha/4)$ are disjoint. Now choose T, $0 < T < \tau$, such that $\varrho(x, xt) \leqq \alpha/8$ for $0 \leqq t \leqq T$. By continuity of π there is a $\delta > 0$ such that $\varrho(x, y) < \delta$ implies $\varrho(xt, yt) < \alpha/8$ for $0 \leqq t \leqq \tau$. Notice in particular that if $\varrho(x, y) < \delta$, then

$$\varrho(x, yt) \leqq \varrho(x, xt) + \varrho(xt, yt) < \alpha/8 + \alpha/8 = \alpha/4 \text{ if } 0 \leqq t \leqq T,$$

and $\varrho(x\tau, y\tau) < \alpha/4$.

Now for sufficiently large n we have $T_n < T$ and $d(x, x_n) < \delta$. Hence $\varrho(x, x_n t) < \alpha/4$ for $0 \leqq t \leqq T_n < T$. And as γ_n is periodic of period T_n, we have $\varrho(x, x_n t) < \alpha/4$ for all $t \in R$. This is impossible, because we must have $\varrho(x, x_n\tau) \geqq \varrho(x, x\tau) - \varrho(x\tau, x_n\tau) > \alpha - \alpha/4 = (3/4)\,\alpha$. This contradiction proves that the point $x \in M$ is a rest point.

As an application of the above lemma we have

3.8 Theorem. Let $M \subset R^n$ be a compact positively invariant set, which is homeomorphic to the closed unit ball in R^n. Then M contains a rest point.

Proof. Consider any sequence $\{\tau_n\}$, $\tau_n > 0$, $\tau_n \to 0$. Consider the sequence of transitions $\{\pi_n\}$, $\pi_n = \pi^{\tau_n}$. As M is positively invariant each π_n maps M into itself. Thus by the Brouwer fixed point theorem, M contains a fixed point of each one of the transitions π_n. Let $x_n \in M$ be a fixed point of π_n. Then since $x_n = \pi_n(x_n) = x_n\tau_n$, the trajectory $\gamma(x_n) = \gamma_n$ is a rest point or a periodic trajectory with a period τ_n, and as M is positively invariant $\gamma_n \subset M$. By the above lemma, M contains a rest point, and the theorem is proved.

We are now in a position to prove the following important result.

3.9 Theorem. Let $M \subset R^n$ be a compact set which is a weak attractor. Let the region of weak attraction $A_\omega(M)$ of M be homeomorphic to R^n. Then M contains a rest point. In particular, when $A_\omega(M) = R^n$ (i.e., M is a global weak attractor), then M contains a rest point.

Proof. By Theorem 1.25 $D^+(M)$ is an asymptotically stable compact set with $A(D^+(M)) = A_\omega(M)$. Let $\Phi(x)$ be any function for the asymptotically stable set $D^+(M)$ as in Theorem 3.1, and consider a set P_α for $\Phi(x)$. Then P_α is compact, positively invariant, and is a retract of $A_\omega(M)$. As $A_\omega(M)$ is homeomorphic to R^n we can choose a compact set B, $P_\alpha \subset B \subset A_\omega(M)$, where B is homeomorphic to the closed unit ball in R^n. Then P_α is a retract of B. Thus P_α has the fixed point property, as B has by the Brouwer fixed point theorem. Since P_α is positively invariant, the transition π^τ maps P_α into P_α for each $\tau \geqq 0$. Thus for each fixed $\tau > 0$, π^τ has a fixed point in P_α, i.e., corresponding to any $\tau > 0$ there is an $x_\tau \in P_\alpha$ such that $\pi^\tau(x_\tau) = x_\tau$. Thus the orbit $\gamma(x_\tau)$ is periodic and has a period τ, moreover $\gamma(x_\tau) \subset P_\alpha$, because $\gamma(x_\tau) = \gamma^+(x_\tau) \subset P_\alpha$. We have thus shown that, corresponding to any sequence $\{\tau_n\}$, $\tau_n > 0$, $\tau_n \to 0$, there is a sequence of closed orbits $\{\gamma_n\}$, $\gamma_n = \gamma(x_{\tau_n})$, with γ_n having a period τ_n. This sequence being in P_α, P_α contains a rest point x^* (Lemma 3.7). However, $M \subset D^+(M) \subset P_\alpha \subset A_\omega(M)$. As M is a weak attractor we have $\Lambda^+(x) \cap M \neq \emptyset$ for each $x \in A_\omega(M)$. Thus $\Lambda^+(x^*) \cap M \neq \emptyset$. But $\Lambda^+(x^*) = \{x^*\}$, as x^* is a rest point. Hence $x^* \in M$. The theorem is proved.

For the following corollaries the dynamical system is assumed to be defined on R^n.

3.10 Corollary. If the dynamical system is ultimately bounded, then it contains a rest point.

This is so, because ultimate boundedness is equivalent to the existence of a compact globally asymptotically stable set (Remark 2.14) which by the above theorem contains a rest point.

3.11 Corollary. The region of attraction of a compact minimal weak attractor M cannot be homeomorphic to R^n, unless M is a rest point.

Note, however, that if a rest point $p \in R^n$ is weakly attracting, or attracting, then $A(p)$ need not be homeomorphic to R^n, as the analytic example 1.7.4 shows. As another corollary we have

3.12 Corollary. If M is compact minimal and a global weak attractor, then M is a singleton. Consequently, M consists of a rest point.

Finally we prove Theorem 3.6.

Proof of Theorem 3.6. By Theorem 3.9, M contains a rest point. We may assume without loss of generality that M contains the origin 0 and 0 is a rest point. Consider now the homeomorphism $h: E - \{0\} \to E - \{0\}$ defined by $h(x) = x/||x||^2$, where $||x||$ is the euclidean norm of x. h maps the given dynamical system into a dynamical system on R^n, with 0 becoming a negatively asymptotically stable rest point, and $R^n - M$ is mapped into $A^-(0) - \{0\}$, where $A^-(0)$ is now the region of negative attraction of 0. By the Corollary 3.5, $A^-(0) - \{0\}$ is homeomorphic to $R^n - \{0\}$. Hence the result follows.

We shall now present one example of application of Theorem 3.9.

3.13 *Example.* Consider a dynamical system in R^n with only two rest points x and y, $x \neq y$. Theorem 3.9 shows that y cannot be asymptotically stable with $A(\{y\}) = R^n - \{x\}$, since $\mathscr{C}(\{x\} \cup \{y\})$ is not homeomorphic to $R^n - \{0\}$.

3.14 *Exercises.*

3.14.1 Show that if X is compact and $M \subset X$ a compact global weak attractor, then $D^+(M) = X$. Consequently, if $M \neq X$, then M cannot be stable.

3.14.2 If M is a compact invariant asymptotically stable set in X, then the restriction of the dynamical system (X, R, π) to the set $A(M) - M$ is parallelizable.

4. Stability and Asymptotic Stability of Closed Sets

In this section X will be an arbitrary metric space, unless otherwise restricted. In contrast to the situation in the last three sections, several notions of stability and asymptotic stability may be introduced. The purpose of this section is to develop several notions of stability and attraction and to provide criteria for these in terms of Liapunov functions.

We shall first discuss stability of closed sets.

4.1 **Definition.** A closed set $M \subset X$ will be said to be

4.1.1 *stable*, if for each $\varepsilon > 0$ and $x \in M$, there is a $\delta = \delta(x, \varepsilon) > 0$ such that

$$S(x, \delta)\, R^+ \subset S(M, \varepsilon),$$

4.1.2 *equi-stable*, if for each $x \notin M$, there is a $\delta = \delta(x) > 0$ such that

$$x \notin \overline{S(M, \delta)\, R^+},$$

4.1.3 *uniformly stable*, if for each $\varepsilon > 0$, there is a $\delta = \delta(\varepsilon) > 0$ such that

$$S(M, \delta)\, R^+ \subset S(M, \varepsilon).$$

4.2 **Proposition.** If X is locally compact and M is compact, then M is uniformly stable whenever it is either equi-stable or stable (or both). (The converse holds always.)

Proof. (i) If M is stable, then for a given $\varepsilon > 0$, let $\delta_x > 0$ be a number corresponding to $x \in M$ such that $S(x, \delta_x)\, R^+ \subset S(M, \varepsilon)$. Since $\cup \{S(x, \delta_x) \colon x \in M\}$ is an open cover of M, there is a finite open cover, consisting of $\{S(x_i, \delta_{x_i}) \colon i = 1, 2, \ldots, n;\ x_i \in M\}$. But then there is a $\delta > 0$ such that $S(M, \delta) \subset \cup \{S(x_i, \delta_{x_i}) \colon i = 1, 2, \ldots, n\}$. Notice now that $S(M, \delta)\, R^+ \subset [\cup \{S(x_i, \delta_{x_i}) \colon i = 1, 2, \ldots, n\}]\, R^+ \subset S(M, \varepsilon)$. Thus M is uniformly stable.

(ii) Let M be equi-stable. Since M is compact and X is locally compact, there is a $\delta > 0$ such that $S[M, \delta]$ and (hence) $H(M, \delta)$ are compact. Then for each $x \in H(M, \delta)$, there is a $\delta_x > 0$ such that $x \notin \overline{S(M, \delta_x)\, R^+}$. But then $x \in \mathscr{C}(S_x)$, where $S_x \equiv \overline{S(M, \delta_x)\, R^+}$. Since each $\mathscr{C}(S_x)$ is open, and $H(M, \delta) \subset \cup \{\mathscr{C}(S_x) \colon x \in H(M, \delta)\}$, we have an open cover of the compact set $H(M, \delta)$. Thus there are points $x_1, x_2, \ldots, x_n$ in $H(M, \delta)$ such that $H(M, \delta) \subset \cup \{\mathscr{C}(S_{x_i}) \colon i = 1, 2, \ldots, n\}$. Since $\cup \{\mathscr{C}(S_{x_i}) \colon i = 1, 2, \ldots, n\} = \mathscr{C}(\cap \{S_{x_i} \colon i = 1, 2, \ldots, n\})$ we have $\cap \{S_{x_i} \colon i = 1, 2, \ldots, n\} \subset S(M, \varepsilon)$. If now $\delta = \min \{\delta_{x_1}, \delta_{x_2}, \ldots, \delta_{x_n}\}$, then $S(M, \delta)\, R^+ \subset \cap \{S_{x_i} \colon i = 1, 2, \ldots, n\} \subset S(M, \varepsilon)$. Thus M is uniformly stable.

4.3 *Remark.* Note that part (i) of the above proof did not use the fact that X is locally compact. Further uniform stability implies both stability and equi-stability, but it cannot be asserted that a closed set which is both stable and equi-stable is uniformly stable.

4.4 **Proposition.** If a closed set is either stable or equi-stable, then it is positively invariant.

The proof is simple and is left as an exercise.

We now indicate the connection between various kinds of stability and Liapunov functions.

4.5 **Theorem.** A closed set M is stable if and only if there exists a function $\varphi(x)$ defined on X with the following properties:

4.5.1 $\varphi(x) = 0$ if and only if $x \in M$.

4.5.2 For every $\varepsilon > 0$, there is a $\delta > 0$ such that $\varphi(x) \geqq \delta$ whenever $\varrho(x, M) \geqq \varepsilon$; also for any sequence $\{x_n\}$, $\varphi(x_n) \to 0$ whenever $x_n \to x \in M$.

4.5.3 $\varphi(xt) \leqq \varphi(x)$ for all $x \in X$, $t \geqq 0$.

Proof. (i) *Sufficiency.* Given $\varepsilon > 0$, set $m_0 = \inf\{\varphi(x) : \varrho(x, M) \geqq \varepsilon\}$. By 4.5.2, $m_0 > 0$. Then for $x \in M$ find $\delta > 0$ such that $\varphi(y) < m_0$ for $y \in S(x, \delta)$. This is also possible by 4.5.1 and 4.5.2. We claim that $S(x, \delta) R^+ \subset S(M, \varepsilon)$. For otherwise there is $y \in S(x, \delta)$ and $t \geqq 0$ such that $\varrho(yt, M) = \varepsilon$. But then $\varphi(yt) \leqq \varphi(y) < m_0$ on one hand by 4.5.3 and also $\varphi(yt) \geqq \inf\{\varphi(x) : \varrho(x, M) \geqq \varepsilon\}$, as $\varrho(yt, M) = \varepsilon$, i.e., $\varphi(yt) \geqq m_0$. This contradiction proves the result.

(ii) *Necessity.* Let M be stable. Define

$$\varphi(x) = \sup\left\{\frac{\varrho(xt, M)}{1 + \varrho(xt, M)} : t \geqq 0\right\}.$$

This $\varphi(x)$ is defined on X, and has all the properties required in the theorem. The verification is left to the reader. The theorem is proved.

4.6 *Remark.* Condition 4.5.2 in the above theorem is equivalent to the requirement that if $\{x_n\}$ is any sequence such that $x_n \to x \in M$, then $\varphi(x_n) \to 0$, and there is a continuous strictly increasing function $\alpha(\mu)$, defined for $\mu \geqq 0$, such that $\varphi(x) \geqq \alpha(\varrho(x, M))$.

4.7 **Theorem.** A closed set $M \subset X$ is equi-stable if and only if there is a function $\varphi(x)$ defined on X such that

4.7.1 $\varphi(x) = 0$ for $x \in M$, $\varphi(x) > 0$ for $x \notin M$,

4.7.2 for every $\varepsilon > 0$ there is a $\delta > 0$ such that $\varphi(x) \leqq \varepsilon$ if $\varrho(x, M) \leqq \delta$, and

4.7.3 $\varphi(xt) \leqq \varphi(x)$ for $x \in X$, $t \geqq 0$.

Proof. (i) *Sufficiency.* Let $x \notin M$. Set $\varrho(x, M) = \varepsilon$. Then by 4.7.2 there is a $\delta > 0$ such that $\varphi(x) \leqq \varepsilon/2$ for $\varrho(x, M) \leqq \delta$. Then indeed $\overline{S(M, \delta) R^+} \subset \overline{S(M, \varepsilon/2)}$, by 4.7.3. Hence $x \notin \overline{S(M, \delta) R^+}$.

(ii) *Necessity.* Set for each $x \notin M$,

$$\varphi(x) = \sup\{\delta > 0\colon x \notin \overline{S(M, \delta) R^+}\},$$

and $\varphi(x) = 0$ for $x \in M$. This $\varphi(x)$ has all the desired properties, the verification of which is left to the reader. Note that $\varphi(x) \leqq \varrho(x, M)$.

4.8 *Remark.* Condition 4.7.2 in the above theorem is equivalent to the existence of a continuous strictly increasing function $\alpha(r)$, $\alpha(0) = 0$, such that $\varphi(x) \leqq \alpha(\varrho(x, M))$.

4.9 **Theorem.** A closed set M is uniformly stable if and only if there is a function $\varphi(x)$ defined on X such that

4.9.1 for every $\varepsilon > 0$ there is a $\delta > 0$ such that $\varphi(x) \geqq \delta$ whenever $\varrho(x, M) \geqq \varepsilon$,

4.9.2 for every $\varepsilon > 0$ there is a $\delta > 0$ such that $\varphi(x) \leqq \varepsilon$ whenever $\varrho(x, M) \leqq \delta$,

4.9.3 $\varphi(xt) \leqq \varphi(x)$ for $x \in X$, $t \geqq 0$.

Proof. We leave the details to the reader, but remark that in the proof of necessity one may choose either of the functions given in the necessity proofs of Theorems 4.5 and 4.7.

4.10 *Remark.*

4.10.1 The theorems above differ from the usual theorems on stability in that the existence of the functions is shown in all of X rather than in a small neighborhood of M. Indeed for sufficiency the functions need be defined on just a neighborhood of M.

4.10.2 It should perhaps be emphasized that in contrast to the situation in section 2, the functions shown to exist in the last three theorems need not be continuous. To further emphasize this point, we give an example.

4.11 *Example.* Consider a planar dynamical system given by the differential system in cartesian coordinates

4.11.1 $$\dot{x}_1 = x_2, \qquad \dot{x}_2 = \sin^2\left(\frac{\pi}{x_1^2 + x_2^2}\right) x_2 - x_1.$$

The phase portrait (Fig. 4.11.3) consists of a rest point P (the origin of coordinates) a sequence $\{\gamma_n\}$ of periodic trajectories,

4.11.2 $$\gamma_n = \{(x_1, x_2)\colon x_1^2 + x_2^2 = 1/n\},$$

and between any two consecutive γ_n the trajectories are spirals which approach the outer circle as $t \to +\infty$. The singleton $\{P\}$ is stable, but there are no other compact stable sets. No continuous function satisfying conditions of Theorem 4.5, 4.7, or 4.9 exists. For if so, then such a function must necessarily be constant on each γ_n (prove this!) and its value on any γ_n must be less than or equal to that on γ_{n+1}. Consequently its value at P must be greater or equal to that on any γ_n.

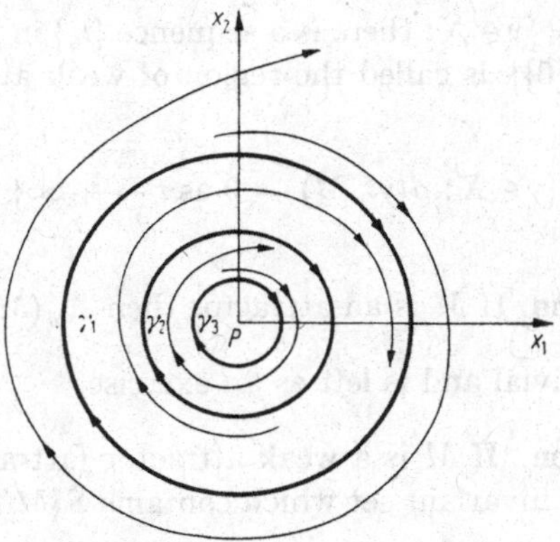

Fig. 4.11.3

We shall now discuss asymptotic stability of closed sets and its relation with the Liapunov functions.

4.12 **Definition.** A closed set $M \subset X$ will be said to be

4.12.1 a semi-weak attractor, if for each $x \in M$, there is a $\delta_x > 0$, and for each $y \in S(x, \delta_x)$ there is a sequence $\{t_n\}$ in R, $t_n \to +\infty$, such that $\varrho(yt_n, M) \to 0$,

4.12.2 a semi-attractor, if for each $x \in M$, there is a $\delta_x > 0$, such that for each $y \in S(x, \delta_x)$, $\varrho(yt, M) \to 0$ as $t \to +\infty$,

4.12.3 a weak attractor, if there is a $\delta > 0$ and for each $y \in S(M, \delta)$, there is a sequence $\{t_n\}$ in R, $t_n \to +\infty$, such that $\varrho(yt_n, M) \to 0$,

4.12.4 an attractor, if there is a $\delta > 0$ such that for each $y \in S(M, \delta)$, $\varrho(yt, M) \to 0$ as $t \to +\infty$,

4.12.5 a uniform attractor, if there is an $\alpha > 0$, and for each $\varepsilon > 0$ there is a $T = T(\varepsilon) > 0$, such that $x[T, +\infty) \subset S(M, \varepsilon)$ for each $x \in S[M, \alpha]$,

4.12.6 an equi-attractor, if it is an attractor, and if there is an $\lambda > 0$ such that for each ε, $0 < \varepsilon < \lambda$, and $T > 0$, there exists a $\delta > 0$ with the property that $x[0, T] \cap S(M, \delta) = \emptyset$ whenever $\varepsilon \leqq \varrho(x, M) \leqq \lambda$,

4.12.7 semi-asymptotically stable, if it is stable and a semi-attractor,

4.12.8 asymptotically stable, if it is uniformly stable and is an attractor,

4.12.9 uniformly asymptotically stable, if it is uniformly stable and a uniform attractor.

4.13 Definition. For any $M \subset X$, the set

4.13.1 $A_\omega(M) = \{y \in X\colon$ there is a sequence $\{t_n\}$ in R, $t_n \to +\infty$, such that $\varrho(yt_n, M) \to \{0\}\}$ is called the region of weak attraction of M,

and the set

4.13.2 $A(M) = \{y \in X\colon \varrho(yt, M) \to 0$ as $t \to +\infty\}$ is called the region of attraction of M.

4.14 Proposition. If M is an attractor then $A_\omega(M) \equiv A(M)$.

The proof is trivial and is left as an exercise.

4.15 Proposition. If M is a weak attractor (attractor), then $A_\omega(M)$ $(A(M))$ is an open invariant set which contains $S(M, \delta)$ for some $\delta > 0$.

The proof is simple and is left as an exercise.

4.16 Theorem. If a compact set M is a semi-weak attractor (semi-attractor), then it is weak attractor (attractor).

The proof is again simple and left as an exercise.

We now discuss the existence of Liapunov functions for various kinds of asymptotic stability.

4.17 Theorem. A closed set M is semi-asymptotically stable, if and only if there exists a function $\varphi(x)$ defined on X which has the following properties:

4.17.1 For each $y \in M$, $\varphi(x)$ is continuous in some neighborhood $S(y, \delta_y)$ of y,

4.17.2 $\varphi(x) = 0$ for $x \in M$, $\varphi(x) > 0$ for $x \notin M$,

4.17.3 there is a strictly increasing function $\alpha(\mu)$, $\alpha(0) = 0$, defined for $\mu \geqq 0$, such that

$$\varphi(x) \geqq \alpha(\varrho(x, M)),$$

4.17.4 $\varphi(xt) \leqq \varphi(x)$ for all $x \in X$, $t \geqq 0$, and for each $y \in M$, there is a $\delta_y > 0$ such that if $x \notin M$, $x \in S(y, \delta_y)$, then $\varphi(xt) < \varphi(x)$ for $t > 0$ and $\varphi(xt) \to 0$ as $t \to +\infty$.

Proof. (i) *Sufficiency.* Stability follows from Theorem 4.5. The semi-attractor property follows from 4.17.3 and 4.17.4.

(ii) *Necessity.* Consider the function

$$\varphi(x) = \sup\left\{\frac{\varrho(xt, M)}{1+\varrho(xt, M)} : t \geqq 0\right\}.$$

This has all the properties 4.17.1 — 4.17.4 except that it may not be strictly decreasing along trajectories originating in any neighborhood of points of M. Before proving this we complete our construction. We define

$$\omega(x) = \int_0^\infty \varphi(x\tau)\, e^{-\tau}\, d\tau.$$

This $\omega(x)$ has all the properties 4.17.1 — 4.17.4 except possibly 4.17.3. The construction is now completed by setting

$$\Phi(x) = \varphi(x) + \omega(x).$$

To see for example that for each $y \in M$, there is a $\delta_y > 0$ such that $\varphi(x)$ is continuous in $S(y, \delta_y)$, we need prove that $\varphi(x)$ is continuous in an open set containing M. If $A(M)$ is the region of attraction of M, then $A(M)$ is invariant, and indeed $\mathscr{I}(A(M))$ is also invariant and open and contains M. We now define the set $W_\varepsilon = \{x \in \mathscr{I}(A(M)) : \gamma^+(x) \subset S(M, \varepsilon)\}$. This W_ε is open positively invariant, and contains an open set containing M, and has the important property that if $x \in \mathscr{I}(A(M))$, then there is a $T > 0$ such that $xT \in W_\varepsilon$. Now let $x \in \mathscr{I}(A(M))$ and let $\varrho(x, M) = \lambda$. There is a $T > 0$ such that $xT \in W_{\lambda/4}$. Since $W_{\lambda/4}$ is open we can find a neighborhood $S(xT, \sigma) \subset W_{\lambda/4}$. Then $S(xT, \sigma)(-T) = N$ is a neighborhood of x, and indeed $N \subset \mathscr{I}(A(M))$. We can thus choose an $\eta > 0$ such that $\eta < \lambda/4$, and $S(x, \eta) \subset N$. Then if $y \in S(x, \eta)$,

$$\varphi(x) - \varphi(y) = \sup\left\{\frac{\varrho(xt, M)}{1+\varrho(xt, M)} : t \geqq 0\right\} - \sup\left\{\frac{\varrho(yt, M)}{1+\varrho(yt, M)} : t \geqq 0\right\}$$
$$= \sup\left\{\frac{\varrho(xt, M)}{1+\varrho(xt, M)} : 0 \leqq t \leqq T\right\} - \sup\left\{\frac{\varrho(yt, M)}{1+\varrho(yt, M)} : 0 \leqq t \leqq T\right\},$$

and so

$$|\varphi(x) - \varphi(y)| \leqq \sup\left\{\left|\frac{\varrho(xt, M)}{1+\varrho(xt, M)} - \frac{\varrho(yt, M)}{1+\varrho(yt, M)}\right| : 0 \leqq t \leqq T\right\}$$
$$= \sup\left\{\left|\frac{\varrho(xt, M) - \varrho(yt, M)}{(1+\varrho(xt, M))\,(1+\varrho(yt, M))}\right| : 0 \leqq t \leqq T\right\}$$
$$\leqq \sup\{|\varrho(xt, M) - \varrho(yt, M)| : 0 \leqq t \leqq T\}$$
$$\leqq \sup\{\varrho(xt, yt) : 0 \leqq t \leqq T\}.$$

By the continuity axiom the right hand side tends to zero as $y \to x$, hence $\varphi(x)$ is continuous in $\mathscr{I}(A(M))$. The rest of the observations on $\varphi(x)$, $\omega(x)$ are easy to verify and are left as an exercise.

4.18 Theorem. Let M be a closed set. Then M is asymptotically stable if and only if there is a function $\varphi(x)$ defined in X with the following properties:

4.18.1 $\varphi(x)$ is continuous in some neighborhood of M which contains the set $S(M, \delta)$ for some $\delta > 0$,

4.18.2 $\varphi(x) = 0$ for $x \in M$, $\varphi(x) > 0$ for $x \notin M$,

4.18.3 there exist strictly increasing functions $\alpha(\mu)$, $\beta(\mu)$, $\alpha(0) = \beta(0) = 0$, defined for $\mu \geqq 0$, such that

$$\alpha(\varrho(x, M)) \leqq \varphi(x) \leqq \beta(\varrho(x, M)),$$

4.18.4 $\varphi(xt) \leqq \varphi(x)$ for all $x \in X$, $t > 0$, and there is a $\delta > 0$ such that if $x \in S(M, \delta) - M$ then $\varphi(xt) < \varphi(x)$ for $t > 0$, and $\varphi(xt) \to 0$ as $t \to +\infty$.

The proof follows exactly the same lines as that of the previous theorem and is left as an exercise. We note, however, that in the proof of necessity, since $A(M)$ is open and invariant, and $A(M) \supset S(M, \delta)$ for some $\delta > 0$, the functions $\varphi(x)$ and $\omega(x)$ can be taken as being defined and continuous on $A(M)$. In the present case $\varphi(x)$ will have the property 4.18.3, whereas $\omega(x)$ may not satisfy the left inequality in 4.18.3 although it will satisfy the right inequality. Thus $\Phi(x) = \varphi(x) + \omega(x)$ will have all the desired properties.

We shall now prove the following very important theorem, which in the case of asymptotic stability of a closed invariant set M, describes the flow in the set $A(M) - M$.

4.19 Theorem. If a closed invariant set $M \subset X$ is asymptotically stable, then for each $x \in A(M)$, $J^+(x) \subset M$, and for each $x \in A(M) - M$, $J^-(x) \cap A(M) = \emptyset$.

Proof. Let, if possible, $x^* \in A(M)$ and $y \in J^+(x^*)$, $y \notin M$. Set $\varrho(y, M) = \alpha\ (> 0)$. Since M is uniformly stable, there is a $\delta > 0$ such that $\gamma^+(S(M, \delta)) \subset S(M, \alpha/2)$. Since $x^* \in A(M)$, there is a $T > 0$ such that $x^*T \in S(M, \delta)$. Since $S(M, \delta)$ is open, there is an $\eta > 0$ such that $S(x^*T, \eta) \subset S(M, \delta)$. Now $N \equiv S(x^*T, \eta)(-T)$ is a neighborhood of x^* such that for each $x \in N$, $xT \in S(x^*T, \eta)$ and consequently $x[T, +\infty) \subset S(M, \alpha/2)$. Now since $y \in J^+(x^*)$, there exist sequences $\{x_n\}$ in X and $\{t_n\}$ in R^+, $t_n \to +\infty$, such that $x_n \to x^*$, $x_n t_n \to y$. We may assume without loss of generality, that $x_n \in N$, and $t_n \geqq T$. But then $x_n t_n \in S(M, \alpha/2)$. Thus if $x_n t_n \to y$, we must have $\varrho(y, M) \leqq \alpha/2$. This is a contradiction, as $\varrho(y, M) = \alpha$. Thus $J^+(x^*) \subset M$. The second statement follows from the fact that if $y \in J^-(x)$, then $x \in J^+(y)$. Now let $x \in A(M) - M$, and assume that $y \in J^-(x) \cap A(M)$. Then we have $y \in A(M)$, $x \in J^+(y)$, $x \notin M$, which has already been ruled out.

4.20 Corollary. Let a closed invariant set M be asymptotically stable. Let $A(M) - M$ (or in particular the space X) be locally compact and contain a countable dense subset. Then the invariant set $A(M) - M$ is parallelizable.

The proof follows from the above theorem, and IV, 2.6.

4.21 *Remark.* The considerations in section 2 show that for X locally compact and $M \subset X$ a compact invariant asymptotically stable set, if $\varphi(x)$ is a function satisfying the conditions of Theorem 2.2 in a neighborhood N of M, then the sets $\{x \in S[M, \varepsilon]: \varphi(x) = \alpha\}$, for fixed $\varepsilon > 0$ such that $S[M, \varepsilon] \subset N$ and α sufficiently small, represent sections of the parallelizable flow in $A(M) - M$. How far the same method of construction can be extended to non-compact closed sets, depends naturally on whether the flow in $A(M) - M$ is parallelizable.

We shall now prove that uniform asymptotic stability of a closed set $M \subset X$ implies that the flow in $A(M) - M$ is parallelizable, even when the subspace $A(M) - M$ of X is assumed to be neither locally compact nor separable.

4.22 Proposition. Let $M \subset X$ be a closed invariant uniformly asymptotically stable set with $A(M)$ as its region of attraction. Then $A(M) - M$ is parallelizable.

Proof. Since M is asymptotically stable, we can find a function $\varphi(x)$ defined on $A(M)$ and having the properties given in Theorem 4.18. Since M is uniformly asymptotically stable, there is an $\alpha > 0$ such that $S[M, \alpha] \subset A(M)$, and such that for any $\varepsilon > 0$, there is a $T > 0$ with the property that $\gamma^+(xT) \subset S(M, \varepsilon)$ for every $x \in S[M, \alpha]$. Now let

$$m_0 = \inf\{\varphi(x): \varrho(x, M) = \alpha\}.$$

Indeed $m_0 > 0$. Consider now any set

$$S_\eta = \{x \in S[M, \alpha]: \varphi(x) = \eta\}.$$

We claim that if $\eta < m_0$, then S_η is a section of the flow in $A(M) - M$ with the property that there is a unique continuous function $\tau(x)$, mapping $A(M) - M$ into R such that for each $x \in A(M) - M$, $x\tau(x) \in S_\eta$. The existence of such a section indeed shows that the flow on $A(M) - M$ is parallelizable (IV, 2.4). To see that S_η has the properties enunciated above, we consider the set $P_\eta = \{x \in S[M, \alpha]: \varphi(x) \leqq \eta\}$. Indeed $P_\eta \subset A(M)$, and $P_\eta \supset M$. We note now that any trajectory in $A(M) - M$ can intersect S_η at most at one point. This is so because if any trajectory in $A(M) - M$ has two points x_1, x_2 on S_η, then we may assume that $x_2 = x_1 t$ where $t > 0$. But then since

$S_\eta \cap M = \emptyset$, $\varphi(x_2) = \varphi(x_1 t) < \varphi(x_1)$, which contradicts the definition of S_η. To see that every trajectory γ in $A(M) - M$ intersects S_η we note first that if $x \notin P_\eta$, there is a $t > 0$ such that $xt \in P_\eta$. But then $x[0, t] \cap \partial P_\eta \neq \emptyset$. However, $S_\eta \equiv \partial P_\eta$. If $x \in P_\eta$, then we claim that there is a $t \leqq 0$ such that $xt \in S_\eta$. For otherwise $\gamma^-(x) \subset \mathscr{I} P_\eta$. In this case we can set $\delta = \inf\{\varrho(xt, M): t \leqq 0\}$, and $\delta > 0$ (otherwise M will be unstable). If now $T > 0$ is such that $y \in S[M, \alpha]$ implies $\varrho(yT, M) < \delta/2$ (such $T > 0$ exists by uniform asymptotic stability), then $x(-T) = y \in P_\eta \subset S[M, \alpha]$, but $yT = x \notin S(M, \delta/2)$. This contradiction shows that every trajectory in $A(M) - M$ intersects S_η exactly once. We now define the function $\tau\colon A(M) - M \to R$ by the requirement that $x\tau(x) \in S_\eta$ for $x \in A(M) - M$. Then $\tau(x)$ is uniquely defined and is continuous. The continuity follows in the same way as in the proof of Theorem 2.9. We have thus proved our proposition.

We shall now prove the following theorem.

4.23 Theorem. A closed invariant set M is uniformly asymptotically stable and equi-attracting with an open set N containing $S(M, \delta)$ for some $\delta > 0$ as its region of attraction, if and only if there exists a continuous function $\varphi(x)$ defined on N with the following properties:

4.23.1 $\varphi(x) = 0$ for $x \in M$, $\varphi(x) > 0$ for $x \notin M$;

4.23.2 there exist strictly increasing continuous functions $\alpha(r)$, $\beta(r)$, $\alpha(0) = \beta(0) = 0$ such that

$$\alpha(\varrho(x, M)) \leqq \varphi(x) \leqq \beta(\varrho(x, M));$$

4.23.3 there is a sequence of closed sets $\{E_n\}$ such that $\mathscr{I}(E_{n+1}) \supset E_n \supset S(M, \delta_n)$ for some $\delta_n > 0$; $\cup\{E_n\colon n = 1, 2, 3, \ldots\} = N$, and this sequence has the property that for any $\alpha > 0$ there is an integer n_0 such that $\varphi(x) > \alpha$ for $x \notin E_{n_0}$;

4.23.4 $\varphi(xt) = e^{-t}\varphi(x)$ for $x \in N$, $t \in R$.

Proof. We shall not give complete details as the arguments are similar to those used earlier. For the proof of sufficiency we remark that 4.23.3 and 4.23.4 imply that N is invariant. 4.23.1, 4.23.2, and 4.23.4 ensure uniform stability, as well as attraction, and show that $N \subset A(M)$. Since N is an invariant neighborhood, it follows that $N = A(M)$. Uniform attraction and equi-attraction follow from 4.23.2 and 4.23.4. To prove necessity, we consider the region of attraction $A(M)$ and define a $\varphi(x)$ on $A(M)$ as in Theorem 4.18. We then consider a section S_η of the flow in $A(M) - M$ defined by this $\varphi(x)$ (Proposition 4.22), with the corresponding continuous map $\tau: A(M) - M \to R$. Lastly we define

$$\varphi(x) = e^{\tau(x)} \quad \text{for} \quad x \in A(M) - M,$$

and

$$\varphi(x) = 0 \quad \text{for} \quad x \in M.$$

This $\varphi(x)$ is easily shown to have all the properties 4.23.1—4.23.4. Note that to get the sequence $\{E_n\}$, we set $E_0 = \{x: \varphi(x) \leqq 1\}$. Then define $E_n = E_0[-n, 0]$. These sets are closed and have the required properties.

Setting $\Phi(x) = -\dfrac{\varphi(x)}{1+\varphi(x)}$, we obtain the following very important corollary.

4.24 Corollary. A closed invariant set M is uniformly asymptotically stable and equi-attracting, with an open set N containing $S(M, \delta)$ for some $\delta > 0$ as its region of attraction, if and only if there exists a continuous function $\Phi(x)$ defined on N with the following properties:

4.24.1 $-1 < \Phi(x) < 0$ for $x \in N - M$,

4.24.2 $\Phi(x) \to 0$ as $\varrho(x, M) \to 0$,

4.24.3 for any $\varepsilon > 0$ there is a $\delta > 0$ such that $\Phi(x) \leqq -\varepsilon$ for $\varrho(x, M) \geqq \delta$,

4.24.4 $\Phi(x) \to -1$ as $x \to y \in \partial N$,

4.24.5 $\left.\dfrac{d\Phi(xt)}{dt}\right|_{t=0} = -\Phi(x)\,(1 + \Phi(x))$.

We shall now give a theorem along the lines of the Theorem 4.18 for the case of uniform asymptotic stability.

4.25 Theorem. Let the space X be locally compact and separable. Then a closed set $M \subset X$ is uniformly asymptotically stable, with an open set N containing $S(M, \delta)$ for some $\delta > 0$ as its region of attraction, if and only if there exists a continuous function $\varphi(x)$ defined on N and having the following properties:

4.25.1 $\varphi(x) = 0$ for $x \in M$, $\varphi(x) > 0$ for $x \notin M$,

4.25.2 there exist continuous strictly increasing functions $\alpha(r)$, $\beta(r)$, $\alpha(0) = \beta(0)$, such that

$$\alpha(\varrho(x, M)) \leqq \varphi(x) \leqq \beta(\varrho(x, M)),$$

4.25.3 there exists a sequence of closed sets $\{E_n\}$, $E_n \subset \mathscr{I} E_{n+1}$, $\bigcup_{n=1}^{\infty} E_n = N$, such that given any $\alpha > 0$ there is an n_0 such that $\varphi(x) > \alpha$ if $x \notin E_{n_0}$, and on every E_n, $\varphi(x)$ is bounded,

4.25.4 $\varphi(xt) \leqq e^{-t}\varphi(x)$.

The conditions can easily be shown to be sufficient. To prove the necessity we need the following lemma.

4.26 Lemma. Let $f(r, x)$ be a function from $(0, 1]\times X \to [0, +\infty)$, where X is locally compact separable metric space. Let $f(r, x)$ be bounded on every compact subset of $(0, 1]\times X$. Then there exist two functions $H(r)$ and $G(x)$ defined on $(0, 1]$ and X respectively, which are bounded on compact subsets of $(0, 1]$ and X respectively (and may even be chosen continuous), such that

$$f(r, x) \leqq H(r) \cdot G(x).$$

Proof. Since X is locally compact and separable we can find a sequence of compact sets U_n such that $U_n \subset U_{n+1}$ and $X = \bigcup_{n=1}^{\infty} U_n$. We now define $H(r) = \sup\{f(r, x) + 1 \colon x \in U_{n_0}$, where $1 + 1/r \geqq n_0 \geqq 1/r\}$, and $G(x) = \sup\left\{\frac{f(r, x)}{H(r)} : 1 \geqq r > 0\right\}$. The above defined $H(r)$ and $G(x)$ have the required properties. Indeed $H(r)$ is defined here as a step function.

Proof of Necessity of Theorem 4.25. Let $A(M)$ be the region of attraction of the set M. Since M is an attractor, there is an $\alpha > 0$ such that $S(M, \alpha) \subset A(M)$. We may choose $\alpha \leqq 1$. For each $r \in (0, \alpha)$ define $T(r, x) = \inf\{\tau > 0 \colon xt \in S(M, r)$ for $t \geqq \tau\}$. We assert that $T(r, x)$ is bounded on any compact set $K \subset A(M)$ and for fixed r. To prove this we need to show that for each compact $K \subset A(M)$, and $r > 0$, there exists a $T > 0$ such that $Kt \subset S(M, r)$ for $t \geqq T$. We note first that by stability of M, there is a $\delta > 0$ such that $y \in S(M, \delta)$ implies $\gamma^+(y) \subset S(M, r)$. For $x \in A(M)$, choose $T(x)$ such that $xT(x) \in S(M, \delta)$. Since $S(M, \delta)$ is open, we can choose a $\sigma > 0$ such that $\overline{S(xT(x), \sigma)}$ is compact and contained in $S(M, \delta)$. Then its inverse image

$$N_x = \overline{S(xT(x), \sigma)}\ (-T(x))$$

is a compact neighborhood of x. N_x has, moreover, the property that $N_x T(x) \subset S(M, \delta)$. Thus we have in fact shown that for each $x \in A(M)$, there exists a $T(x)$ and a $\varrho(x) > 0$ such that $y \in S(x, \varrho(x))$ implies $yt \in S(M, r)$ for $t \geqq T(x)$. Consider now the open cover $\{S(x, \varrho(x)) \colon x \in K\}$ of the compact set K. By the Heine-Borel theorem, there exist a finite number of sets, say, $S(x_1, \varrho(x_1)), \ldots, S(x_n, \varrho(x_n))$ which cover K. We can now choose $T = \max(T(x_1), \ldots, T(x_n))$. Then $x \in K$ implies $xt \in S(M, r)$ for $t \geqq T$.

For any given integer $n > 1/\alpha$, define

$$\varphi_n(x) = \sup\{\varrho(x\tau, S(M, 1/n)) \cdot \exp(\tau) \colon \tau \geqq 0\}.$$

We assert that $\varphi_n(x)$ is continuous on $A(M)$. To see this, note that for $\varrho > 0$ such that $\overline{S(x, \varrho)}$ is a compact subset of $A(M)$, there exists a $T > 0$ such that

$$\varrho(y\tau, S(M, 1/n)) = 0$$

for $y \in S(x, \varrho)$ and $\tau \geqq T$. Therefore, if $y \in S(x, \varrho)$ we have

$$\begin{aligned} |\varphi_n(x) - \varphi_n(y)| &= |\sup\{\varrho(x\tau, S(M, 1/n)) \cdot \exp(\tau) \colon 0 \leqq \tau \leqq T\} \\ &\quad - \sup\{\varrho(y\tau, S(M, 1/n)) \cdot \exp(\tau) \colon 0 \leqq \tau \leqq T\}| \\ &\leqq \exp(T) \cdot \sup\{|\varrho(x\tau, S(M, 1/n)) - \varrho(y\tau, S(M, 1/n))| \colon 0 \leqq \tau \leqq T\}. \end{aligned}$$

This implies that

$$|\varphi_n(x) - \varphi_n(y)| \leqq \exp(T) \cdot \sup\{\varrho(x\tau, y\tau) \colon 0 \leqq \tau \leqq T\}.$$

Using the continuity axiom we conclude that the right hand side tends to zero as $\varrho(x, y) \to 0$. Thus $\varphi_n(x)$ is continuous on $A(M)$. This $\varphi_n(x)$ has further the following important property

$$\varphi_n(xt) \leqq \exp(-t)\, \varphi_n(x), \qquad \text{for} \quad t > 0.$$

To see this, note that for $t \geqq 0$

$$\begin{aligned} \varphi_n(xt) &= \sup\{\varrho(x(t + \tau), S(M, 1/n)) \exp(\tau) \colon \tau \geqq 0\} \\ &= \sup\{\varrho(x\tau, S(M, 1/n)) \exp(\tau - t) \colon \tau \geqq t\} \\ &= \exp(-t) \cdot \sup\{\varrho(x\tau, S(M, 1/n)) \exp(\tau) \colon \tau \geqq t\} \\ &\leqq \exp(-t) \cdot \varphi_n(x) \text{ as } t \geqq 0. \end{aligned}$$

We now note that

$$\varphi_n(x) = \sup\{\varrho(x\tau, S(M, 1/n)) \exp(\tau) \colon 0 \leqq \tau \leqq T(1/n, x)\}$$

as $\varrho(x\tau, S(M, 1/n)) = 0$ for $\tau \geqq T(1/n, x)$. Thus

$$\varphi_n(x) \leqq \exp(T(1/n, x)) \sup\{\varrho(x\tau, S(M, 1/n)) \colon \tau \geqq 0\}.$$

Since the function $\exp(T(1/n, x))$ has the properties of the function $f(r, x)$ of Lemma 4.26, we can choose a function $H(r)$, such that $\varphi_n(x)/H(1/n)$ is uniformly bounded on each compact subset $K \subset A(M)$.

We now define

$$\Phi(x) = \sum_{n=n_0}^{\infty} \varphi_n(x)/H(1/n)\, n!, \text{ where } n_0 \geqq 1/\alpha.$$

Then $\Phi(x)$ is continuous on $A(M)$ and

$$\Phi(xt) \leqq e^{-t}\, \Phi(x).$$

Note that $\Phi(x) = 0$ for $x \in M$, and $\Phi(x) > 0$ for $x \notin M$. By uniform asymptotic stability, $\Phi(x) \to 0$ if $\varrho(x, M) \to 0$; thus there is a strictly increasing function $\beta(r)$, $\beta(0) = 0$ such that

$$\Phi(x) \leqq \beta(\varrho(x, M)).$$

Further if $\varrho(x, M) \geqq \varepsilon > 0$, then for sufficiently large n, $\varphi_n(x) \geqq \delta_n > 0$ for some δ_n. And hence $\Phi(x) > \delta > 0$. Thus there is a strictly increasing function $\alpha(r)$, $\alpha(0) = 0$ such that $\Phi(x) \geqq \alpha(\varrho(x, M))$. We now choose $k > 0$ such that

$$k < \inf\{\Phi(x) : \varrho(x, M) = \alpha\}.$$

Consider the sets

$$P_k = \{x \in A(M) : \Phi(x) < k\} \cap S[M, \alpha],$$

and

$$S_k = \{x \in A(M) : \Phi(x) = k\} \cap S[M, \alpha].$$

Then as shown in Theorem 4.23, S_k is the section of the flow in $A(M)$ consisting of all those trajectories in $A(M)$ which are not in M. For each $x \in A(M)$, $x \notin P_k$, we can define $\tau(x)$ by the requirement that $x\tau(x) \in S_k$. Then $\tau(x)$ is continuous. Now define

$$\varphi(x) = \Phi(x) \quad \text{for} \quad x \in P_k,$$

and

$$\varphi(x) = k e^{\tau(x)} \quad \text{for} \quad x \notin P_k.$$

This $\varphi(x)$ has all the properties required in the theorem as may easily be verified.

4.27 *Remark.* Note that if in the above theorem we assume M to be invariant and construct $\varphi(x)$ as in Theorem 4.23, then this $\varphi(x)$ need not satisfy the property 4.25.2 in the above theorem. Indeed if that were the case, then construction of the $\varphi_n(x)$ would be superfluous and then uniform asymptotic stability would imply equi-attraction, which is indeed not the case.

4.28 *Exercises.*

4.28.1 Give proofs or provide the missing details in the proofs of Proposition 4.4, Theorems 4.5, 4.7, 4.9, Propositions 4.14, 4.15, Theorems 4.16, 4.17, 4.18, 4.20, 4.23, 4.24, 4.25.

4.28.2 Show that in a locally compact metric space X, a compact set M is stable if and only if there exists a lower semi-continuous function $\varphi(x)$ defined on some neighborhood N of M and having the properties: (a) $\varphi(x) = 0$ for $x \in M$, $\varphi(x) > 0$ for $x \notin M$, (b) $\varphi(x)$ is continuous on M, and (c) $\varphi(xt) \leqq \varphi(x)$ if $x[0, t] \subset N$.

4.28.3 Give examples to show that the various concepts introduced in this section are indeed distinct concepts.

4.28.4 In the following figures, trajectories of certain dynamical systems defined in the euclidean plane R^2 are shown. The x_1-axis in each case has an appropriate stability or attractor property, but no stronger property is implied. Such a property is written just below the figure. Construct examples of differential equations which conform to the behavior indicated in each figure.

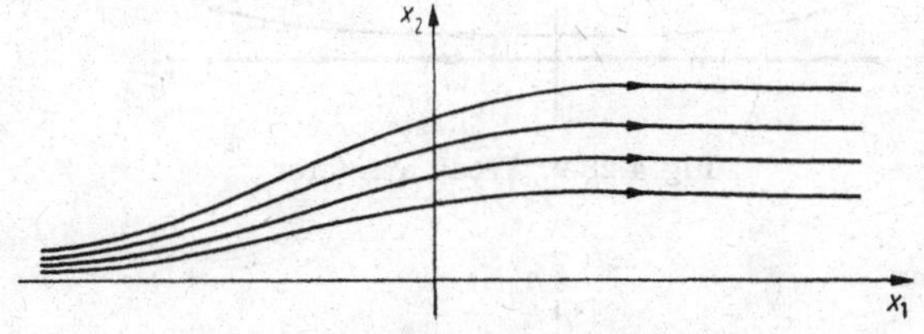

Fig. 4.28.5. Stable but not equistable

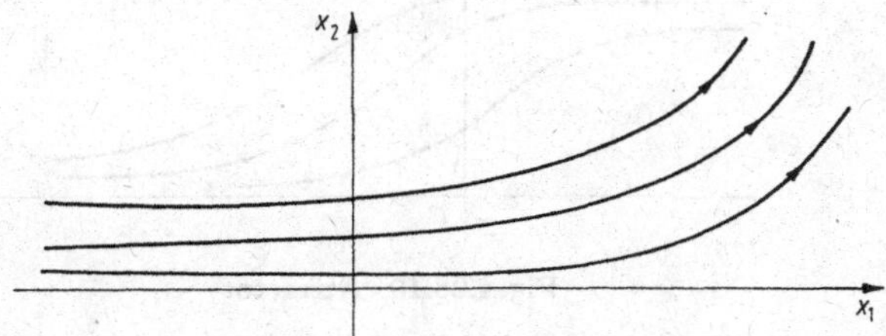

Fig. 4.28.6. Equistable but not stable

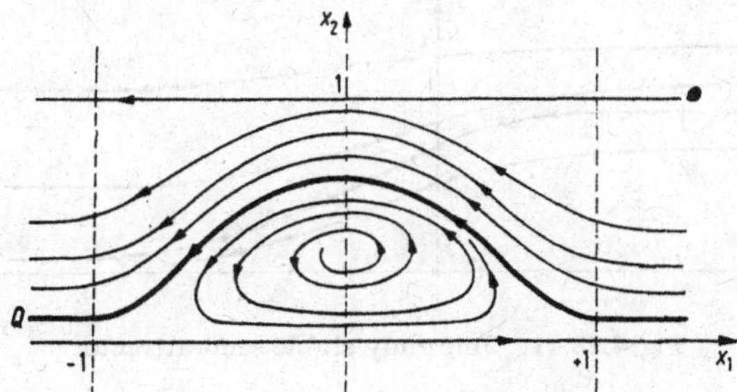

Fig. 4.28.7. Semi-weak attractor

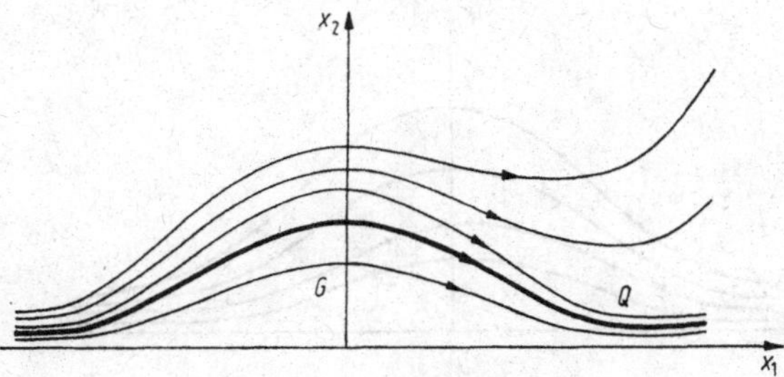

Fig. 4.28.8. Semi-attractor

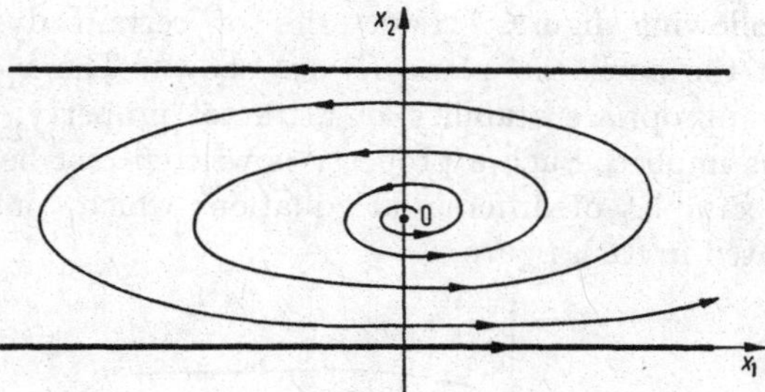

Fig. 4.28.9. Weak attractor

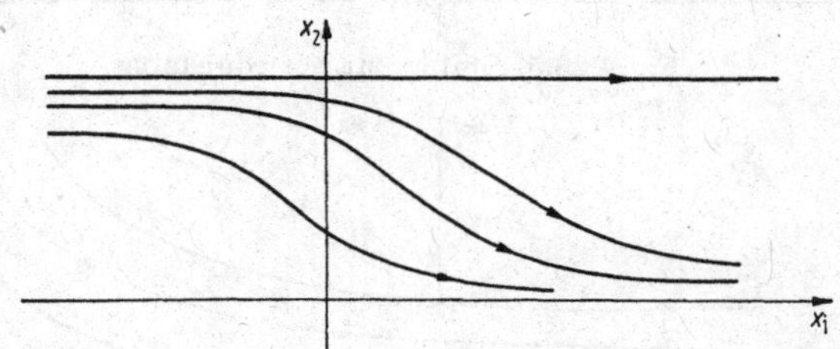

Fig. 4.28.10. Attractor

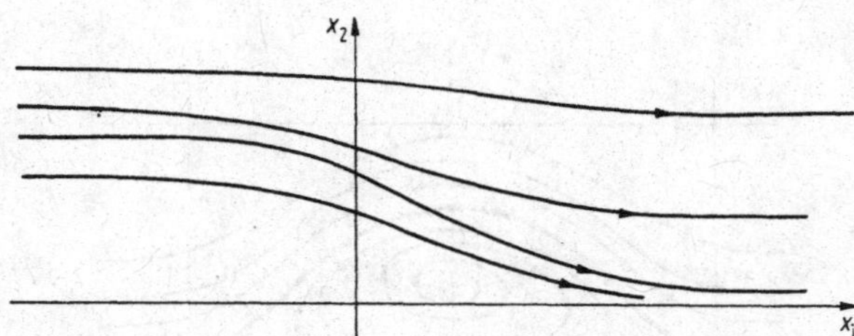

Fig. 4.28.11. Uniformly stable semi-attractor

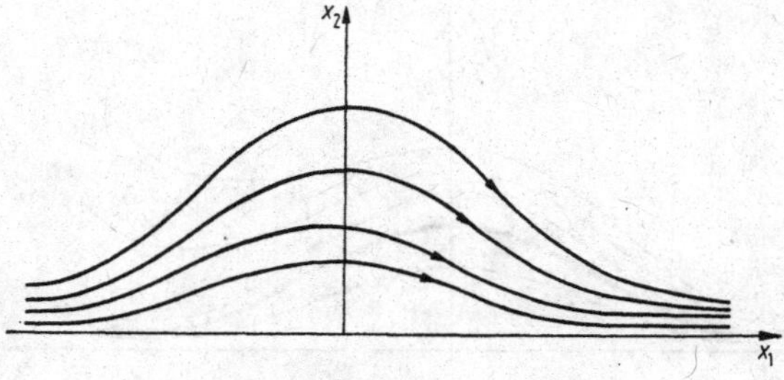

Fig. 4.28.12. Stable attractor

4.28.13 In the euclidean plane, consider the system

$$\dot{x}_1 = 1, \dot{x}_2 = 0 \qquad \text{for } x_1 \leqq 0,$$

$$\dot{x}_1 = 1, \dot{x}_2 = -\frac{2x_1x_2}{(1+x_1^2)} \quad \text{for } x_1 \geqq 0.$$

The solution through any point (x_1^0, x_2^0) has the form

$$x_1 = t + x_1^0, x_2 = \frac{1+(x_1^0)^2}{1+(t+x_1^0)^2} x_2^0 \text{ for } t \geqq -x_1^0,$$

and

$$x_1 = t + x_1^0, \quad x_2 = x_2^0 \quad \text{for } t \leqq -x_1^0.$$

The x_1-axis is a uniformly stable attractor, but is not a uniform attractor.

4.29 *Remark.* The notions in this section are described in terms of positive semi-trajectories and/or using sequences $\{xt_n\}$ with $t_n \to +\infty$ (on a positive semi-trajectory $\gamma^+(x)$). It is clear that notions dual to these may be obtained by using negative semi-trajectories and/or using sequences $\{xt_n\}$ with $t_n \to -\infty$ on a negative semi-trajectory $\gamma^-(x)$. The dual notions, whenever needed, will be described by using the adjective negative before the appropriate term describing that notion. One should also note that some of these negative notions have in the literature been termed as instability notions.

5. Relative Stability Properties

In what follows the space X is assumed to be locally compact. In this section we shall use the concept of relative prolongation $D^+(M, U)$ and of relative prolongational limit set $J^+(M, U)$ (Definition II, 4.10) of a set $M \in X$, relative to the set $U \subset X$. These concepts will allow us to characterize the properties of relative stability and attraction of a compact set.

5.1 **Definition.** Given a compact set $M \subset X$ and a set $U \subset X$, the set M is said to be

5.1.1 stable, relatively to the set U, if given an $\varepsilon > 0$ there exists $\delta > 0$, such that

$$\gamma^+(S(M, \delta) \cap U) \subset S(M, \varepsilon);$$

5.1.2 a weak attractor relative to U, if

$$\Lambda^+(x) \cap M \neq \emptyset, \text{ for each } x \in U;$$

5.1.3 an attractor relative to U, if

$$\Lambda^+(x) \neq \emptyset, \Lambda^+(x) \subset M, \text{ for each } x \in U;$$

5.1.4 a uniform attractor relative to U, if

$$J^+(x, U) \neq \emptyset, \; J^+(x, U) \subset M, \text{ for each } x \in U;$$

5.1.5 asymptotically stable, relatively to U if M is a uniform attractor relative to U and it is positively invariant.

5.2 *Remark.* If in Definition 5.1 U is neighborhood of M, then the stability, weak attraction, attraction, uniform attraction and asymptotic stability of M, relative to U, reduces to the stability, weak attraction, attraction, uniform attraction and asymptotic stability of M, respectively, as defined in Definition 1.5.

5.3 **Theorem.** A compact set $M \subset X$ is stable relatively to $U \subset X$ if and only if $M \supset D^+(M, U)$.

Proof. (i) *Sufficiency.* Let $M \supset D^+(M, U)$ and let, if possible, M be not stable relative to U. Then there is an $\varepsilon > 0$; a sequence $\{x_n\}$ in U, $x_n \to x \in M$, and a sequence $\{t_n\}$, $t_n \geqq 0$, such that $\varrho(x_n t_n, M) = \varepsilon$. We may assume that $H(M, \varepsilon)$ is compact. Thus the sequence $\{x_n t_n\}$ may be assumed to converge to a point $y \in H(M, \varepsilon)$. Then $y \in D^+(x, U) \subset D^+(M, U)$, but $y \notin M$. This contradiction proves sufficiency.

(ii) *Necessity.* Let M be relatively stable with respect to U. Then clearly $D^+(M, U) \subset S[M, \varepsilon]$ for arbitrary $\varepsilon > 0$. Hence $D^+(M, U) \subset \cap \{S[M, \varepsilon] : \varepsilon > 0\} = M$. This proves the theorem.

5.4 *Remark.* In the above theorem or definitions, the set U need not contain the point x or the set M. If, however, $U \supset M$, then in the above theorem we may replace condition $M \supset D^+(M, U)$ by $M = D^+(M, U)$. Further, in case U is a neighborhood of M, one obtains Theorem 1.12.

We have now a theorem for the characterization of relative uniform attraction. Its proof follows that of the analogous proposition 1.2 and it is left as an exercise to the reader.

5.5 **Theorem.** A necessary and sufficient condition for a compact set M to be a uniform attractor relative to $U \subset X$ is that for each $\varepsilon > 0$ and each compact set $K \subset U$ we can find a real number $T(K, \varepsilon) > 0$, such that $Kt \subset S(M, \varepsilon)$ for each $t \geqq T$.

We shall now prove a theorem (5.10) on relative weak attraction, which is an extension of the weak attractor theorem (1.25). Its proof will be based upon the two following Lemmas 5.6 and 5.8.

5.6 Lemma. Let $M \subset X$ be a compact set such that $A_\omega(M) \neq M$. Let $U \subset A_\omega(M)$ be a set with the following properties:

5.6.1 U is closed and positively invariant.

5.6.2 $U \cap M \neq \emptyset$.

Then the set $D^+(M, U)$ is non-empty and compact and in addition, given $\varepsilon > 0$ we can find $T > 0$ such that

5.6.3 $D^+(M, U) \subset S[M, \varepsilon] \cdot [0, T]$.

Proof. Choose $\varepsilon > 0$ so small that the set $V = H(M, \varepsilon) \cap U$ is non-empty and compact. Let

$$\tau_x = \operatorname{Inf} \{t \in R^+ \colon xt \in S(M, \varepsilon)\}$$

and

$$T = \operatorname{Sup} \{\tau_x \colon x \in V\}.$$

Clearly $T < +\infty$. If not there would exist a sequence $\{x^n\} \subset V$ such that $\tau_{x^n} \to +\infty$. Since V is compact there exists a point $x \in V$ such that $x^n \to x$ and $x\tau \in S(M, \varepsilon)$, $\tau > 0$. Then $\tau_{x^n} < \tau$ for a sufficiently large n; which is absurd.

Let now

$$y \in D^+(M, U) - S[M, \varepsilon].$$

Then there exist sequences $\{x^n\} \subset U$ and $\{t_n\} \subset R^+$ such that $x^n \to x \in M$ and $x^n t_n \to y$.

Notice that from the hypothesis 5.6.1, $\gamma^+(x^n) \subset U$, $x^n \in S(M, \varepsilon)$ and $x^n t_n \in \mathscr{C}\{S(M, \varepsilon)\}$ for sufficiently large values of n. Hence there exists a sequence $\{\tau_n\} \subset R^+$ such that $x^n \tau_n \in H(M, \varepsilon) \cap U = V$ and $x^n t \notin S[M, \varepsilon]$ for $t \in (\tau_n, t_n]$. Since $x^n t_n = x^n \tau_n\ (t_n - \tau_n)$ and $x^n \tau_n \in V$, it follows that $x^n t_n \in V[0, T] \subset S[M, \varepsilon] \cdot [0, T]$.

5.7 *Remark.* The hypothesis 5.6.1 is necessary for Lemma 5.6 to be true. Consider for that the planar flow shown in Fig. 5.7.1, where P is a rest point.

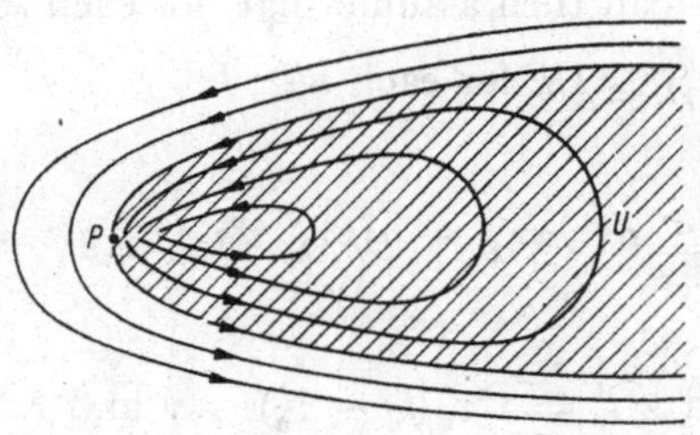

Fig. 5.7.1

In this case the region of weak attraction of P is not closed and if $U = A_\omega(P)$, the set $D^+(M, U)$ is not compact. Consider next the planar flow shown in Fig. 5.7.2.

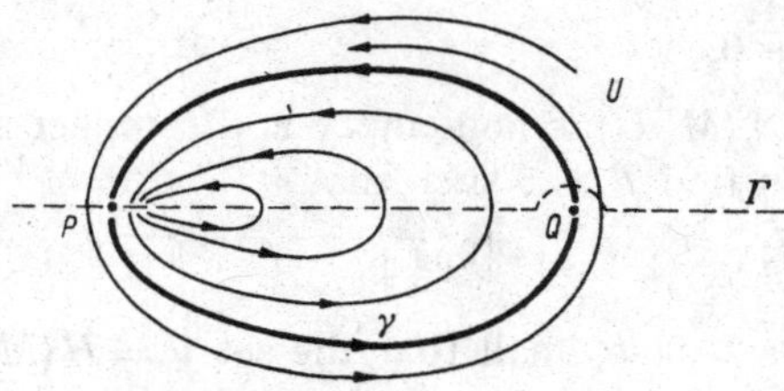

Fig. 5.7.2

In this flow the points P and Q are rest points, and $A_\omega(P) = R^2 - \{Q \cup \gamma\}$. Let U be the closed set above the dotted line Γ, this set is closed, contained in $A_\omega(P)$, but it is not positively invariant. In this case we have $Q \in D^+(P, U)$ and therefore Lemma 5.6 does not hold.

The hypothesis 5.6.2 on the other hand ensures that Lemma 5.6 is not trivial. In fact since U is closed, $U \cap M \neq \emptyset$ would imply that $D^+(M, U) \neq \emptyset$.

5.8 Lemma. Let $x \in X$ and $U \subset X$ be positively invariant, then $z \in \Lambda^+(x)$ implies

$$J^+(x, U) \subseteq D^+(z, U).$$

Proof. If $x \notin \bar{U}$ the theorem is trivially true. Assume then that $x \in \bar{U}$. If $z \in \Lambda^+(x)$ and $y \in J^+(x, U)$ there exist then the following sequences:

5.8.1 $\{\tau_n'\} \subset R^+,\ \tau_n' \to +\infty,\ \{x\tau_n'\} \subset \bar{U},\ x\tau_n' \to z.$

5.8.2 $\{x^n\} \subset U,\ x^n \to x.$

5.8.3 $\{t_n'\} \subset R^+,\ t_n' \to +\infty,\ \{x^n t_n'\} \subset \bar{U},\ x^n t_n' \to y.$

We can assume, possibly by choosing suitable subsequences that $t_n' - \tau_n' > 0$ for each n. For k fixed, consider the sequence $\{x^n\tau_k'\} \subseteq U$. We have $x^n\tau_k' \to x\tau_k' \in U$. We can then assume that, for each k,

5.8.4 $\varrho(x^n\tau_k', x\tau_k') \leqq 1/k$ for each $n \geqq k$.

Now

5.8.5 $\varrho(z, x^n\tau_n') \leqq \varrho(z, x\tau_k') + \varrho(x\tau_k', x^n\tau_n') \leqq \varrho(z, x\tau_k') + 1/n \to 0$

as $n \to +\infty$.

On the other hand $x^n t_n' = x^n\tau_n'(t_n' - \tau_n') \to y$ and $\{x^n\tau_n'\} \subset U,\ x^n\tau_n' \to z$, $\{t^n - \tau_n'\} \subset R^+$. Thus $y \in D^+(z, U)$.

5.9 *Remark.* In Lemma 5.8 the set U does not have to be closed. On the other hand the positive invariance of U is essential. This can be shown from the following planar flow (Fig. 5.9.1), in which $z \in \Lambda^+(x)$ and $D^+(z, U) = \emptyset$, while $J^+(x, U) \neq \emptyset$.

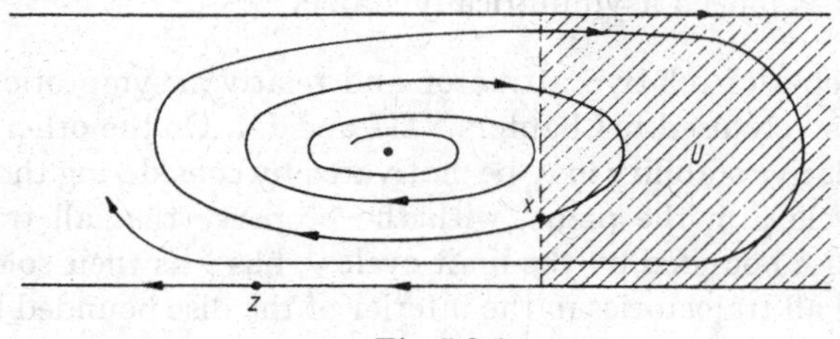

Fig. 5.9.1

We shall proceed next to the proof of the theorem on the prolongation of a relative weak attractor.

5.10 **Theorem.** Let $M \subset X$ be compact and such that $A_\omega(M) - M \neq \emptyset$. Let $U \subseteq A_\omega(M)$ be a set with the following properties:

5.10.1 U is closed and positively invariant.

5.10.2 $U \cap M \neq \emptyset$.

Then the set $D^+(M, U)$ is compact and asymptotically stable relatively to U.

Proof. From Lemma 5.6 we have that $D^+(M, U)$ is a compact set, which is contained in $U \subseteq A_\omega(M)$. We shall see that $J^+[D^+(M, U), U] \subseteq D^+(M, U)$. Let $x \in D^+(M, U) \subset U$, then $\Lambda^+(x) \cap M \neq \emptyset$. Let $z \in \Lambda^+(x) \cap M$. From Lemma 5.8 it follows that $J^+(x, U) \subseteq D^+(z, U) \subseteq D^+(M, U)$. Thus $D^+(M, U)$ is a uniform attractor relative to U. Since $D^+(M, U)$ is also positively invariant the theorem is proved.

5.11 *Remark.* From what was previously seen, it follows that a necessary and sufficient condition for a compact set $M \subset X$ to be asymptotically stable relatively to a closed and positively invariant set $U \subset X$ with $M \subset U$, is that M is a stable weak attractor relative to U.

5.12 **Theorem.** Let $M \subset X$ be compact and positively invariant and let $\tilde{M} \subset M$ be the largest invariant set contained in M. Then $\tilde{M}$ is a stable attractor, relative to M.

Proof. For any $x \in M$, $\Lambda^+(x) \neq \emptyset$ and compact, since $\overline{\gamma^+(x)} \subset M$. Again $\Lambda^+(x) \subset \tilde{M}$, because $\Lambda^+(x)$ is an invariant subset of M. Hence $\tilde{M}$ is an attractor, relative to M. To see that it is stable, one easily shows that $J^+(\tilde{M}, M)$ is an invariant subset of M; then by definition of $\tilde{M}$, $D^+(\tilde{M}, M) \subset \tilde{M}$ holds, so that $\tilde{M}$ is stable relative to M by 5.3.

5.13 *Remark.* In the case in which in Theorem 5.12 we have that $\tilde{M} \subset \mathscr{I}M$, M is indeed asymptotically stable.

The concepts of relative attractor and relative asymptotic stability will be extensively used in Chapters VIII and IX. On the other hand the concept of relative stability may be motivated by considering the example of a limit cycle γ in the plane, with the property that all trajectories outside the disc bounded by the limit cycle γ, has γ as their sole positive limit set, and all trajectories in the interior of the disc bounded by γ tend to the equilibrium point 0.

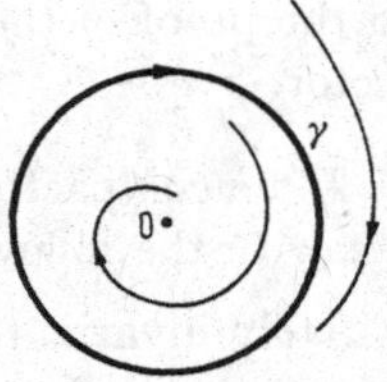

Fig. 5.13.1

Notice that if U is the complement of the disc bounded by γ, then γ is relatively stable with respect to U. Notice also that if γ is an asymptotically stable limit cycle, then γ is stable with respect to every component of $R^2 - \gamma$. These considerations lead to the following definition and theorem.

5.14 **Definition.** Let $M \subset X$ be compact. We say that M is *component-wise stable* if M is relatively stable with respect to every component of $X - M$.

We have then

5.15 **Theorem.** Let a compact set $M \subset X$ be positively stable. Then M is component-wise stable.

The proof is obvious and is omitted.

The converse of Theorem 5.15 is in general not true. To see this, we consider a simple example.

5.16 *Example.* Let $X \subset R^2$ be given by $X = \{(x, y) \in R^2: y = 1/n,\ n$ any integer, or $y = 0\}$. The space X is a metric space with the distance between any two points being the euclidean distance between the points in R^2. We define a dynamical system on X by the differential equations

5.16.1 $\dot{x} = |y|, \quad \dot{y} = 0.$

Then the set $\{(0, 0)\} \subset X$ is component-wise stable, but is not stable.

The question now arises, as to when the converse of Theorem 5.15 is true. For this purpose the following definition is convenient.

5.17 Definition. Let $M \subset X$ be compact. We shall say that the pair (M, X) is stability-additive if the converse of Theorem 5.15 holds for every dynamical system defined on X which admits M as an invariant set.

In this connection the following theorems are important.

5.18 Theorem. The pair (M, X) is stability-additive if $X - M$ has a finite number of components.

5.19 Theorem. The pair (M, X) is stability-additive if $X - M$ is locally connected.

The proof of Theorem 5.18 is immediate and is left as an exercise. (It is indeed a corollary of Theorem 5.19.)

5.20 *Proof of Theorem 5.19.* Let M be a compact invariant set for a given dynamical system on X and let M be component-wise stable. Since M is locally compact, there is an $\varepsilon > 0$ such that $S[M, \varepsilon]$, and hence also $H(M, \varepsilon)$, is compact. We claim that only a finite number of components of $X - M$ can intersect $H(M, \varepsilon)$. For otherwise, if an infinite number of components of $X - M$ intersect $H(M, \varepsilon)$, then we may choose a sequence of points $\{x_n\}$ in $H(M, \varepsilon)$ such that no two points of the sequence are in the same component. Since $H(M, \varepsilon)$ is compact we may assume that $x_n \to x \in H(M, \varepsilon)$. Since X is locally connected, there is a neighborhood of x, say N, such that N is a subset of a component of $X - M$. Now there is an integer n_0 such that $x_n \in N$ for $n \geqq n_0$ and hence all x_n for $n \geqq n_0$ belong to the same component of $X - M$, which is a contradiction. Now notice that every component of $X - M$ is an invariant set. If $C_1, C_2, \ldots, C_p$ are the components of $X - M$ which intersect $H(M, \varepsilon)$, then by component-wise stability we have positive numbers $\delta_1, \delta_2, \ldots, \delta_p$ such that $\gamma^+(S(M, \delta_i) \cap C_i) \subset S(M, \varepsilon)$, $i = 1, 2, \ldots, p$. If now $\delta = \min(\delta_1, \delta_2, \ldots, \delta_p)$ we get $\gamma^+(S(M, \delta)) \subset S(M, \varepsilon)$, i.e., M is stable. To see this last assertion, note that if $x \in S(M, \delta)$, then either $x \in C_i$ for some $i = 1, 2, \ldots, p$ and hence $\gamma^+(x) \subset S(M, \varepsilon)$, or x is an element of a component of $X - M$ which does not intersect $H(M, \varepsilon)$ and hence is contained in $S(M, \varepsilon)$, or $x \in M$. In either of the last two cases $\gamma^+(x) \subset S(M, \varepsilon)$. The theorem is proved.

6. Stability of a Motion and Almost Periodic Motions

In this section we shall assume throughout that the metric space X is complete.

The concept of almost periodicity is intermediate between that of periodicity and recurrence, and the concept of stability of motion plays a central role in its study. We therefore first introduce the concept of stability of a motion.

6.1 Definition. A motion π_x is said to be positively (Liapunov) stable in a subset N of X if for any $\varepsilon > 0$, there is a $\delta > 0$ such that $y \in N \cap S(x, \delta)$ implies $\varrho(xt, yt) < \varepsilon$ for $t \in R^+$.

Any motion π_x is called *negatively stable*, or *stable in both directions* in a subset N of X, if the above condition is satisfied with $t \in R^+$ replaced by $t \in R^-$ or $t \in R$, respectively.

If in the above definition N is a neighborhood of x, then the qualifier "in the subset N of X" will be deleted. Thus a motion π_x is positively stable if given $\varepsilon > 0$, there is a $\delta > 0$ such that $y \in S(x, \delta)$ implies $\varrho(xt, yt) < \varepsilon$ for $t \in R^+$.

6.2 *Exercise.* Show that a motion π_x is positively stable if and only if every motion π_{xt}, $t \in R$, is positively stable.

6.3 Definition. If $A \subset B \subset X$, then the motions through A (i.e., motions π_x with $x \in A$) will be called *uniformly positively stable, uniformly negatively stable*, or *uniformly stable in both directions* in B, if given any $\varepsilon > 0$, there is a $\delta > 0$ such that $\varrho(xt, yt) < \varepsilon$ for $t \in R^+$, $t \in R^-$, or $t \in R$ respectively whenever $x \in A$, $y \in B$, and $\varrho(x, y) < \delta$.

6.4 *Exercise.* Show that if A is a compact subset of B, then the motions through A are uniformly positively stable in B, whenever each motion through A is positively stable in B.

We now introduce the concept of almost periodicity.

6.5 Definition. A motion π_x is said to be *almost periodic* if for every $\varepsilon > 0$ there exists a relatively dense subset of numbers $\{\tau_n\}$, called displacements, such that

$$\varrho(xt, x(t + \tau_n)) < \varepsilon$$

for all $t \in R$ and each τ_n.

It is obvious that periodic motions and rest points are special cases of almost periodic motions. That every almost periodic motion is recurrent follows from Theorem III, 3.12 and we leave this as an exercise. Later in this section we shall consider examples to show that not every recurrent motion is almost periodic, and that an almost periodic motion need not be periodic.

The following theorems show how stability is deeply connected with almost periodic motions. First observe the following lemma.

6.6 Lemma. If a motion π_x is almost periodic, then every motion π_y with $y \in \gamma(x)$ is almost periodic with the same set of displacements $\{\tau_n\}$ corresponding to a given $\varepsilon > 0$.

Proof. Indeed for any $\varepsilon > 0$ there is a set of displacements $\{\tau_n\}$ such that

$$\varrho(xt, x(t+\tau_n)) < \varepsilon \text{ for } t \in R, \text{ and each } \tau_n.$$

If $y \in \gamma(x)$, then there is a $\tau \in R$ such that $y = x\tau$, or $x = y(-\tau)$. The above inequality together with the group axiom then gives

$$\varrho(y(t-\tau), y(t-\tau-\tau_n)) < \varepsilon \text{ for } t \in R.$$

Setting $t - \tau = s$, we see that

$$\varrho(ys, y(s+\tau_n)) < \varepsilon \text{ for } s \in R \text{ and each } \tau_n,$$

as τ is fixed. This proves the lemma.

6.7 Theorem. Let the motion π_x be almost periodic and let $\overline{\gamma(x)}$ be compact. Then

6.7.1 every motion π_y with $y \in \overline{\gamma(x)}$ is almost periodic with the same set of displacements $\{\tau_n\}$ for any given $\varepsilon > 0$, but with the strict inequality $<$ replaced by $\leqq$;

6.7.2 the motion π_x is stable in both directions in $\gamma(x)$.

Proof of 6.7.1. For any $y \in \overline{\gamma(x)}$, there is a sequence $\{x_n\} \subset \gamma(x)$ such that $x_n \to y$. By Lemma 6.6, for any $\varepsilon > 0$ there is a set of displacements $\{\tau_n\}$ such that $\varrho(x_n t, x_n(t+\tau_m)) < \varepsilon$ for all $t \in R$, $x_n \in \{x_n\}$, and $\tau_m \in \{\tau_n\}$. Now keeping $t \in R$, and $\tau_m \in \{\tau_n\}$ fixed but arbitrary and proceeding to the limit we get $\varrho(yt, y(t+\tau_m)) \leqq \varepsilon$ for all $t \in R$ and $\tau_m \in \{\tau_n\}$. This completes the proof of 6.7.1.

Proof of 6.7.2. Given $\varepsilon > 0$, let $\{\tau_n\}$ be a set of displacements corresponding to $\varepsilon/3$ for the almost periodic motion π_x, and let $T > 0$ be such that $\{\tau_n\} \cap [t-T, t+T] \neq \emptyset$ for $t \in R$. Then by 6.7.1, $\varrho(yt, y(t+\tau_n)) \leqq \varepsilon/3$ for all $y \in \overline{\gamma(x)}$, $t \in R$, and each τ_n. By the continuity axiom, for $\varepsilon/3 > 0$ and $T > 0$ as above, there is a $\delta > 0$ such that $\varrho(y, z) < \delta$

implies $\varrho(yt, zt) < \varepsilon/3$ for all $|t| \leqq T$, whenever $\{y, z\} \subset \overline{\gamma(x)}$ as this last set is compact. Now for any $y \in \overline{\gamma(x)}$ and $\varrho(x, y) < \delta$, we have for any $t \in R$

$$\varrho(xt, yt) \leqq \varrho(xt, x(t+\tau_n)) + \varrho(x(t+\tau_n), y(t+\tau_n)) + \varrho(y(t+\tau_n), yt) < \varepsilon/3 + \varepsilon/3 + \varepsilon/3 = \varepsilon,$$

because for any $t \in R$ we can choose τ_n such that $|t + \tau_n| \leqq T$. This proves the theorem completely.

6.8 Corollary. If M is a compact minimal set, and if one motion in M is almost periodic, then every motion in M is almost periodic.

6.9 Corollary. If M is a compact minimal set of almost periodic motions, then the motions through M are uniformly stable in both directions in M.

The above corollary follows from 6.7.2, and from Exercise 6.4. We now investigate when a recurrent motion is almost periodic.

6.10 Theorem. If a motion π_x is recurrent and stable in both directions in $\gamma(x)$, then it is almost periodic.

Proof. We have indeed that given $\varepsilon > 0$ there is a $\delta > 0$ such that $\varrho(xt, yt) < \varepsilon$ for all $t \in R$, whenever $\{x, y\} \subset \gamma(x)$ and $\varrho(x, y) < \delta$. Further, by recurrence of π_x (Corollary III, 3.10 and Theorem III, 3.12), there is a relatively dense set $\{\tau_n\}$ of displacements such that

$$\varrho(x, x\tau_n) < \delta \text{ for each } \tau_n.$$

From the above two results we conclude

$$\varrho(xt, x(t+\tau_n)) < \varepsilon \text{ for } t \in R$$

and each τ_n. The theorem is proved.

A stronger result is the following:

6.11 Theorem. If a motion π_x is recurrent and positively stable in $\gamma(x)$, then it is almost periodic.

Proof.

(a) By positive stability of π_x in $\gamma(x)$, we have given $\varepsilon > 0$ there is a $\delta > 0$ such that $\varrho(x, x\tau) < \delta$ implies $\varrho(xt, x(t+\tau)) < \varepsilon/2$ for all $t \in R^+$.

(b) By recurrence of π_x, there is a relatively dense set $\{\tau_n\}$ such that $\varrho(x, x\tau_n) < \delta/2$ for each τ_n.

(c) By the continuity axiom, for any τ_n there is a $\sigma > 0$ such that $\varrho(x, y) < \sigma$ implies $\varrho(x\tau_n, y\tau_n) < \delta/2$. Now let $t \in R$ and τ_n be arbitrary but fixed. Then by recurrence of π_x, there is a $\tau < t$ such that $\varrho(x, x\tau) < \min(\sigma, \delta)$. Then by (c) $\varrho(x\tau_n, x(\tau + \tau_n)) < \delta/2$, so that

$$\varrho(x, x(\tau + \tau_n)) \leqq \varrho(x, x\tau_n) + \varrho(x\tau_n, x(\tau + \tau_n)) < \delta/2 + \delta/2 = \delta.$$

Hence by (a), since $t - \tau > 0$, we get $\varrho(x(t - \tau), x(t + \tau_n)) < \varepsilon/2$. Further, $\varrho(xt, x(t - \tau)) = \varrho(x\tau(t - \tau), x(t - \tau)) < \varepsilon/2$ by (a), as $\varrho(x, x\tau) < \min(\sigma, \delta) \leqq \delta$, and $t - \tau > 0$. Thus we get

$$\varrho(xt, x(t + \tau_n)) \leqq \varrho(xt, x(t - \tau)) + \varrho(x(t - \tau), x(t + \tau_n)) < \varepsilon/2 + \varepsilon/2 = \varepsilon.$$

This shows that π_x is almost periodic and the theorem is proved.

6.12 Theorem. If the motions in $\gamma(x)$ are uniformly positively stable in $\gamma(x)$ and are negatively Lagrange stable, then they are almost periodic.

Proof. It is sufficient to prove that the motion π_x is recurrent, as the rest follows from the last theorem.

By negative Lagrange stability of π_x we conclude that $\Lambda^-(x)$ is compact, and indeed $\Lambda^-(x) \subset \overline{\gamma(x)}$. Since $\Lambda^-(x)$ is also invariant, there is a compact minimal set M, $M \subset \Lambda^-(x)$. If π_x is not recurrent, then $M \neq \overline{\gamma(x)}$, and in particular $x \notin M$. Let $\varrho(x, M) = \alpha > 0$. We will show, that every motion π_y, $y \in \Lambda^-(x)$, is positively stable in $\gamma(x)$. To this end, given $\varepsilon > 0$, there is a $\delta = \delta(\varepsilon) > 0$ (by uniform positive stability in $\gamma(x)$ of motions in $\gamma(x)$), such that $\{x_n, x_m\} \subset \gamma(x)$, $\varrho(x_n, x_m) < \delta$, implies $\varrho(x_n t, x_m t) < \varepsilon$ for $t \geqq 0$. Now for $y \in \Lambda^-(x)$, there is a sequence $\{t_n\}$ in R^-, $t_n \to -\infty$, such that $xt_n \to y$. There is then an integer N such that $n \geqq N$, $m \geqq N$ imply $\varrho(xt_n, xt_m) < \delta$ and $\varrho(xt_n, y) < \delta$, and consequently $\varrho(xt_n(t), xt_m(t)) < \varepsilon$ for $t \geqq 0$. Keeping, in this last inequality, $t \in R^+$ and $n \geqq N$ arbitrary but fixed and letting $m \to \infty$ we get $\varrho(yt, xt_n(t)) \leqq \varepsilon$ for $t \geqq 0$ whenever $\varrho(y, xt_n) < \delta$. Choosing now $\varepsilon = \alpha/2$ and $t = -t_n$, we see that $\varrho(y(-t_n), x) \leqq \alpha/2$. Since $y(-t_n) \in M$, this contradicts the assumption that $\varrho(x, M) = \alpha$. The theorem is proved.

The remaining portion of this section will be devoted to finding conditions under which a limit set $\Lambda^+(x)$ is compact and minimal, and further when such a set consists of almost periodic motions only. For this, the following definition is useful.

6.13 Definition. A semi-trajectory $\gamma^+(x)$ is said to uniformly approximate its limit set $\Lambda^+(x)$, if given any $\varepsilon > 0$, there is a $T = T(\varepsilon) > 0$ such that $\Lambda^+(x) \subset S(x[t, t + T], \varepsilon)$ for each $t \in R^+$.

6.14 Theorem. Let the motion π_x be positively Lagrange stable. Then the limit set $\Lambda^+(x)$ is minimal if and only if the semi-trajectory $\gamma^+(x)$ uniformly approximates $\Lambda^+(x)$.

Proof. The set $\Lambda^+(x)$ is non-empty, compact and invariant. Now let $\gamma^+(x)$ uniformly approximate $\Lambda^+(x)$. If $\Lambda^+(x)$ is not minimal, then there are points $y, z \in \Lambda^+(x)$ such that $z \notin \overline{\gamma(y)}$ (otherwise $\overline{\gamma(y)} = \Lambda^+(x)$ for each $y \in \Lambda^+(x)$, and $\Lambda^+(x)$ is minimal; Theorem III, 3.4). Let $\varrho(z, \overline{\gamma(y)}) = \varepsilon > 0$. By uniform approximation there is a $T > 0$ such that $S(x[t, t+T], \varepsilon/2) \supset \Lambda^+(x)$ for $t \geqq 0$. Further, there is a $\delta > 0$ such that $\varrho(y, w) < \delta$ implies $\varrho(yt, wt) < \varepsilon/2$ for $|t| \leqq T$. Since $y \in \Lambda^+(x)$, there is a point $x_1 \in \gamma^+(x)$ such that $\varrho(x_1, y) < \delta$. And because $S(x[t, t+T], \varepsilon/2) \supset \Lambda^+(x)$, we have in particular $S(x_1[0, T], \varepsilon/2) \supset \Lambda^+(x)$. Thus there is a point $x_2 \in x_1[0, T]$ such that $\varrho(z, x_2) < \varepsilon/2$. If then $x_2 = x_1\tau$, where $0 \leqq \tau \leqq T$, then $\varrho(x_1\tau, y\tau) < \varepsilon/2$, so that $\varrho(z, y\tau) \leqq \varrho(z, x_2) + \varrho(x_2, y\tau) < \varepsilon/2 + \varepsilon/2 = \varepsilon$, as $x_2 = x_1\tau$. This is a contradiction, as $\varrho(z, \overline{\gamma(y)}) = \varepsilon$. Thus $\Lambda^+(x)$ is minimal. Now let $\Lambda^+(x)$ be minimal, so that for any $y \in \Lambda^+(x)$, $\overline{\gamma(y)} = \Lambda^+(x)$. Now assume, if possible, that $\gamma^+(x)$ does not uniformly approximate $\Lambda^+(x)$. Then there is an $\varepsilon > 0$, a sequence of intervals $\{(t_n, \tau_n)\}$, and a sequence $\{y_n\} \subset \Lambda^+(x)$ such that $t_n \to +\infty$, $(\tau_n - t_n) \to +\infty$, $y_n \to y\ (\in \Lambda^+(x))$, and $y_n \notin S(x[t_n, \tau_n], \varepsilon)$. We may also assume that $\varrho(y_n, y) < \varepsilon/3$ for all n. Then for arbitrary n

$$\varrho(y, x[t_n, \tau_n]) \geqq \varrho(y_n, x[t_n, \tau_n]) - \varrho(y_n, y) > \varepsilon - \varepsilon/3 = 2\varepsilon/3.$$

Consider now the sequence of points $\{x_n\}$, where

$$x_n = x\left(\frac{t_n + \tau_n}{2}\right) = xt'_n.$$

Clearly $t'_n \to +\infty$. Since $\overline{\gamma^+(x)}$ is compact, we may assume that $xt'_n \to z\ (\in \Lambda^+(x))$. Since $\Lambda^+(x)$ is minimal, $\overline{\gamma(z)} = \Lambda^+(x)$, so that there is a $T \in R$ such that $\varrho(zT, y) < \varepsilon/3$. By the continuity axiom we can choose a $\sigma > 0$ such that $\varrho(zT, wT) < \varepsilon/3$ whenever $\varrho(z, w) < \sigma$. Now choose N large enough such that $\varrho(z, x_N) < \sigma$, and $(\tau_N - t_N)/2 > |T|$. Then

$$x_N T = x(t'_N + T) \in x[t_n, \tau_n],$$

and hence $\varrho(y, x_N T) > 2\varepsilon/3$. On the other hand $\varrho(x_N T, zT) < \varepsilon/3$, so that

$$\varrho(y, x_N T) \leqq \varrho(x_N T, zT) + \varrho(zT, y) < \varepsilon/3 + \varepsilon/3 = 2\varepsilon/3.$$

This contradiction proves the result.

The following theorem gives a sufficient condition for a positive limit set $\Lambda^+(x)$ to be a minimal set of almost periodic motions. No necessary and sufficient condition is known as yet.

6.15 Theorem. Let the motion π_x be positively Lagrange stable, and let the motions in $\gamma^+(x)$ be uniformly positively stable in $\gamma^+(x)$. If moreover $\gamma^+(x)$ uniformly approximates $\Lambda^+(x)$, then $\Lambda^+(x)$ is a minimal set of almost periodic motions.

Proof. By Theorem 6.14, $\Lambda^+(x)$ is a compact minimal set. In view of Theorem 6.11 we need only prove that every motion through $\Lambda^+(x)$ is positively stable in $\gamma^+(x)$.

By uniform positive stability of motions in $\gamma^+(x)$, we have given $\varepsilon > 0$, there is a $\delta > 0$ such that $\varrho(xt_1, xt_2) < \delta$ implies $\varrho(x(t_1 + t), x(t_2 + t)) < \varepsilon/3$ for $t \geqq 0$. Let $\{y, z\} \subset \Lambda^+(x)$ and $\varrho(y, z) < \delta/3$. Let $\tau > 0$ be arbitrary. We wish to estimate $\varrho(y\tau, z\tau)$. By the continuity axiom there is a $\sigma > 0$ such that $\varrho(y, w) < \sigma$, $\varrho(z, u) < \sigma$ imply $\varrho(y\tau, w\tau) < \varepsilon/3$ and $\varrho(z\tau, u\tau) < \varepsilon/3$. If $\zeta = \min[\sigma, \delta/3]$. There are $t_1 > 0$ and $t_2 > 0$ such that $\varrho(xt_1, y) < \zeta$ and $\varrho(xt_2, y) < \zeta$. Thus

$$\varrho(xt_1, xt_2) \leqq \varrho(xt_1, y) + \varrho(y, z) + \varrho(z, xt_2) < \zeta + \delta/3 + \zeta \leqq \delta.$$

Consequently $\varrho(x(t_1 + \tau), x(t_2 + \tau)) < \varepsilon/3$, and $\varrho(y\tau, x(t_1 + \tau)) < \varepsilon/3$ and also $\varrho(z\tau, x(t_2 + \tau)) < \varepsilon/3$. The last three inequalities yield

$$\varrho(y\tau, z\tau) \leqq \varrho(y\tau, x(t_1 + \tau)) + \varrho(x(t_1 + \tau), x(t_2 + \tau)) + \varrho(x(t_2, \tau), z\tau)$$
$$< \varepsilon/3 + \varepsilon/3 + \varepsilon/3 = \varepsilon.$$

This shows in fact that the motions through $\Lambda^+(x)$ are uniformly positively stable in $\Lambda^+(x)$. The theorem is proved.

We now give a simple example of an almost periodic motion which is neither a rest point nor a periodic motion.

6.16 *Example*. Consider a dynamical system defined on a torus by differential equations (see Example III, 2.7).

$$\frac{d\varphi}{dt} = 1, \quad \frac{d\theta}{dt} = \alpha,$$

where α is irrational. For any point P on the torus $\overline{\gamma(P)}$ = the torus, and since α is irrational, no trajectory is periodic. The torus thus is a minimal set of recurrent motions. To see that the motions are almost periodic, we note first that if $P_1 = (\varphi_1, \theta_1)$, $P_2 = (\varphi_2, \theta_2)$, then $\varrho(P_1, P_2) = \sqrt{(\varphi_1 - \varphi_2)^2 + (\theta_1 - \theta_2)^2}$, where the values of $\varphi_1 - \varphi_2$ and $\theta_1 - \theta_2$ are taken as the smallest in absolute value of the differences (mod 1). Now any motion on the torus is given by $\varphi = \varphi_0 + t$, $\theta = \theta_0 + \alpha t$. Then for the motions through P_1 and P_2 we have

$$\varrho(P_1 t, P_2 t) = \sqrt{(\varphi_1 - \varphi_2)^2 + (\theta_1 - \theta_2)^2} = \varrho(P_1, P_2).$$

Thus the motions are uniformly stable in both directions in the torus. Thus by Theorem 6.10 the torus is a minimal set of almost periodic motions.

Notes and References

A definition of stability (of motion) was given in a precise form by A. M. Liapunov [1] (see also Notes and References of Chapter VIII). The concept of stability of equilibrium was previously discussed by Lagrange and by Dirichlet. Lagrange gave also a criteria for stability of equilibrium points based upon the consideration of the potential energy: If the potential energy has a minimum value at a given rest point, then that rest point is stable. It is interesting to notice that the energy is a particular case of a Liapunov function.

The first complete development of stability theory of dynamical systems is due to V. I. Zubov [1] (see also R. Bass [2] and S. Lefschetz [2]). In a sense, in the theory of stability of dynamical systems one has achieved a synthesis between the point of view of Liapunov and that of Poincaré, in fact Liapunov was interested in local (stability) properties of n-dimensional systems and Poincaré in global properties of planar systems. In the theory of stability, the concept of "region of asymptotic stability" and of "global asymptotic stability" is of rather recent origin and it was developed in the early fifties in the Soviet Union by M. Aizermann, E. Barbashin and N. N. Krasovskii. Most of the machinery needed for these developments is already present in the book by Nemytskii and Stepanov [1]. Results in more general framework than the one presented here can be found in the monographs by G. P. Szegö and G. Treccani (for the case of flows without uniqueness) by N. P. Bhatia and O. Hajek (for the case of local semiflows) and in the papers by D. Bushaw [2], J. Clay [2], T. Ura and I. Kimura [2], and E. Roxin [3, 5, 6].

Section 1. Notice that the discovery of "unstable attractors" for dynamical systems is rather recent and it is due to P. Mendelson [2]. On the other hand in the case of non-autonomous differential equations unstable attractors were known and were called quasi-asymptotically stable sets (see H. Antosiewicz [1]). Weak attractors were introduced by N. P. Bhatia [2]. The concept of uniform attractors is used by S. Lefschetz [2] and extensively used in the paper by N. P. Bhatia, A. Lazer and G. P. Szegö where the characterization presented in Exercise 7.29.2 is used as a definition. The same definition is also used in the monograph by N. P. Bhatia and G. P. Szegö [1] where the characterization 1.5.3 is derived. The complete diagram 1.17 is due to N. P. Bhatia [7]. Theorem 1.22 is essentially due to Desbrow [1] who proved it for a connected, locally compact, locally connected, metrizable space. The weak attractor theorem (1.25) is due to N. P. Bhatia and it was previously proved for the case of an attractor by J. Auslander, N. P. Bhatia and P. Seibert [1].

Section 2. This section contains results of Bhatia [5]. Some remarks are in order. Earlier results in this direction, for example those of Zubov [1], Auslander and Seibert [2] and Bhatia [1], used essentially the same methods as used for the well-developed theory in the case of ordinary differential equations. For results on ordinary differential equations see, for example, A. M. Liapunov, I. G. Malkin, E. A. Barbashin, N. N. Krasovskii, J. Kurzweil, I. Vrkoc, K. P. Persidskii, S. K. Persidskii, V. I. Zubov, J. Massera, H. Antosiewicz, T. Yoshizawa, W. Hahn. The basic feature of the results in this section is that Liapunov functions are shown to exist on the whole region of attraction instead of on a sufficiently small neighborhood in earlier results. The functions, in general, have sufficient properties to allow the derivation of theorems on global asymptotic stability and ultimate boundedness as corollaries. Indeed J. Auslander and P. Seibert [2] established formally the long suspected duality between stability and boundedness in locally compact separable metric spaces.

Theorem 2.16 is essentially due to V. I. Zubov [1]. The proof given here is simpler than the original because here we deal only with the case of compact sets while V. I. Zubov considers the more general case of closed sets.

Theorems 2.2, 2.9, 2.10, 2.12, 2.15 are due to N. P. Bhatia [5].

Section 3. The results presented in this section are due to N. P. Bhatia, A. Laser and G. P. Szegö [1] and to N. P. Bhatia and G. P. Szegö [2]. For the case of the flow defined by differential equations similar results are proved by F. Wilson [2].

Most of the results given here could have also been proved by means of the asymptotic fixed point theorem, due to G. S. JONES [3]. For the case of ordinary differential equations on the plane some results in the same spirit (identification of the canonical region, i.e., of the regions in which the stability properties are invariant) are given by L. MARKUS [5].

The important Corollary 3.12 can also be proved directly as follows.

Direct Proof of Corollary 3.12. By Theorem 1.25, $D^+(M)$ is compact and globally asymptotically stable. Let $x_0 \in E^n$ be arbitrary but fixed. Choose $\alpha > 0$ sufficiently large so that $D^+(M) \subset S(x_0, \alpha)$. Choose further $\varepsilon > 0$ sufficiently small such that $S(D^+(M), \varepsilon) \subset S[x_0, \alpha]$. By Theorem 1.16, $D^+(M)$ is uniformly attracting. Hence (1.2.3) there is a $T > 0$ such that $xt \in S(D^+(M), \varepsilon) \subset S[x_0, \alpha]$, whenever $x \in S[x_0, \alpha]$ and $t \geqq T$. For each $\tau \geqq T$ define the map $\pi_\tau: X \to X$ by $\pi_\tau(x) = \pi(x, \tau)$. Then π_τ is continuous and $\pi_\tau(S[x_0, \alpha]) \subset S[x_0, \alpha]$. Thus by the Brouwer fixed point theorem $S[x_0, \alpha]$ contains a fixed point of the map π_τ, i.e., there is an $x \in S[x_0, \alpha]$ such that $\pi_\tau(x) = x = \pi(x, \tau)$. Hence $\pi(x, t) = \pi(\pi(x, \tau), t) = \pi(x, t + \tau)$ for all $t \in R$, and so $\gamma(x)$ is a periodic trajectory. Notice that $x \in M$, for otherwise, if $x \notin M$, then $\gamma(x) \cap M = \emptyset$, for M is invariant. On the other hand, since $\gamma(x)$ is a periodic trajectory, we have $\gamma(x) \equiv \Lambda^+(x)$, and as $x \in A_\omega(M)$, we must have $\Lambda^+(x) \cap M \neq \emptyset$, i.e., $\gamma(x) \cap M \neq \emptyset$. This contradiction proves that $\gamma(x) \subset M$. Since $\overline{\gamma(x)} = \gamma(x)$, we must have $\gamma(x) \equiv M$, as M is minimal. Thus M is a periodic trajectory with period τ. If now M is not a rest point, then it will have a least period τ_0, and all other periods must be the numbers $m\tau_0$, where m is an integer. However, we have in fact shown that all numbers $\tau \geqq T$ are periods of M. This is a contradiction and so M is a rest point, and the theorem is proved.

Section 4. Theorems for the characterization of the stability properties of closed sets are given by N. P. BHATIA whose presentation we essentially follow. Other characterizations were given by V. I. ZUBOV [1], T. YOSHIZAWA, while sufficient conditions were given by G. P. SZEGÖ and G. GEISS [1].

The notion of equistability used here was introduced in the monograph by N. P. BHATIA and G. P. SZEGÖ together with Theorem 4.19. This theorem seems to lead the way for the use of the theory of parallelizable flows in the study of problems of asymptotic stability of closed sets and of the existence of Liapunov functions. The exposition in this section of this topic is more general than the one of H. ANTOSIEWICZ and J. DUGUNDJI.

Theorem 4.24 is an improvement of an analogous theorem due to V. I. ZUBOV [1].

Section 5. This section contains some new results (5.5—5.10), but the remaining parts are adopted from T. URA [5]. For the case of differential equations some different results on relative stability properties were derived by U. D'AMBROSIO and V. LAKSHMIKANTHAM [1], A. A. KAYANDE and V. LAKSHMIKANTHAM [1] and by V. LAKSHMIKANTHAM [4].

Section 6. This section brings to a completion the classification of compact minimal sets, viz., a rest point, a periodic trajectory, the closure of an almost periodic trajectory, and the closure of a recurrent trajectory. The relationship between almost periodicity and stability is clarified.

The notion of an almost periodic function is due to H. BOHR and Theorem 6.7 is due to S. BOCHNER.

A. A. MARKOV [3] showed the relationship between stability of motion and almost periodicity (Theorems 6.11, 6.12).

Definition 6.13 and the following material is due to V. NEMYTSKII. In this connection one may also see the paper of DEYSACH and SELL [1] on the existence of almost periodic motions.

Chapter VI

Flow near a Compact Invariant Set

This chapter is devoted to a rather deep analysis of the flow in the vicinity of a compact invariant set of a locally compact phase space. The main result was motivated by a desire to find if every neighborhood of a rest point contained a semi-trajectory different than the one through the rest point (see II, 2.6).

1. Description of Flow near a Compact Invariant Set

Our principal result is the following

1.1 **Theorem.** Let the phase space X be compact and $M \subset X$ a compact invariant subset of X. Then one of the four conditions given below holds.

1.1.1 M is positively asymptotically stable,

1.1.2 M is negatively asymptotically stable,

1.1.3 there exist $x \notin M$, $y \notin M$ such that $\emptyset \neq \Lambda^+(x) \subset M$, and $\emptyset \neq \Lambda^-(y) \subset M$,

1.1.4 every neighborhood U of M contains an $x \notin M$ with $\gamma(x) \subset U$. (Note that since M is invariant, $\gamma(x) \cap M = \emptyset$.)

Proof. Since X is compact, $\Lambda^+(x) \neq \emptyset$, $\Lambda^-(x) \neq \emptyset$ for every $x \in X$. If M is an open subset of X then 1.1.1 and 1.1.2 hold trivially. Let U be an arbitrary open neighborhood of M. Then for any $x \in U - M$ one of the following conditions must hold.

1.1.5 $\Lambda^+(x) \subset M$,

1.1.6 $\Lambda^+(x) \subset U$ but 1.1.5 does not hold,

1.1.7 $\Lambda^+(x) \cap \mathscr{C}(U) \neq \emptyset$.

Note now that if 1.1.6 holds, then U contains the invariant set $\Lambda^+(x) - M$ which is non-empty. A similar analysis will show that for every $x \in U$ one of the following conditions holds.

1.1.8 $\Lambda^-(x) \subset M$,

1.1.9 $\Lambda^-(x) \subset U$ but 1.1.8 does not hold,

1.1.10 $\Lambda^-(x) \cap \mathcal{C}(U) \neq \emptyset$,

and if 1.1.9 holds then U contains the invariant set $\Lambda^-(x) - M$ which is non-empty. To see that at least one of the conditions 1.1.1–1.1.4 must hold for M, we assume that 1.1.4 and 1.1.3 do not hold. Then we claim that there exists an open neighborhood U of M such that $\mathcal{C}(U)$ is either a weak attractor or a negative weak attractor with the region of weak attraction (positive or negative) coinciding with $\mathcal{C}(M)$. To see this note that if 1.1.4 does not hold then there is a neighborhood U of M such that for no $x \in U - M$, 1.1.6 or 1.1.9 holds. If now for some $x \in U - M$, the condition 1.1.5 holds, then 1.1.8 cannot hold for any $y \notin M$ for then 1.1.3 will be fulfilled. Then 1.1.10 holds for every $x \in U - M$ and consequently for every $x \notin M$, and $\mathcal{C}(U)$ is a weak negative attractor. Similarly, if 1.1.8 holds for some $x \in U - M$, then $\mathcal{C}(U)$ will be a weak attractor. To complete the proof, it is now sufficient to show that if $\mathcal{C}(U)$ is a weak attractor with $A_\omega(M) = \mathcal{C}(M)$, then M is negatively asymptotically stable, and if $\mathcal{C}(U)$ is a negative weak attractor with its region of negative weak attraction coinciding with $\mathcal{C}(M)$, then M is asymptotically stable. Because of the dual nature of the concepts we need only prove one of the foregoing assertions.

So let $\mathcal{C}(U)$ be a weak attractor with $A_\omega(\mathcal{C}(U)) = \mathcal{C}(M)$. Then by the weak attractor Theorem V, 1.25, $D^+(\mathcal{C}(U))$ is compact and asymptotically stable with $A_\omega(D^+(\mathcal{C}(U))) = A(\mathcal{C}(U)) = \mathcal{C}(M)$. Since $D^+(\mathcal{C}(U)) \subset \mathcal{C}(M)$, the open set $\mathcal{C}(D^+(U)) \equiv V$ is an open neighborhood of M. We claim that for each $x \in V$, $J^-(x) \subset M$. This follows from the fact if for some $x \in V$, $J^-(x) \not\subset M$, then there is a $y \notin M$, with $y \in J^-(x)$. But then $x \in J^+(y)$ by Theorem II, 4.9. Consequently, there is a $y \in A(\mathcal{C}(U)) = \mathcal{C}(M)$ with $J^+(y) \not\subset D^+(\mathcal{C}(U))$. Since the latter set is uniformly attracting by Theorem V, 1.16, we have arrived at a contradiction. Hence M is a negative uniform attractor, and since it is invariant, it is negatively asymptotically stable. This completes the proof.

The above theorem, though proved for compact spaces X, carries over to locally compact phase spaces. To see this one may use the result in Exercise II, 4.8.4. Then the result holds for the dynamical system $\tilde{\pi}$ on X. But since $\tilde{\pi} = \pi$ on $\tilde{X}$, the result holds in X. Thus we have

1.2 **Corollary.** Theorem 1.1 holds for locally compact spaces X.

1.3 *Exercises.* Let X be locally compact and M a compact invariant subset of X.

1.3.1 If there is a relatively compact neighborhood U of M such that for each $x \in M$ either $\Lambda^+(x) \cap \mathcal{C}(U) \neq \emptyset$, or there is a $T > 0$ with $\gamma^+(xT) \subset \mathcal{C}(U)$, then M is negatively asymptotically stable. Show that the converse holds also.

1.3.2 If M is asymptotically stable, and $A(M)$ is relatively compact, then $\mathcal{C}(A(M))$ is negatively asymptotically stable.

1.3.3 If M is asymptotically stable, then for each $x \notin M$, $J^-(x) \subset \mathcal{C}(A(M))$ holds.

1.3.4 Let M be stable. Then for every $x \notin M$, either $\Lambda^+(x) \neq \emptyset$ and $\Lambda^+(x) \subset M$, or $\Lambda^+(x) \cap M = \emptyset$.

1.3.5 If M is stable but not asymptotically stable, then every neighborhood U of M contains a compact invariant set N disjoint from M.

1.3.6 If M is a weak attractor, then $\Lambda^-(x) \cap M \neq \emptyset$ for every $x \in D^+(M)$.

1.3.7 Give an example of a weak attractor for which condition 1.1.3 of Theorem 1.1 holds, but not the condition 1.1.4.

1.3.8 Prove that if x is a rest point, then every neighborhood U of x contains a $y \neq x$ with either $\gamma^+(y)$ or $\gamma^-(y)$ contained in U.

1.3.9 Show that the results in this section hold for closed invariant sets M with compact boundary ∂M.

2. Flow near a Compact Invariant Set (Continued)

In Theorem 1.1, the condition 1.1.3 may be fulfilled in several ways. For example one may have points $x \notin M$, $y \notin M$, such that $\emptyset \neq J^+(x) \subset M$, $\emptyset \neq J^-(y) \subset M$. Similarly, condition 1.1.4 of Theorem 1.1 may be fulfilled in several ways. For example every neighborhood U of M contains a non-empty compact invariant set disjoint from M. It is therefore desirable to try to develop more detailed descriptions of the flow near a compact invariant set than the one contained in Theorem 1.1. We shall indicate one such generalization of Theorem 1.1 here.

2.1 **Theorem.** Let the phase space X be compact and $M \subset X$ a compact invariant subset of X. Then one of the following holds. (Here $A_u^+(M)$ will denote the set $A_u(M)$ defined in V, 1.1.3, and $A_u^-(M)$ the corresponding negative concept.)

2.1.1 $A_u^+(M) \cup A_u^-(M) \cup M$ is a neighborhood of M. (Thus M is either positively or negatively asymptotically stable, or there is a neighborhood U of M such that either $\emptyset \neq J^+(x) \subset M$ or $\emptyset \neq J^-(x) \subset M$ holds for each $x \in U - M$.)

2.1.2 There exist $x, y \notin M$ such that

$$\emptyset \neq \Lambda^+(x) \subset M, \ \emptyset \neq \Lambda^-(y) \subset M.$$

2.1.3 Every neighborhood U of M contains an $x \notin M$ such that $\gamma(x) \subset U$ and

$$\Lambda^+(x) \cap M \neq \emptyset, \quad \Lambda^-(x) \cap M \neq \emptyset,$$

but $J^+(x) \cup J^-(x) \not\subset M$.

2.1.4 Every neighborhood U of M contains an $x \notin M$ such that $\gamma(x) \subset U - M$.

Proof. Suppose that 2.1.1 and 2.1.2 do not happen. Then by Theorem 1.1 condition 1.1.4 of Theorem 1.1 holds. Thus every neighborhood U of M contains an x with $\gamma(x) \subset U - M$. Notice first that in fact we can choose x such that $\gamma(x) \subset U$, because for any neighborhood U of M there is a compact neighborhood V of M with $V \subset U$ and, by the above requirement, there is an $x \in V$ with $\gamma(x) \subset V - M$. Now compactness of V ensures $\gamma(x) \subset V \subset U$. Note also that in this case $\gamma(x)$ is compact and therefore both $\Lambda^+(x)$ and $\Lambda^-(x)$ are non-empty compact sets. If now $\Lambda^+(x) \cap M = \emptyset$ or $\Lambda^-(x) \cap M = \emptyset$, then 2.1.4 holds by choosing a $y \in \Lambda^+(x)$ in the first, or a $y \in \Lambda^-(x)$ in the second case (thus ensuring $\gamma(y) \subset \Lambda^+(x) \subset U - M$ for example in the first case). If therefore 2.1.4 does not hold we must have an $x \in U - M$, with $\Lambda^+(x) \cap M \neq \emptyset$, $\Lambda^-(y) \cap M \neq \emptyset$ and $\gamma(x) \subset U - M$. However, since 2.1.2 does not hold we cannot have both $\Lambda^+(x) \subset M$ and $\Lambda^-(x) \subset M$, and therefore 2.1.3 holds. This proves the theorem.

Notes and References

The problem of the qualitative behavior of the flow defined in a neighborhood of an isolated rest point of a planar differential equation of analytic type was one of the first dealt by H. Poincaré [3]. In this situation he proves that, if the differential equation has the form $\dot{x} = Ax + g(x)$, $x \in R^2$, $g(0) = 0$ and it is such that the power series expansion of $g(x)$ in the neighborhood $x = 0$ begins with terms of degree at least two, then, if the linear system $\dot{x} = Ax$ has eigenvalues with non-zero real parts, the qualitative behavior of the nonlinear system and that of the linear system in a sufficiently small neighborhood of the rest point $x = 0$ coincide. A classification of the possible behaviors of linear systems was also developed (see A. Andronov and C. Chaikin [1], N. Minorsky [1]). Classifications for more general planar cases are given by S. Lefschetz [1] and S. Barocio [1, 2].

The same problem was studied in the case of the n-th order ordinary differential equation by D. GROBMAN [4], A. I. PEROV [1] and by P. HARTMAN [4]. A complete presentation of this problem can be found in the book by HARTMAN [1, chapter 9].

Recently V. V. NEMYTSKII [12] proposed a new classification for the case of a differential equation on the plane. This classification was also used by N. P. PAPUSH. Recently M. B. KUDAEV [2] extended the Nemytskii classification to higher order singular points and proposed a method [3], based upon the consideration of the higher derivatives of a Liapunov function (see also Chapter VIII) for classifying the flow.

The qualitative behavior of differential equations on manifolds or near manifolds was in addition investigated by F. HAAS [1—4], A. DENJOY [1—5], F. B. FULLER [1], T. SAITO [1, 2], C. L. SIEGEL [1], S. STERNBERG [1], W. KAPLAN [6], J. K. HALE and A. STOKES and by A. SCHWARTZ [1, 2].

For the case of the second order differential equation the global problem of the classification of the flow was studied by I. BENDIXSON. He posed and gave a solution to the problem of the possible qualitative behaviors of the solutions in a neighborhood of a rest point. For the case of a dynamical system defined in a locally compact space T. URA and I. KIMURA [1] were able to propose a description. A similar description (Theorem 1.1) was with an independent proof, proposed by N. P. BHATIA [7]. The more advanced result (Theorem 2.1) was proved by N. P. BHATIA [9] in the more general framework of dynamical system in a locally compact Hausdorff space. Other classification theorems for dynamical systems are due to T. SAITO [3, 4]. Classification theorems were also proved for the case of local semi-dynamical systems by N. P. BHATIA [8] and by N. P. BHATIA and O. HAJEK [1] and for the case of flows without uniqueness by G. P. SZEGÖ and G. TRECCANI [1, 2].

For further research in the direction of the results of this chapter, one may find the work of K. S. SIBIRSKII [1, 2] to be useful and interesting. SIBIRSKII introduces the notion of $\psi - (\beta-)$ limit point, which essentially guarantees that the limit set of a point x has precisely one minimal subset. G. D. BIRKHOFF [1, p. 204] used this last idea to define semi-asymptotic central motions. The notion of ambits in Topological Dynamics as introduced by W. A. GOTTSCHALK [12] is also based on a similar idea.

It should be noted that in Theorem 1.1, if M is compact but not open, then the cases 1.1.1 and 1.1.2 cannot occur simultaneously. In fact, if 1.1.1 occurs, then none of 1.1.2—1.1.4 occurs. On the other hand 1.1.3 and 1.1.4 may occur simultaneously. This is so, because the cases 1.1.1 and 1.1.2 characterize the flow through all points of a neighborhood of M, whereas the cases 1.1.3 and 1.1.4 require the existence of some points with a given behavior in each neighborhood. It will be useful to develop description of flows containing as many cases of the nature of 1.1.1 and 1.1.2 as possible and reduce or eliminate the number of cases of the type 1.1.3 or 1.1.4. In Theorem 2.1 we have a description containing three cases of the former (all grouped under 2.1.1) and three cases of the latter type.

The problem of the classification of the flow via Liapunov-type functions will again be discussed in Chapters VIII and IX.

Chapter VII

Higher Prolongations

The first positive prolongation and the first positive prolongational limit set have been shown to be useful in characterizing various concepts in dynamical systems. Notable applications are the characterization of stability of a compact set in a locally compact metric space (Theorem V, 1.12) and the characterization of a dispersive flow (Theorem IV, 1.8). The first positive prolongation may be thought of as an extension of the positive semi-trajectory. For example consider a dynamical system (R^2, R, π) which is geometrically described by Fig. 0.1.

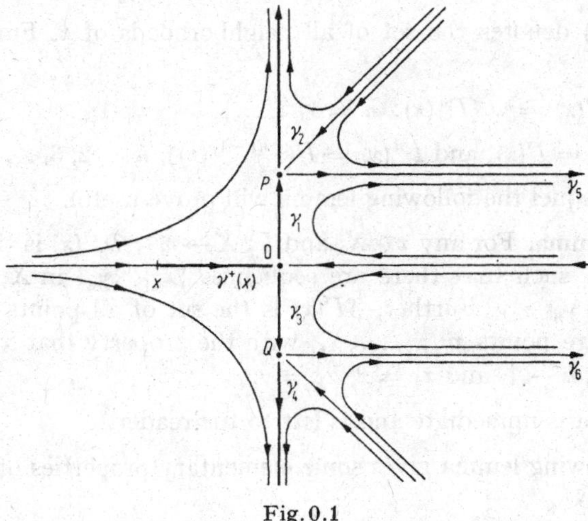

Fig. 0.1

The first positive prolongation of the point x in the figure consists of the semi-positive trajectory $\gamma^+(x)$, the equilibrium points 0, P, and Q, and the trajectories $\gamma_1, \gamma_2, \gamma_3$, and γ_4. In a way, to get the prolongation of a point we might find ourselves arguing that we move along the positive semi-trajectory and approach the equilibrium point 0. So we transfer to the point 0. From 0 we transfer to a trajectory which leaves 0, e.g., we can transfer to γ_1 or γ_3. If we transfer to γ_1, then we approach the equilibrium point P. So we transfer to P, and thence to a trajectory leaving P, and so on. If indeed this procedure were laid down to define the prolongation of a point, then notice that we would have to include the trajectories γ_5 and γ_6 in the prolongation of x. The definition of a prolongation, however, excludes the trajectories γ_5 and γ_6 from the prolongation of x. If, however, we wished to include these in a prolongation, then either we must change the definition of prolongation, or in a sense introduce other prolongations which will do precisely what we did with the intuitive reasoning above. Just as the first positive prolongation is in fact a meaningful extension of the positive trajectory, the second and higher prolongations which will be presently introduced, will be shown to be meaningful extensions of the first prolongation.

1. Definition of Higher Prolongations

The description of higher prolongations is facilitated by the introduction of two operations $\mathcal{S}$ and $\mathcal{D}$ on the class of maps from X into 2^X.

1.1 *The Operators $\mathcal{S}$ and $\mathcal{D}$.* If $\Gamma: X \to 2^X$, we define $\mathcal{D}\Gamma$ by

1.1.1 $$\mathcal{D}\Gamma(x) = \cap \{\overline{\Gamma(U)}: U \in \mathcal{N}(x)\},$$

where $\mathcal{N}(x)$ denotes the set of all neighborhoods of x. Further, $\mathcal{S}\Gamma$ is defined by

1.1.2 $\mathcal{S}\Gamma(x) = \cup \{\Gamma^n(x): n = 1, 2, \ldots\}$,

where $\Gamma^1(x) = \Gamma(x)$, and $\Gamma^n(x) = \Gamma(\Gamma^{(n-1)}(x))$, $n = 2, 3, \ldots$.

In the sequel the following lemma will prove useful.

1.2 **Lemma.** For any $x \in X$, and $\Gamma: X \to 2^X$, $\mathcal{D}\Gamma(x)$ is the set of all points $y \in X$ such that there are sequences $\{x_n\}$, $\{y_n\}$ in X, $y_n \in \Gamma(x_n)$, and $x_n \to x$, $y_n \to y$. Further, $\mathcal{S}\Gamma(x)$ is the set of all points $y \in X$ such that there are points $x_1, x_2, \ldots, x_k$, with the property that $x_{i+1} \in \Gamma(x_i)$, $i = 1, 2, \ldots, k-1$, and $x_1 = x$, $x_k = y$.

The proof is immediate and is left to the reader.

The following lemma gives some elementary properties of the operators $\mathcal{D}$ and $\mathcal{S}$.

1.3 **Lemma.**

1.3.1 $\mathcal{D}^2 = \mathcal{D}$, and $\mathcal{S}^2 = \mathcal{S}$. Thus $\mathcal{D}$ and $\mathcal{S}$ are idempotent operators.

1.3.2 If $M \subset X$ is compact, then

$$\mathcal{D}\Gamma(M) = \cup \{\mathcal{D}\Gamma(x): x \in M\}$$

is closed.

1.3.3 If $v = \varphi(x)$ is a continuous real-valued function on X, such that $\varphi(y) \leqq \varphi(x)$, whenever $y \in \Gamma(x)$, then $\varphi(y) \leqq \varphi(x)$ whenever $y \in \mathcal{D}\Gamma(x) \cup \mathcal{S}\Gamma(x)$.

Proof of 1.3.1. Let $\Gamma: X \to 2^X$. If $y \in \mathcal{D}\Gamma(x)$, then indeed $y \in \mathcal{D}\mathcal{D}\Gamma(x)$. Thus $\mathcal{D}\Gamma(x) \subset \mathcal{D}\mathcal{D}\Gamma(x)$. If $y \in \mathcal{D}\mathcal{D}\Gamma(x)$, then there are sequences $\{x_n\}$, $\{y_n\}$, $y_n \in \mathcal{D}\Gamma(x_n)$, such that $x_n \to x$ and $y_n \to y$. For each x_k, y_k, $y_k \in \mathcal{D}\Gamma(x_k)$, there are sequences $\{x_k^n\}$, $\{y_k^n\}$, $y_k^n \in \Gamma(x_k^n)$, such that $x_k^n \to x_k$, $y_k^n \to y_k$. We may assume $\varrho(x_k, x_k^n)$ and $\varrho(y_k, y_k^n) \leqq 1/k$ for $n \geqq k$ and each k. Now consider the sequences $\{x_n^n\}$, $\{y_n^n\}$. Clearly $y_n^n \in \Gamma(x_n^n)$. Further $x_n^n \to x$, and $y_n^n \to y$. Thus $y \in \mathcal{D}\Gamma(x)$. We have thus proved that $\mathcal{D}\mathcal{D}\Gamma(x) \subset \mathcal{D}\Gamma(x)$. This together with the previous observation shows that $\mathcal{D}\mathcal{D}\Gamma(x) = \mathcal{D}\Gamma(x)$. Hence $\mathcal{D}^2 = \mathcal{D}$. The proof of $\mathcal{S}^2 = \mathcal{S}$ is even simpler and is left as an exercise.

Proof of 1.3.2. Let $\{y_n\}$ be a sequence in $\mathcal{D}\Gamma(M)$ such that $y_n \to y$. Then there is a sequence $\{x_n\}$ in M such that $y_n \in \mathcal{D}\Gamma(x_n)$. Since M is compact, we may assume that $x_n \to x \in M$. Thus $y \in \mathcal{D}\mathcal{D}\Gamma(x) = \mathcal{D}\Gamma(x) \subset \mathcal{D}\Gamma(M)$. This shows that $\mathcal{D}\Gamma(M)$ is closed.

Proof of 1.3.3. If $y \in \mathcal{D}\Gamma(x)$, then there are sequences $\{x_n\}$, $\{y_n\}$, $y_n \in \Gamma(x_n)$, such that $x_n \to x$, and $y_n \to y$. It is given that $\varphi(y_n) \leqq \varphi(x_n)$. Since φ is continuous we get by proceeding to the limit $\varphi(y) \leqq \varphi(x)$. If $y \in \mathcal{S}\Gamma(x)$, then there are points $x = x_1, x_2, \ldots, x_n = y$ such that $x_{i+1} \in \Gamma(x_i)$, $i = 1, 2, \ldots, n-1$. Hence $\varphi(y) = \varphi(x_n) \leqq \varphi(x_{n-1}) \leqq \cdots \leqq \varphi(x_2) \leqq \varphi(x_1) = \varphi(x)$. This completes the proof of the lemma.

1.4 Definition. A map $\Gamma: X \to 2^X$ will be called transitive if $\mathcal{S}\Gamma = \Gamma$.

1.5 Exercise.

1.5.1 A map $\Gamma: X \to 2^X$, is transitive if and only if $\Gamma^2 = \Gamma$.

1.5.2 Given $\Gamma: X \to 2^X$, $\mathcal{S}\Gamma$ is transitive.

1.6 Definition. A map $\Gamma: X \to 2^X$ will be called a *cluster map* if $\mathcal{D}\Gamma = \Gamma$.

1.7 Definition. A map $\Gamma: X \to 2^X$ will be called a $c - c$ map if it has the following property: For any compact set $K \subset X$ and $x \in K$, one has either $\Gamma(x) \subset K$, or $\Gamma(x) \cap \partial K \neq \emptyset$.

1.8 Theorem. Let the space X be locally compact. Let Γ be a $c - c$ map. Then if $\Gamma(x)$ is compact, it is connected.

Proof. If $\Gamma(x)$ is compact, but not connected, then we can write $\Gamma(x) = M_1 \cup M_2$, where M_1, M_2 are non-empty compact disjoint sets.

Since X is locally compact, we can choose compact neighborhoods U_1, U_2 of M_1, M_2 respectively such that $U_1 \cap U_2 = \emptyset$. Note that $\Gamma(x) \cap U_1 \neq \emptyset$, but $U_1 \not\supset \Gamma(x)$, as $M_2 \cap U_1 = \emptyset$. However, $\partial U_1 \cap \Gamma(x) = \emptyset$, contradicting the fact that $\Gamma(x)$ is a $c - c$ map. Thus $\Gamma(x)$ is connected.

1.9 Theorem. Let the space X be locally compact. Let Γ be a $c - c$ map. If $M \subset X$ is compact, then $\mathcal{D}\Gamma(M) = M$ if and only if for each neighborhood U of M, there is a neighborhood W of M such that $\overline{\Gamma(W)} \subset U$.

Proof. (i) *Sufficiency.* Note that for any $c - c$ map Γ, $x \in \Gamma(x)$. Hence $\mathcal{D}\Gamma(M) \supset M$ always. Consider now a sequence of closed neighborhoods $\{U_n\}$ of M, such that $\cap U_n = M$. Then there is a sequence of neighborhoods $\{W_n\}$ such that $\overline{\Gamma(W_n)} \subset U_n$. Then $\mathcal{D}\Gamma(M) \subset \cap \overline{\Gamma(W_n)} \subset \cap U_n = M$. Thus we have proved that $\mathcal{D}\Gamma(M) = M$, and sufficiency is proved.

(ii) *Necessity.* Indeed assume, if possible, that there is a neighborhood U of M, such that for every neighborhood W of M, $\Gamma(W) \not\subset U$. We may assume without loss of generality that U is compact (because X is locally compact). Then there is a sequence $\{x_n\}$, $x_n \to x \in M$, and a sequence $\{y_n\}$, $y_n \in \Gamma(x_n)$ such that $y_n \notin U$, $n = 1, 2, \ldots$. Indeed we may assume that $x_n \in U$, $n = 1, 2, \ldots$. But then since Γ is a $c - c$ map and $\Gamma(x_n) \not\subset U$, we must have $\Gamma(x_n) \cap \partial U \neq \emptyset$. Consequently, there is a sequence $\{z_n\}$, $z_n \in \Gamma(x_n) \cap \partial U$, $n = 1, 2, \ldots$. Since ∂U is compact, we may assume that $z_n \to z \in \partial U$. But then $z \in \mathcal{D}\Gamma(x) \subset \mathcal{D}\Gamma(M)$. But $z \notin M$, a contradiction, as $\mathcal{D}\Gamma(M) = M$. This proves necessity and the theorem is proved.

We remark now that $c - c$ maps have also the following properties, which help to build families of $c - c$ maps.

1.10 Lemma.

1.10.1 Let $\{\Gamma_\alpha\}$, $\alpha \in A$, be a family of $c - c$ maps. Then $\Gamma = \cup \Gamma_\alpha$ is a $c - c$ map.

1.10.2 If Γ_1, Γ_2 are $c - c$ maps, then so is the composition map $\Gamma = \Gamma_1 \circ \Gamma_2$.

1.10.3 If Γ is a $c - c$ map, then so are $\mathcal{S}\Gamma$ and $\mathcal{D}\Gamma$.

Proof of 1.10.1. Let K be a compact set, and $x \in K$. If $\Gamma(x) \not\subset K$, we must prove that $\Gamma(x) \cap \partial K \neq \emptyset$. Indeed if $\Gamma(x) \not\subset K$, then there is at least one map, say Γ_α, such that $\Gamma_\alpha(x) \not\subset K$. Since $x \in K$, and Γ_α is a $c - c$ map, we must have $\Gamma_\alpha(x) \cap \partial K \neq \emptyset$. Thus $\Gamma(x) \cap \partial K \neq \emptyset$, and so Γ is a $c - c$ map.

Proof of 1.10.2. We need consider the case that $\Gamma_2(x) \subset \mathcal{I}(K)$, where $x \in K$, and K compact. This is so, because $\Gamma_1 \circ \Gamma_2(x) = \Gamma_1(\Gamma_2(x)) \supset \Gamma_2(x)$.

If $\Gamma_1 \circ \Gamma_2(x) \not\subset K$, and $\Gamma_2(x) \subset \mathcal{I}(K)$, then there is a $y \in \Gamma_2(x)$ such that $\Gamma_1(y) \not\subset K$. But then $\Gamma_1(y) \cap \partial K \neq \emptyset$, and since $\Gamma_1(y) \subset \Gamma_1(\Gamma_2(x)) = \Gamma_1 \circ \Gamma_2(x)$, we have $\Gamma_1 \circ \Gamma_2(x) \cap \partial K \neq \emptyset$. Thus $\Gamma = \Gamma_1 \circ \Gamma_2$ is a $c - c$ map.

Proof of 1.10.3. If Γ is a $c - c$ map, then $\mathcal{S}\Gamma$ is indeed a $c - c$ map by the assertions 1.10.1 and 1.10.2. To show that $\mathcal{D}\Gamma$ is a $c - c$ map, let $x \in K$, where K is compact. If $\mathcal{D}\Gamma(x) \not\subset K$, we may assume without loss of generality that $x \in \mathcal{I}(K)$. If $\mathcal{D}\Gamma(x) \not\subset K$, then there is a sequence $\{x_n\}$ in K, $x_n \to x$, and a sequence $\{y_n\}$, $y_n \notin K$, $y_n \to y \notin K$, $y_n \in \Gamma(x_n)$. Since $\Gamma(x_n) \not\subset K$, and $x_n \in K$, there is a sequence $\{z_n\}$, $z_n \in \Gamma(x_n)$, and $z_n \in \partial K$. Since ∂K is compact, we may assume that $z_n \to z \in \partial K$. But then $z \in \mathcal{D}\Gamma(x)$, so that $\mathcal{D}\Gamma(x) \cap \partial K \neq \emptyset$. This proves that $\mathcal{D}\Gamma$ is a $c - c$ map.

We now prove the following interesting theorem.

1.11 Theorem. Let X be locally compact and M a compact subset of X. Let Γ be a $c - c$ map which is moreover a transitive map as well as a cluster map. Then $\Gamma(M) = M$ if and only if there exists a fundamental system of compact neighborhoods $\{U_n\}$ of M such that $\Gamma(U_n) = U_n$.

Proof. Let $\{W_n\}$ be a fundamental system of neighborhoods of M. Since $\mathcal{D}\Gamma(M) = \Gamma(M) = M$, by Theorem 1.9, we have compact neighborhoods K_n of M such that $\Gamma(K_n) \subset W_n$. Now notice that $\Gamma(K_n)$ are closed and may be considered as compact. Setting $\Gamma(K_n) = U_n$, we get a fundamental system of compact neighborhoods of M such that $\Gamma(U_n) = U_n$. This is so because $\Gamma(U_n) = \Gamma(\Gamma(K_n)) = \Gamma(K_n) = U_n$, as Γ is transitive. Indeed $U_n \subset \Gamma(U_n)$ so that equality follows.

We shall now apply the theory constructed above to a dynamical system (X, R, π).

1.12 Definition. *The Higher Prolongations.* Consider the map $\gamma^+ : X \to 2^X$ which defines the positive semi-trajectory through each point $x \in X$. Then since $\gamma^+(x)$ is connected, γ^+ is a $c - c$ map. Further, notice that $\gamma^+(\gamma^+(x)) = \gamma^+(x)$, so that γ^+ is a transitive map, i.e., $\mathcal{S}\gamma^+ = \gamma^+$. We now set $\mathcal{D}\mathcal{S}\gamma^+ \equiv \mathcal{D}\gamma^+ = D_1^+$, and call $D_1^+(x)$ as the first positive prolongation of x. This is clearly the same as defined in Chapter II, where it is denoted by D^+. Indeed D_1^+ is a cluster map as $\mathcal{D}$ is idempotent, but D_1^+ is not transitive as simple examples will show. We, therefore, consider the map $\mathcal{D}\mathcal{S}D_1^+$ and denote it by D_2^+ and call $D_2^+(x)$ as the second prolongation of x. Having defined D_n^+, we define $D_{n+1}^+ = \mathcal{D}\mathcal{S}D_n^+$, and call $D_n^+(x)$ to be the n-th prolongation of x. This defines a prolongation of x for any positive integer n. Using transfinite induction, we define a prolongation $D_\alpha^+(x)$ of x for any ordinal number α as follows: If α is a successor ordinal, then having defined $D^+_{\alpha-1}$, we set $D^+_\alpha = \mathcal{D}\mathcal{S}D^+_{\alpha-1}$. If α is not a successor ordinal, then having defined D^+_β for every $\beta < \alpha$, we set $D^+_\alpha = \mathcal{D} \cup \{\mathcal{S}D^+_\beta : \beta < \alpha\}$. This defines for each $x \in X$, a prolongation $D^+_\alpha(x)$ for every ordinal α.

Notice that each of the map considered above is a $c - c$ map. Moreover, the prolongations $D^+_\alpha(x)$ for any ordinal α are closed.

We give below some properties of these prolongations.

1.13 Theorem. Let Ω be the first uncountable ordinal number. Then

1.13.1 $D^+_\Omega = \cup \{D^+_\alpha : \alpha < \Omega\}$,

1.13.2 D^+_Ω is a transitive map.

Proof. Recall that $D^+_\Omega = \mathcal{D} \cup \{\mathcal{S}D^+_\alpha : \alpha < \Omega\}$, as Ω is not a successor ordinal. Let for any $x \in X$, $y \in D^+_\Omega(x)$. Then there are sequences $\{x_n\}$, $\{y_n\}$, $x_n \to x$, $y_n \to y$, and $y_n \in \Gamma_{\alpha_n}(x_n)$, where $\Gamma_{\alpha_n} = \mathcal{S}D^+_{\alpha_n}$ and α_n is some ordinal number, $\alpha_n < \Omega$. Let β be an ordinal number such that $\alpha_n < \beta < \beta + 1 < \Omega$ (such ordinals exist). Then indeed $y_n \in \mathcal{S}D^+_\beta(x_n)$ for each n, so that $y \in \mathcal{D}\mathcal{S}D^+_\beta(x) = D^+_{\beta+1}(x)$. Consequently $y \in \Gamma(x)$ where Γ stands for $\cup \{D^+_\alpha : \alpha < \Omega\}$. Indeed for any $\alpha < \Omega$, $D^+_\alpha(x) \subset D^+_\Omega(x)$, so that $\Gamma(x) \subset D^+_\Omega(x)$. This proves 1.13.1. To prove that D^+_Ω is transitive, we need show only that $D^+_\Omega \circ D^+_\Omega = D^+_\Omega$. Suppose that $y \in D^+_\Omega(x)$, and $z \in D^+_\Omega(y)$, so that $z \in D^+_\Omega \circ D^+_\Omega(x)$. Then there is a $\beta < \Omega$ such that $y \in D^+_\beta(x)$, and $z \in D^+_\beta(y)$. But then $z \in \mathcal{S}D^+_\beta(x) \subset \mathcal{D}\mathcal{S}D^+_\beta(x) = D^+_{\beta+1}(x) \subset D^+_\Omega(x)$. Hence $\mathcal{S}D^+_\Omega = D^+_\Omega$. The theorem is proved.

1.14 Corollary. If $\alpha > \Omega$, then $D^+_\alpha = D^+_\Omega$. This follows by induction as D^+_Ω is transitive, and indeed also a cluster map.

2. Absolute Stability

We shall now define a host of stability concepts with the help of the higher prolongations introduced above.

2.1 Definition. Let X be locally compact. Let M be a compact subset of X. The set M will be called stable of order α, or α-stable, if $D^+_\alpha(M) = M$. If M is stable of order α, for every ordinal number α, then M is said to be *absolutely stable*.

2.2 *Remark.* Stability of order 1, is the same thing as stability defined in V, 1.5, as is evident from Theorem V, 1.12. Note also that absolute stability is the same as stability of order Ω, where Ω is the first uncountable ordinal, as is clear from Theorem 1.13 and Corollary 1.14.

One may easily prove the following theorem.

2.3 Theorem. A compact set $M \subset X$, where X is locally compact, is α-stable if and only if for every neighborhood U of M, there exists a neighborhood W of M, such that $D_\alpha^+(W) \subset U$.

We now give a few simple examples to illustrate that the various higher prolongations introduced above are indeed different concepts.

2.4 *Example.* Consider a dynamical system defined on the real line. The points of the form $\pm \frac{n}{1+n}$, $n = 0, 1, 2, \ldots$, are equilibrium points, and so are the points -1, and $+1$. Between any two successive (isolated) equilibrium points p, q, such that $p < q$, there is a single trajectory which has q as its positive limit point, and p as its negative limit point. There is a single trajectory with -1 as its sole positive limit point, and it has no negative limit points, and there is a single trajectory with $+1$ as its negative limit point, and it has no positive limit points (see Fig. 2.4.1). For any point P such that $P < -1$, $D_1^+(P) = \{x \in R\colon P \leqq x \leqq -1\}$, $D_2^+(P) = D_1^+(P)$, but $D_3^+(P) = \{x \in R\colon P \leqq x \leqq +1\}$, and $D_4^+(P) = \{x \in R\colon P \leqq x\}$. Note also, that if P is the point -1, then $D_1^+(P) = P$, $D_2^+(P) = \{x \in R\colon -1 \leqq x \leqq 1\}$, $D_3^+(P) = \{x\colon x \geqq -1\}$.

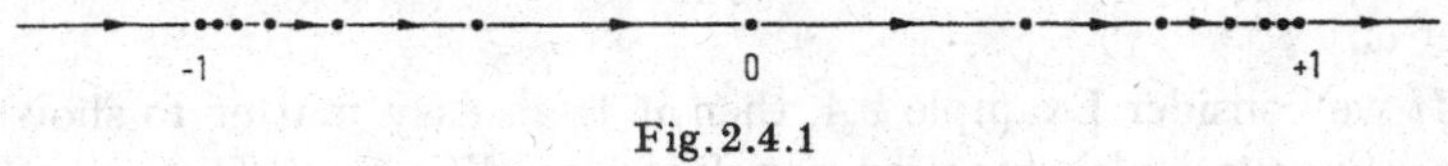

Fig. 2.4.1

2.5 *Example.* Consider again a flow on the real line, such that we have the equilibrium points as in the above example, and, moreover, between any two such successive equilibrium points, say q and p, there are two sequences of equilibrium points, say $\{p_n\}$, and $\{q_n\}$, $\cdots q_n \leqq q_{n-1} \cdots \leqq q_1 \leqq p_1 \leqq p_2 \leqq \cdots$, $p_n \to p$, and $q_n \to q$. Then direction of motion on a trajectory between any two equilibrium points is again from left to right, as in the previous example. In this case, if we consider the point $P = -1$, then indeed $D_1^+(P) = P$, $D_2^+(P) = P$, but $D_3^+(P) = \{x \in R\colon -1 \leqq x \leqq +1\}$, and $D_4^+(P) = \{x \in R\colon -1 \leqq x\}$.

Proceeding in this fashion it is easy to see that we can construct examples on the real line in which a point is stable of order n (n integer), but is not stable of order $n + 1$.

2.6 *Example.* We now give an example of a dynamical system defined on the real line, in which an equilibrium point is stable of every integral order n, but not stable of order ω, where ω is the first uncountable ordinal. To obtain such an example we consider a sequence of points $\{P_n\}$, $P_n \to 0$, $P_1 > P_2 > P_3 > \cdots > 0$. To the right of P_1, we introduce a sequence $\{P_1^n\}$, such that $P_1^1 > P_1^2 > \cdots > P_1$, and $P_1^n \to P_1$. Between

P_1 and P_2, we first introduce a sequence $\{P_{2n}\}$, $P_1 > P_{21} > P_{22} > P_{23} > \cdots > P_2$, $P_{2n} \to P_2$. Then for each P_{2k}, we introduce a sequence $\{P_{2k}^n\}$ between P_{2k} and $P_{2(k-1)}$, such that $P_{2k}^1 > P_{2k}^2 > \cdots > P_{2k}$, and $P_{2k}^n \to P_{2k}$. Between P_2 and P_3 we first introduce a sequence of points as between P_1 and P_2, and then between any two successive points we introduce a monotone decreasing sequence converging to the point on the left. We now proceed inductively. Having introduced a suitable sequence between say P_{n-1} and P_n, we introduce a sequence between P_n and P_{n+1} similar to the one introduced between P_{n-1} and P_n, then between each pair of successive points of this sequence, we introduce a monotonic decreasing sequence converging to the point on the left. Now we are ready to introduce the dynamical system on the real line. Each point of the countable set of points introduced on the line is an equilibrium point. There are no other equilibrium points, and the motion between any two successive equilibrium points is from left to right. It is easy to see that each point $\{P_n\}$ of the first sequence introduced above has the following property. P_1 is not stable, P_2 is stable of order 1, but not stable of order 2, P_3 is stable of order 2 (and hence also of order 1), but is not stable of order 3, P_{n+1} is stable of order n but not stable of order $n + 1$. The point 0 is stable of every integral order n, but is not stable of order ω.

If we consider Example 2.4, then it is an easy matter to show that no *continuous* scalar function satisfying conditions of Theorem V, 4.9 exists for the uniformly stable equilibrium point -1. An example in the plane, e.g., Example V, 4.11 was previously used to establish the same thing. In fact, even the point 0 in Example 2.6 which is stable of every integral order n is such that no *continuous* function satisfying conditions of Theorem V, 4.9 can exist for this point. The question obviously arises, as to what are the implications of the existence of a continuous function satisfying the conditions of Theorem V, 4.9 for a given closed set M. The answer for a compact set M in locally compact spaces X is given by the following theorem.

2.7 Theorem. Let X be locally compact, and let $M \subset X$ be compact. Then the following are equivalent:

2.7.1 There is a real-valued function $\varphi(x)$ satisfying conditions of Theorem V, 4.9 which is continuous in some neighborhood of M,

2.7.2 M possesses a fundamental system of absolutely stable compact neighborhoods,

2.7.3 M is absolutely stable.

We shall need the following lemma, whose proof is immediate from the definitions.

2.8 Lemma. Let $\varphi(x)$ be a real-valued function satisfying conditions of Theorem V, 4.9. If M is compact, and the space X is locally compact, then the set $\{U_\alpha: \alpha > 0\}$ is a fundamental system of neighborhoods of M, where $U_\alpha = \{x \in X: \varphi(x) \leqq \alpha\}$.

Proof of Theorem 2.7. 2.7.1 implies 2.7.2. Let U be a compact neighborhood of M. Let $m_0 = \min\{\varphi(x): x \in \partial U\}$. Then $m_0 > 0$, and $\{U_\alpha: 0 < \alpha < m_0\}$, where $U_\alpha = \{x \in U: \varphi(x) \leqq \alpha\}$, is a fundamental system of compact, positively invariant neighborhoods of M. We will now show that each U_α is absolutely stable. We shall show this by using 1.3.3. To do this, we consider the function $\Phi(x)$ defined on X, by means of $\Phi(x) = \varphi(x)$ for $x \in U_{m_0}$, and $\Phi(x) = m_0$ for $x \notin U_{m_0}$. This is a continuous function which is decreasing along the trajectories. For $0 < \alpha < \beta < m_0$, U_β is indeed a compact neighborhood of U_α. Since $\Phi(x)$ is decreasing along the trajectories, we get for $y \in D_\Omega^+(x)$, $\Phi(y) \leqq \Phi(x)$. If $D_\Omega^+(U_\alpha) \neq U_\alpha$, then there is a $\beta > \alpha$ such that $D_\Omega^+ U_\alpha \not\subset U_\beta$. Since D_Ω^+ is a $c - c$ map, there is an $x \in U_\alpha$, and a $y \in D_\Omega^+(x) \cap \partial U_\beta$. On one hand, therefore, $\Phi(y) \leqq \Phi(x) \leqq \alpha$, and, on the other hand, $\Phi(y) = \beta > \alpha$. This contradiction shows that $D_\Omega^+(U_\alpha) = U_\alpha$, i.e., each U_α is absolutely stable.

2.7.2 implies 2.7.3. This is immediate.

2.7.3 implies 2.7.1. Using Theorem 1.11 (since D_Ω^+ is a $c - c$ map which is moreover a transitive as well as a cluster map) we first construct a fundamental system of absolutely stable neighborhoods $U_{1/2^n}$, $n = 0, 1, 2, \ldots$, such that $(U_{1/2^n}) \subset \mathscr{I}(U_{1/2^{n-1}})$.

We now extend this system of absolutely stable compact neighborhoods to one defined over the diadic rationals, i.e., numbers of the type $\alpha = j/2^n$, $n = 0, 1, 2, \ldots$; $j = 1, 2, \ldots, 2^n$, in such a way that (a) the compact neighborhood corresponding to any diadic rational is absolutely stable, (b) if $\alpha < \beta$ are diadic rationals, then $U_\alpha \subset \mathscr{I}(U_\beta)$, (c) $M = \cap\{U_\alpha: \alpha$ diadic rational$\}$. Indeed this is possible by using Theorem 1.11. Now if $x \in U_1$, define $\varphi(x) = \inf\{\alpha: x \in U_\alpha, \alpha$ diadic rational$\}$. Clearly $\varphi(x) = 0$ if and only if $x \in M$. If $t > 0$, then $\varphi(xt) \leqq \varphi(x)$. This is so, because if $x \in U_\alpha$, then since U_α is positively invariant, we have $xt \in U_\alpha$, hence $\varphi(xt) \leqq \varphi(x)$. Finally, to see that $\varphi(x)$ is continuous on U_1, we assume that this is not true. Then there is an $x \in U_\alpha$, and a sequence $\{x_n\}$ in U_1, $x_n \to x$, such that $\varphi(x_n) \to \alpha \neq \varphi(x) = \alpha_x$. If $\alpha < \alpha_x$, then we can choose diadic rationals α_1, α_2, such that $\alpha < \alpha_1 < \alpha_2 < \alpha_x$. Then for large x_n, $x_n \in U_{\alpha_1}$, whereas $x \notin U_{\alpha_2}$. Since U_{α_1} is closed, and $x_n \to x$, $x \in U_{\alpha_1}$. This is a contradiction as $U_{\alpha_1} \subset \mathscr{I}(U_{\alpha_2})$. If again $\alpha > \alpha_x$, then choose diadic rationals α_1, α_2, such that $\alpha > \alpha_1 > \alpha_2 > \alpha_x$. Then $x_n \notin U_{\alpha_1}$ for large n, whereas $x \in U_{\alpha_2}$. But $U_{\alpha_2} \subset \mathscr{I}(U_{\alpha_1})$, which contradicts $x_n \to x$. This completes the proof of the theorem.

2.9 *Remark. Prolongations and Stability of Closed Sets.* Although Theorem V, 1.2 gives an excellent characterization of Liapunov stability of compact sets in locally compact spaces, a similar characterization is not available for closed (non-compact) sets, or in general metric spaces. Indeed we defined several concepts of stability of closed sets in section V, 4 and it appears that if we are to reach at a characterization we must first change the definition of prolongation for non-compact sets.

The following lemma gives an insight into what may be done.

2.10 **Lemma.** If the set $M \subset X$ is compact, then

$$D_1^+(M) = \cap \{\overline{\gamma^+(S(M, \delta))}: \delta > 0\}.$$

The proof is elementary and is left as an exercise. We only recall that $D_1^+(M)$ is by definition the set $\cup \{D_1^+(x): x \in M\}$.

It is now to be noted that $D_1^+(M)$ need not even be closed for closed sets M, which are not compact. And further, in general, if for any closed set M, we have $D_1^+(M) = M$, then the set M need not be weakly stable, or equistable. We now introduce the following definition.

2.11 **Definition.** Given any non-empty set M in X, we shall call the set $\cap \{\overline{\gamma^+(S(M, \delta))}: \delta > 0\}$ as the uniform (first) (positive) prolongation of M and denote it by $D_u^+(M)$.

Lemma 2.10 says that if M is compact, then $D_u^+(M) = D_1^+(M)$.

The uniform prolongation has further the following properties.

2.12 **Lemma.** For any non-empty set $M \subset X$,

2.12.1 $D_u^+(M)$ is closed and positively invariant,

2.12.2 $D_u^+(M) = \{y \in X$: there are sequences $\{x_n\}$ in X and $\{t_n\}$ in R^+ such that $\varrho(x_n, M) \to 0$ and $x_n t_n \to y\}$,

2.12.3 $D_u^+(M) \supset \overline{MR^+}$.

The proofs are immediate consequences of the definition.

The uniform prolongation is useful in characterizing the equi-stability of a closed set.

2.13 **Theorem.** A closed set $M \subset X$ is equi-stable if and only if $D_u^+(M) = M$.

This is an immediate consequence of the definitions and we leave the details to the reader. We note that Theorem V, 1.2 falls as a corollary of this theorem.

3. Generalized Recurrence

In Chapter II we introduced the first positive prolongation and the first positive prolongational limit set, and we studied some of their properties. We introduced the higher prolongations in section 1. We shall now introduce also the higher prolongational limit sets and study some of the properties. We shall then use these to characterize a notion of generalized recurrence.

3.1 Definition. The first positive prolongational limit set $J_1^+(x)$ of any point $x \in X$ is defined by $J_1^+(x) = \{y \in X$: there are sequences $\{x_n\}$ in X and $\{t_n\}$ in R such that $x_n \to x$, $t_n \to +\infty$, and $x_n t_n \to y\}$. In Chapter II this set was simply denoted by $J^+(x)$. Using now the operators $\mathcal{S}$ and $\mathcal{D}$ introduced in section 1, we define for any $x \in X$

$$J_2^+(x) = \mathcal{D}\mathcal{S}J_1^+(x),$$

and if α is any ordinal number, and J_β^+ has been defined for all $\beta < \alpha$, we set

$$J_\alpha^+ = \mathcal{D}(\cup \{\mathcal{S}J_\beta^+ : \beta < \alpha\}).$$

We have immediately the following lemma as a consequence of the definition. In the sequel we denote J_α^+ simply by J_α.

3.2 Lemma. If $\alpha > 1$, then $y \in J_\alpha(x)$ if and only if there are sequences $\{x_n\}$, $\{y_n\}$, $y_n \in J_{\beta_n}^{k_n}(x_n)$, $x_n \to x$, $y_n \to y$, where β_n are ordinal numbers less than α, and k_n are positive integers. (Recall that for any map Γ: $X \to 2^X$, $\Gamma^n = \Gamma \circ \Gamma^{n-1}$, where $\Gamma^1 = \Gamma$.)

We leave the proof to the reader. It is also to be noted that:

3.3 Lemma. For any ordinal $\alpha > 1$, $y \in D_\alpha^+(x)$ if and only if there are sequences $\{x_n\}$, $\{y_n\}$ in X such that $x_n \to x$, $y_n \to y$, and $y_n \in D_{\beta_n}^{k_n}(x_n)$, where for each n, β_n is an ordinal less than α and k_n is a positive integer. (In this lemma and hereafter D_α^+ is simply written as D_α to facilitate the use of upper indices.)

The following lemma now expresses some elementary properties of prolongations and prolongational limit sets.

3.4 Lemma. For any $x \in X$, and any ordinal α,

3.4.1 $J_\alpha(x)$ is closed and invariant,

3.4.2 $J_\alpha(xt) = J_\alpha(x)\,t = J_\alpha(x)$, for all $t \in R$,

3.4.3 $D_\alpha(x) = \gamma^+(x) \cup J_\alpha(x)$,

3.4.4 $D_\alpha(x)$ is closed and positively invariant,

3.4.5 if the space X is locally compact, then $D_\alpha(x)$ and $J_\alpha(x)$ are connected, whenever they are compact (if one is compact, then so is the other), and if $D_\alpha(x)$ $(J_\alpha(x))$ is not compact it does not possess any compact components.

Proof of 3.4.1. $J_1(x)$ has been proved to be closed and invariant in Chapter II. $J_\alpha(x)$ is closed by construction. To prove invariance, let $J_\beta(x)$ be invariant for all $\beta < \alpha$. Let $y \in J_\alpha(x)$, and $t \in R$. Let $x_n \to x$, $y_n \to y$, $y_n \in J_{\beta_n}^{k_n}(x_n)$, where $\beta_n < \alpha$ and k_n is a positive integer. Then by the induction hypothesis $y_n t \in J_{\beta_n}^{k_n}(x_n)$. Since $y_n t \to yt$, we have $yt \in J_\alpha(x)$, and the result follows.

Proof of 3.4.2. $J_\alpha(x)\, t = J_\alpha(x)$ is a trivial consequence of invariance of $J_\alpha(x)$. To see $J_\alpha(xt) = J_\alpha(x)\, t$, note that $J_1(xt) = J_1(x)\, t$ (Lemma V, 1.4). Now assume that $J_\beta(xt) = J_\beta(x)\, t$ for all $\beta < \alpha$. Let $y \in J_\alpha(xt)$. Let $y_n \in J_{\beta_n}^{k_n}(x_n t)$ (where $\beta_n < \alpha$, and k_n positive integers) such that $x_n t \to xt$ and $y_n \to y$. Now $x_n \to x$, and $y_n(-t) \in J_{\beta_n}^{k_n}(x_n t)\,(-t) = J_{\beta_n}^{k_n}(x_n)$, by the induction hypothesis. Since $y_n(-t) \to y(-t)$, so $y(-t) \in J_\alpha(x)$, and $y \in J_\alpha(x)\, t$. Hence $J_\alpha(xt) \subset J_\alpha(x)\, t$. Now $J_\alpha(x)\, t = J_\alpha(xt(-t))\, t \subset J_\alpha(xt)\,(-t)\, t = J_\alpha(xt)$. This proves 3.4.2.

Proof of 3.4.3. In Chapter II we proved that $D_1(x) = \gamma^+(x) \cup J_1(x)$. Now assume that the result is true for all $\beta < \alpha$. Notice that if $y' \in D_\beta^k(x')$, then by 3.4.2 $y' \in \gamma^+(x')$, or $y' \in J_\beta^m(x')$, where $m \leqq k$. Now if $y \in D_\alpha(x)$, let $x_n \to x$, $y_n \to y$, $y_n \in D_{\beta_n}^{k_n}(x_n)$ (where $\beta_n < \alpha$, k_n positive integers). If $y_n \in J_{\beta_n}^{l_n}(x_n)$ $(l_n \leqq k_n)$ for infinitely many n, then $y \in J_\alpha(x)$. If $y_n \in \gamma^+(x_n)$ for infinitely many n, then $y \in D_1(x) = \gamma^+(x) \cup J_1(x)$. In either case $y \in \gamma^+(x) \cup J_\alpha(x)$. Thus $D_\alpha(x) \subset \gamma^+(x) \cup J_\alpha(x)$. Since $\gamma^+(x) \cup J_\alpha(x) \subset D_\alpha(x)$ is obvious, we have $D_\alpha(x) = \gamma^+(x) \cup J_\alpha(x)$. This completes the proof of 3.4.3.

Proof of 3.4.4. Positive invariance is an immediate consequence of 3.4.3, and $D_\alpha(x)$ is closed by definition.

Proof of 3.4.5. The proof of this statement may easily be constructed by the method adopted for the proof of a similar statement about $\Lambda^+(x)$ and about $D_1(x)$ and $J_1(x)$ (II, 3.6, 4.4, 4.7). This we leave to the reader.

3.5 *Exercise.* Show that for any ordinal α

$$J_\alpha(x) = \cap \{D_\alpha(xt) : t \in R\}.$$

We now recall some of the notions of recurrence that have occurred earlier, namely, a rest point, a periodic trajectory (or periodic point), a positively or negatively Poisson stable motion (or point), a non-wandering point. We recall that these concepts are respectively equivalent to $x = xt$ for all $t \in R$, $xt = x(t + T)$ for all $t \in R$, $x \in \Lambda^+(x)$ or $x \in \Lambda^-(x)$, and $x \in J_1^+(x)$ which is equivalent to $x \in J_1^-(x)$.

Now let $\mathcal{V}$ denote the class of real-valued continuous functions f on X such that $f(xt) \leqq f(x)$, for all $x \in X$ and all $t > 0$.

3.6 **Definition.** Let $\mathcal{R}$ denote the set of all points $x \in X$ such that $f(xt) = f(x)$, for all $f \in \mathcal{V}$, and all $t \geqq 0$. $\mathcal{R}$ will be called the *generalized recurrent set.*

We have immediately

3.7 **Lemma.** $\mathcal{R}$ includes the non-wandering points in X.

Proof. Let $x \in J_1(x)$. Let $t > 0$, and $f \in \mathcal{V}$. Then indeed $x \in J_1(xt)$, and there are sequences $x_n \to xt$, $t_n \to +\infty$, and $x_n t_n \to x$. Then indeed $f(x_n t_n) \leqq f(x_n)$, and since f is continuous, we have

$$f(x) \leqq f(xt).$$

As $f(xt) \leqq f(x)$ holds by hypothesis, we get $f(xt) = f(x)$. Thus $x \in \mathcal{R}$.

Now we have

3.8 **Theorem.** $\mathcal{R}$ is closed and invariant.

Proof. That $\mathcal{R}$ is closed is clear. To see invariance, let first $x \in \mathcal{R}$ and $\tau > 0$. Then for any $f \in \mathcal{V}$ $f((x\tau)t) = f(x(\tau + t)) = f(x) = f(x\tau)$. Thus $x\tau \in \mathcal{R}$. Secondly, let $\tau < 0$ and $x\tau \notin \mathcal{R}$. Then there is an $f \in \mathcal{V}$ and a $t_0 > 0$ such that $f((x\tau)\, t_0) < f(x\tau)$. Define now $g \in \mathcal{V}$ by $g(x) = f(x\tau)$ for any $x \in X$. Then $g(xt_0) = f((xt_0)\, \tau) = f((x\tau)\, t_0) < f(x\tau) = g(x)$. This contradicts $x \in \mathcal{R}$, and the theorem is proved.

It is clear that if $f \in \mathcal{V}$, then so are $\tan f$ and $cf + d$, where c and d are real numbers with $c \geqq 0$. This remark and the above theorem yield

3.9 **Lemma.** Let a and b, $a < b$, be real numbers. Set $\mathcal{V}_{a,b} = \{f \in \mathcal{V}: a \leqq f(x) \leqq b$ for all $x \in X\}$. Then $x \in \mathcal{R}$ if and only if $f(xt) = f(x)$ for all $f \in \mathcal{V}_{a,b}$ and all real t.

From now on we shall assume that the space X is locally compact and separable.

The following theorem shows that in the class $\mathcal{V}$ of functions there is a function which is constant along any trajectory in the recurrent set, but is strictly decreasing along any trajectory which is not in the recurrent set.

3.10 Theorem. There is an $f \in \mathcal{V}$ such that

3.10.1 if $x \in \mathcal{R}$, then $f(x) = f(xt)$ for all real t, and

3.10.2 if $x \notin \mathcal{R}$ and $t > 0$, then $f(xt) < f(x)$.

Proof. Let $C(X)$ denote the continuous real-valued functions on X, provided with the topology of uniform convergence on compact sets. Then $C(X)$ contains a countable dense subset and so does $\mathcal{V}' = \mathcal{V}_{-1,1}$. Let $\{f_k\}$, $k = 1, 2, \ldots$, be a countable dense set in $\mathcal{V}'$. Then $x \in \mathcal{R}$ if and only if $f_k(xt) = f_k(x)$ for $k = 1, 2, \ldots$, and real t. Set $g = \sum_{k=1}^{\infty} \frac{1}{2^k} f_k$. Since $|f_k(x)| \leqq 1$, it follows that g is continuous and $|g(x)| \leqq 1$. Thus $g \in \mathcal{V}'$. If $g(xt) = g(x)$, for all $t > 0$, then $f_k(xt) = f_k(x)$ for $k = 1, 2, \ldots$, and so $x \in \mathcal{R}$. If $x \notin \mathcal{R}$, there is a sequence $\{t_n\}$ in R^+ with $t_n \to +\infty$ such that $g(x) > g(xt_1) > g(xt_2) \ldots$. Define $f(x) = \int_0^{\infty} e^{-t} g(xt)\, dt$. Then indeed $f \in \mathcal{V}'$, and f has the properties required in the theorem.

We shall now obtain a characterization of $\mathcal{R}$ by means of the prolongational limit sets. First, the following lemma.

3.11 Lemma. If $f \in \mathcal{V}$ and $y \in D_\alpha(x)$, then

$$f(y) \leqq f(x).$$

This is an immediate consequence of Lemma 1.3.3 and the definition of $D_\alpha(x)$.

3.12 Definition. The set of all points $x \in X$ such that $x \in J_\alpha(x)$ will be denoted by $\mathcal{R}_\alpha$. And we set $\mathcal{R}' = \bigcup \{\mathcal{R}_\alpha : \alpha \text{ an ordinal number}\}$.

The following theorem characterizes $\mathcal{R}$.

3.13 Theorem. $\mathcal{R} = \mathcal{R}'$. That is, $x \in \mathcal{R}$ if and only if $x \in J_\alpha(x)$ for some ordinal α.

For the proof we need the following topological theorem.

3.14 Theorem. Let X be a locally compact separable metric space and let $<$ be a closed quasi-order on X. Let x and y in X such that $x < y$ does not hold. Then there is an f in $C(X)$ such that (i) if $z < z'$, then $f(z) \leqq f(z')$, and (ii) $f(y) < f(x)$.

Proof of Theorem 3.13. We first show that $\mathcal{R}' \subset \mathcal{R}$. Indeed let $x \in J_\alpha(x)$ for some α. Then for any real t, $x \in J_\alpha(xt) = J_\alpha(x)$, and in particular this holds for $t > 0$. Then if $f \in \mathcal{V}$, we have $f(x) \leqq f(xt)$. However, we have

$f(xt) \leqq f(x)$ by definition of f. Thus for each $f \in \mathcal{V}$, $f(xt) = f(x)$ for all $t > 0$, and, therefore, $x \in \mathcal{R}$. This proves $\mathcal{R}' \subset \mathcal{R}$. To prove $\mathcal{R} \subset \mathcal{R}'$, we define a relation $\prec$ on X by $y \prec x$ if and only if $y \in D'(x) = \cup D_\alpha(x)$. Then $\prec$ is a closed quasi-order on X. Observe that $xt \prec x$ whenever $x \in X$, and $t < 0$. If $x \prec y$ but not $y \prec x$, we write $x \prec\prec y$. Note now that if $x \notin \mathcal{R}'$, and $t > 0$, then $xt \prec\prec x$. To see this note that $xt \prec x$. If $x \prec xt$, then $x \in D'(xt)$. Thus $x \in D_\alpha(xt)$ for $t > 0$ and some ordinal α. Then $x \in \gamma^+(xt) \cup J_\alpha(xt)$. Thus either $x \in J_\alpha(xt)$ or $x \in \gamma^+(xt)$. In the second case $\gamma(x)$ is periodic and so $x \in J_1(xt) \subset J_\alpha(xt)$. In any case then $x \in J_\alpha(xt) = J_\alpha(x)$. Thus $x \in \mathcal{R}'$ and this is a contradiction. The rest follows from Theorem 3.14.

3.15 *Remark.* By a quasi-order on X one means a reflexive, transitive, but not necessarily antisymmetric relation.

Notes and References

Section 1. The notion of higher prolongations is due to URA [4] who also showed their close connection with stability and introduced the notion of stability of order α. The exposition here is based on AUSLANDER and SEIBERT [2]. We have followed the enumeration of AUSLANDER and SEIBERT for prolongations. URA's enumeration is different. For example the second prolongation of URA is $D_1^+ \cdot D_1^+$. The prolongation D_2^+ here is what URA labels as D_ω where ω is the first countable ordinal. URA [4, p. 195] also showed that the prolongations introduced here are the only ones which lead to different concepts of stability. The notion of a $c - c$ map is one of the axioms of AUSLANDER and SEIBERT [2], for an abstract prolongation. We show that this is the concept which leads to various properties which are needed for results on stability. Thus articles 1.1 to 1.11 are independent of the notion of a dynamical system. For example 1.9 contains as a particular case URA's characterization of stability given in Theorem V, 1.12. Again Theorem 1.8 shows that the property "for given $x \in X$, $\Lambda^+(x)$, $D^+(x)$, $J^+(x)$ are connected whenever compact" is not dependent on the notion of a dynamical system. Finally, Theorem 1.13 is due to AUSLANDER and SEIBERT [2].

Section 2. The notion of absolute stability is due to URA [4]. The important connection of absolute stability with the existence of continuous Liapunov functions was established by AUSLANDER and SEIBERT [2]. Attempts on the characterization of stability of compact sets in non-locally compact spaces (or of stability of arbitrary sets) using URA's prolongations have not been very fruitful. For more recent work in this direction using the notion of strong attraction see BHATIA [9].

Section 3. This section is almost exclusively a reproduction of results of JOSEPH AUSLANDER [3]. The only exception is the last statement in 3.4.5. Notice that first part of the statement about $D_\alpha(x)$ follows from Theorem 1.8 as D_α is a $c - c$ map. However, the remaining part does need a separate proof. Indeed it is not too difficult to construct examples of $c - c$ maps Γ such that $\Gamma(x)$ is closed but not compact, and $\Gamma(x)$ has a compact component.

Chapter VIII

$\mathscr{C}^1$-Liapunov Functions for Ordinary Differential Equations

1. Introduction

Throughout this chapter we shall assume that the ordinary differential equation

1.1 $\qquad \dot{x} = f(x), \; x \in R^n, \; f\colon R^n \to R^n$

defines a dynamical system (R^n, R, π) in the space R^n, i.e., we shall assume that equation 1.1 satisfies conditions for the existence, uniqueness and extendability to the whole real line of its solutions for all points $x \in R^n$.

This chapter is devoted to the study of sufficient conditions for stability, asymptotic stability and instability of compact invariant sets in the flow (R^n, R, π), defined by the solutions of the ordinary differential equation 1.1. In Chapter V (section 5.1) and in Chapter VI we have defined various possible qualitative properties of the flow in a neighborhood of a compact invariant set. It is now our aim to analyze the qualitative properties of a given differential equation in the neighborhood of a compact connected invariant set $M \subset R^n$, estimate the region $\Gamma \subset R^n$ in which the behavior shown to hold near M does not change, and finally to estimate $\partial\Gamma$ in the case $\Gamma \neq R^n$. These conditions that we derive are based upon the construction of suitable continuously differentiable real-valued functions called Liapunov functions.

For simplicity we have only considered the case of a compact connected set; in view of Theorems V, 1.12 and V, 1.22 this is not a restriction and with some more involved proofs practically the same results apply to the more general case of compact sets.

The theory we present here is essentially a geometric theory whose foundation lies on some ideas introduced by STRAUSS [5] and by LA SALLE [4]. This theory will finally be based upon the so-called extension theorems. These theorems allow an estimate of the region Γ, $M \subset \Gamma \subset R^n$, in a new way, which, in contrast to the classical theorems, does not require an analysis of the global geometrical properties of the level surfaces of Liapunov functions. Liapunov functions have already been applied in section V, 2 for the characterization of asymptotic stability of compact sets for the dynamical system (X, R, π). The Liapunov functions used there were continuous, but not necessarily continuously differentiable. Their application for the solution of the problem was based upon inequalities of the type V, 2.2.2 which for their evaluation required knowledge of the positive semi-trajectories $\gamma^+(x)$ of the dynamical system (X, R, π) for all points x in a neighborhood N of the compact set $M \subset X$. For the case of a given differential equation 1.1, the theory developed in section V, 2 has not much practical use for analyzing stability properties. However, in the case of continuously differentiable Liapunov function $v = \varphi(x)$, the inequality 2.2.2 follows from the inequality

1.2 $$\left.\frac{d\varphi(xt)}{dt}\right|_{t=0} < 0$$

evaluated along the positive semi-trajectories $\gamma^+(x)$ such that $x[0, t) \subset N$. Under the same assumption of $\varphi(x)$, the function $\left.\frac{d\varphi(xt)}{dt}\right|_{t=0}$ can be evaluated without explicit knowledge of the solutions of the equation 1.1 by using the relationship

1.3 $$\left.\frac{d\varphi(xt)}{dt}\right|_{t=0} = \langle \operatorname{grad} \varphi(x), \dot{x}\rangle = \langle \operatorname{grad} \varphi(x), f(x)\rangle.$$

On the other hand if we let

1.4 $$\psi(x) = \langle \operatorname{grad} \varphi(x), f(x)\rangle,$$

by integration we have

1.5 $$\varphi(xt) - \varphi(x) = \int_0^t \psi(x\tau)\, d\tau,$$

from which the inequality V, 2.2.2 follows if $\psi(x) < 0$ for all $x \notin M$. Thus from Theorem V, 2.2 we immediately have

1.6 Theorem. Let $\dot{x} = f(x)$, $x \in R^n$, $f\colon R^n \to R^n$ define a dynamical system in R^n. Let $M \subset R^n$ be a non-empty compact set. If there exists a continuously differentiable real-valued function $v = \varphi(x)$ defined on a neighborhood U of M, such that

1.6.1 $\varphi(x) = 0$, if $x \in M$ and $\varphi(x) > 0$ if $x \notin M$,

1.6.2 $\psi(x) = \langle \operatorname{grad} \varphi(x), f(x)\rangle < 0$ for $x \notin M$.

Then M is asymptotically stable.

2. Preliminary Definitions and Properties

The theory that we shall present is a geometric theory which is based upon the properties of certain sets defined in the space R^n by a real-valued function and a constant. We shall proceed next with definition of these sets and the proof of some of their properties.

2.1 Definition. Let $v = \varphi(x)$ be a real-valued function defined on R^n and assume that there exists a compact, connected set $M \subset X$ such that

2.1.1 $\varphi(x) = 0$ if $x \in M$.

For a real number β, consider the following sets:

2.1.2 $N'(\beta) = \{x \in R^n : \varphi(x) < \beta\}$,

2.1.3 $K'(\beta) = \{x \in R^n : \varphi(x) \leqq \beta\}$,

2.1.4 $H(\beta) = \{x \in R^n : \varphi(x) = \beta\}$,

2.1.5 $L'(\beta) = \{x \in R^n : \varphi(x) \geqq \beta\}$,

2.1.6 $O'(\beta) = \{x \in R^n : \varphi(x) > \beta\}$.

If $N'(\beta) \supset M$, let then

2.1.7 $N(\beta)$ be the component of $N'(\beta)$ containing M, otherwise let $N(\beta) = \emptyset$.

Similarly we denote with $K(\beta)$, $L(\beta)$, $O(\beta)$ the component containing M of the sets $K'(\beta)$, $L'(\beta)$ and $O'(\beta)$, respectively, if such exists.

Let next

2.1.8 $$N_c(\beta) = \begin{cases} \emptyset \text{ if } \overline{N(\beta)} \text{ is not compact,} \\ N(\beta) \text{ if } \overline{N(\beta)} \text{ is compact.} \end{cases}$$

Similarly we shall define $K_c(\beta)$, $L_c(\beta)$ and $O_c(\beta)$ as sets which are empty if the sets $K(\beta)$, $L(\beta)$ and $\overline{O(\beta)}$ are respectively not compact, and equal to $K(\beta)$, $L(\beta)$ and $O(\beta)$ if the sets $K(\beta)$, $L(\beta)$ and $\overline{O(\beta)}$ are compact, respectively.

The sets defined above have the following properties, whose proofs are simple and left as exercises.

2.2 Properties. Let $v = \varphi(x)$ be a continuous real-valued function defined on R^n and let $M \subset R^n$ be a non-empty, compact, connected set such that

2.2.1 $\varphi(x) = 0$ if $x \in M$.

Then the following properties hold:

2.2.2 If $\beta < 0$, then $N(\beta) = N_c(\beta) = \emptyset$, $O(\beta) \neq \emptyset$ and $O(\beta)$ is a neighborhood of M.

2.2.3 If $\beta > 0$, then $O(\beta) = O_c(\beta) = \emptyset$, $N(\beta) \neq \emptyset$ and $N(\beta)$ is a neighborhood of M.

2.2.4 The sets $N(\beta)$ and $O(\beta)$ are open, while the sets $K(\beta)$, $H(\beta)$ and $L(\beta)$ are closed.

2.2.5 $\overline{N(\beta)} \subseteq K(\beta)$, $\overline{O(\beta)} \subseteq L(\beta)$.

2.2.6 $\partial N(\beta) \subseteq H(\beta)$, $\partial O(\beta) \subseteq H(\beta)$.

2.3 **Properties.** Let $v = \varphi(x)$ be a continuously differentiable real-valued function defined on R^n and let $M \subset R^n$ be a non-empty, compact, connected set, such that

2.3.1 $\varphi(x) = 0$ if $x \in M$.

Assume that

2.3.2 $N_c(\beta) \neq \emptyset$

and that

2.3.3 $\operatorname{grad} \varphi(x) \neq 0$ if $x \in \overline{N_c(\beta)} - M$.

Then

2.3.4 $\overline{N_c(\beta)} = K_c(\beta)$

and

2.3.5 there exists a real number $\varepsilon > 0$, such that for all $\tilde{\beta}$ with $\beta \leqq \tilde{\beta} \leqq \beta + \varepsilon$, $N(\tilde{\beta}) = N_c(\tilde{\beta})$.

The proof of these properties is based upon the Dini's implicit function theorem and is left as exercise to the reader.

Clearly in many cases the sets $N_c(\beta)$ and $O_c(\beta)$ may be empty. We shall give next a sufficient condition for $N_c(\beta)$ to be non-empty. A similar theorem, by reversing the sign condition can also be proved for $O_c(\beta)$. From now on we shall discuss mostly properties of $N_c(\beta)$. The analogous properties of $O_c(\beta)$ can be derived and proved by the reader.

2.4 *Remark.* The Definitions 2.1 and the Properties 2.2 and 2.3 can also be applied to the case in which the real valued function $v = \varphi(x)$ is defined in an open neighborhood G of the compact, connected set $M \subset R^n$, satisfying 2.1.1. In this case definitions and properties apply to the sets $N'(\beta)$, etc., which are contained in G.

2.5 **Theorem.** Let $v = \varphi(x)$ be a continuous real-valued function defined in an open neighborhood G of the compact, connected set $M \subset R^n$ such that

2.5.1 $\varphi(x) = 0$ if $x \in M$, $\varphi(x) > 0$ if $x \in G - M$.

Then there exists a real number $\beta > 0$, such that the open set $N_c(\beta)$ is not empty. Furthermore, for every neighborhood $U \subset M$ there is a $\beta > 0$ such that $N_c(\beta) \subset U$.

Proof. Choose $\delta > 0$ such that $S(M, \delta) \subset G$. Let $\nu = \min\{\varphi(x): x \in H(M, \delta/2)\}$. From 2.5.1 it follows that $\nu > 0$. Consider next the set $N'(\nu/2)$. The component of this set which contains M is contained in $S[M, \delta/2]$ and therefore its closure is compact. From definition 2.1.8, it follows that $N_c(\nu/2)$ is not empty. Since this is true for all $\delta' < \delta$, the result follows.

2.6 *Remark.* It is easy to show that Theorem 2.5 does not provide a necessary condition for $N_c(\beta)$ to be non-empty. Consider for instance the following continuous real-valued function defined on the x, y plane:

2.6.1 $v = (x^2 + y^2) \sin 1/(x^2 + y^2)$.

Clearly this function is indefinite (it changes its sign) in each neighborhood of the point (0,0) but the set $N_c(\beta)$ is not empty.

In the classical stability theory much use is made of the concept of definite functions.

2.7 **Definition.** Let $v = \varphi(x)$ be a real-valued function defined in an open neighborhood G of a non-empty, compact, connected set $M \subset R^n$. We shall say that

2.7.1 $\varphi(x)$ is *positive definite* if $\varphi(x) = 0$, if $x \in M$, $\varphi(x) > 0$ for all $x \in G - M$.

2.7.2 $\varphi(x)$ is *negative definite* if $\varphi(x) = 0$, if $x \in M$, $\varphi(x) < 0$ for all $x \in G - M$.

2.7.3 $\varphi(x)$ is *definite* if it is either positive or negative definite.

3. Local Theorems

We shall prove next some theorems providing sufficient conditions for stability, asymptotic stability and instability of compact sets. The proof of the first theorem follows from the following lemma.

3.1 **Lemma.** Consider a dynamical system defined by the ordinary differential equation

3.1.1 $\dot{x} = f(x), \quad x \in R^n, \quad f\colon R^n \to R^n.$

Let $v = \varphi(x)$ and $w = \psi(x)$ be real-valued functions defined in R^n. Assume that

3.1.2 $v = \varphi(x) \in \mathcal{C}^1,$

3.1.3 $\psi(x) = \langle \text{grad}\, \varphi(x), f(x) \rangle \leqq 0$ for all $x \in R^n$.

Then for each real number β, each component of the sets $N'(\beta)$ and $K'(\beta)$ is positively invariant.

Proof. The inequality 3.1.3 implies that $\varphi(x)$ is decreasing along trajectories. Thus $N'(\beta)$ and $K'(\beta)$ are positively invariant and consequently so are their components (see II, 1.14).

In the case in which the real-valued function $v = \varphi(x)$ is defined only on an open set G containing a continuum $M \subset R^n$, the same result as Lemma 3.1 applies to these components of $N'(\beta)$ and $K'(\beta)$ which have a compact closure which is contained in G. In general the following result holds:

3.2 Lemma. Consider a dynamical system defined by the ordinary differential equation

3.2.1 $\dot{x} = f(x), \quad x \in R^n, \quad f\colon R^n \to R^n.$

Let $v = \varphi(x)$ and $w = \psi(x)$ be real-valued functions defined on an open neighborhood G of a continuum $M \subset R^n$. Assume that

3.2.2 $v = \varphi(x) \in \mathcal{C}^1$ on G,

3.2.3 $\psi(x) = \langle \text{grad}\, \varphi(x), f(x) \rangle \leqq 0$, for all $x \in G$.

Let $\tilde{N}(\beta)$ be a component of $N'(\beta)$. Then if $x \in \tilde{N}(\beta)$ and $\tau > 0$ is the smallest positive real number such that $x\tau \in \mathcal{C}(\tilde{N}(\beta))$, then $x\tau \subset \mathcal{C}(G)$. From these it follows that the components of $N'(\beta)$ whose closure lies in G are positively invariant.

We shall next prove the stability theorem.

3.3 Theorem. Consider a dynamical system represented by the differential equation

3.3.1 $\dot{x} = f(x), \quad x \in R^n, \quad f\colon R^n \to R^n.$

Let $v = \varphi(x)$ and $w = \psi(x)$ be real-valued functions defined in an open neighborhood G of a non-empty, compact, connected set $M \subset R^n$. Assume that

3.3.2 $\varphi(x) \in \mathcal{C}^1$ on G,

3.3.3 $\varphi(x) = 0, \quad x \in M,$

3.3.4 $\psi(x) = \langle \operatorname{grad} \varphi(x), f(x) \rangle \leqq 0$ for all $x \in G$,

3.3.5 there exists a real number $\beta > 0$ such that $\emptyset \neq N_c(\beta) \subset G$ and for each neighborhood U of M there exists $\tilde{\beta}$, $0 \leqq \tilde{\beta} < \beta$ such that $N_c(\tilde{\beta}) \subset U$.

Then M is (positively) stable.

Proof. From the hypothesis 3.3.5 we have that for a sufficiently small $\beta > 0$, $N_c(\beta) \subset G$. In addition from Lemma 3.1 we have that for all real numbers $\tilde{\beta}$ with $0 < \tilde{\beta} < \beta$, the set $N_c(\tilde{\beta})$ is positively invariant. Then M is stable.

3.4 **Corollary.** Theorem 3.3 is true if we replace condition 3.3.4 with:

3.4.1 $\psi(x) \geqq 0$ if $x \in G$

and condition 3.3.5 with:

3.4.2 there exists a real number $\beta < 0$ such that $\emptyset \neq O_c(\beta) \subset G$ and for each neighborhood U of M there exists $\tilde{\beta}$, $\beta < \tilde{\beta} < 0$ such that $O_c(\tilde{\beta}) \subset U$.

By applying Theorem 2.5 and the analogous result on $O_c(\beta)$ to Theorem 3.3 and Corollary 3.4 we can immediately prove the following corollary, which requires a sign condition on $\varphi(x)$ which implies the property 3.3.5. This corollary is the classical theorem of stability due to LIAPUNOV.

3.5 **Corollary.** Consider a dynamical system represented by the differential equation

3.5.1 $\dot{x} = f(x), \quad x \in R^n, \quad f\colon R^n \to R^n.$

Let $v = \varphi(x)$ and $w = \psi(x)$ be real-valued functions defined in an open neighborhood G of a compact, connected set $M \in R^n$. Assume that

3.5.2 $\varphi(x) \in \mathscr{C}^1$ if $x \in G$,

3.5.3 $\varphi(x) = 0$ if $x \in M$, $\varphi(x) \neq 0$ if $x \in G - M$,

3.5.4 $\psi(x) = \langle \operatorname{grad} \varphi(x), f(x) \rangle$,

3.5.5 either $\psi(x) \geqq 0$ or $\psi(x) \leqq 0$ if $x \in G - M$,

3.5.6 $\operatorname{sign} \varphi(x) \neq \operatorname{sign} \psi(x)$, for each $x \in G$, with $\psi(x) \neq 0$.

Then M is (positively) stable.

3.6 *Remark.* As shown from the example 2.6.1 condition 3.5.3 is more restrictive than conditions 3.3.5 and 3.4.2. Thus Theorems 3.3 and 3.4 together, cover situations which need not be covered by Corollary 3.5. This can be directly shown by the following example.

3.7 *Example.* Consider the following system on the plane (x, y):

3.7.1 $$\begin{cases} \dot{x} = y, \\ \dot{y} = -x - yr^2(r \sin 1/r - \cos 1/r), \end{cases}$$

where $r = x^2 + y^2$, and the real-valued function

3.7.2 $$v = \varphi(x, y) = (x^2 + y^2) \sin 1/(x^2 + y^2).$$

Then

3.7.3 $$\psi(x, y) = \frac{d\varphi}{dt} = -2y^2r^3 [\sin 1/r - 1/r (\cos 1/r)]^2.$$

This real-valued function satisfies the hypotheses of both Theorem 3.3 and Corollary 3.5. The real-valued function 3.7.2, on the other hand, satisfies the hypotheses of Theorem 3.3, but not of Corollary 3.5.

We shall give next a theorem on asymptotic stability. For its proof the following lemma is useful.

3.8 **Lemma.** Consider the dynamical system represented by the differential equation

3.8.1 $$\dot{x} = f(x), \quad x \in R^n, \quad f\colon R^n \to R^n.$$

Let $v = \varphi(x)$ and $w = \psi(x)$ be real-valued functions defined in an open neighborhood G of a compact, connected set $M \subset R^n$. Assume that

3.8.2 $$v = \varphi(x) \in \mathscr{C}^1$$

and

3.8.3 $$\psi(x) = \langle \operatorname{grad} \varphi(x), f(x) \rangle \leqq 0.$$

If for an $x \in G$, $\Lambda^+(x)$ is a non-empty, compact subset of G, then

3.8.4 $$\psi(y) \equiv 0 \text{ for all } y \in \Lambda^+(x),\ x \in G.$$

Proof. Condition 3.8.3 ensures that $\varphi(x)$ is decreasing along trajectories as long as they remain in G. Compactness of $\Lambda^+(x)$ ensures that there is a $T > 0$, with $xt \in G$, for $t \geqq T$. Then using Lemma V, 2.1 the result follows.

3.9 **Theorem.** Consider the dynamical system represented by the differential equation

3.9.1 $$\dot{x} = f(x), \quad x \in R^n, \quad f\colon R^n \to R^n.$$

Let $v = \varphi(x)$ and $w = \psi(x)$ be real-valued functions defined in an open neighborhood G of a compact, connected, invariant set $M \subset R^n$. Assume that

3.9.2 $$v = \varphi(x) \in \mathscr{C}^1, \quad x \in G,$$

3.9.3 $$\varphi(x) = 0 \text{ for all } x \in M,$$

3.9.4 there exists a real number $\beta > 0$, such that $\overline{N_c(\beta)} \subsetneq G$ is not empty,

3.9.5 $\psi(x) = \langle \text{grad}\, \varphi(x), f(x) \rangle < 0$ for all $x \in G - M$.

Then M is asymptotically stable and

3.9.6 $A(M) \supset N_c(\beta)$.

Proof. Choose $\beta > 0$ in such a way that $\overline{N_c(\beta)} \subset G$. Then from Lemma 3.1, $x \in N_c(\beta)$ implies $\overline{\gamma^+(x)} \subset \overline{N_c(\beta)}$. Now, if M were not an attractor there would exist $y \in N_c(\beta)$, such that $\Lambda^+(y) \neq \emptyset$ and $\Lambda^+(y) \not\subset M$. On the other hand, from Lemma 3.8 it follows that $\psi(x) = 0$ if $x \in \Lambda^+(y)$. From hypothesis 3.9.5 this implies that $\Lambda^+(y) \subset M$. Thus M is an attractor and 3.9.6 holds. Let us now prove that M is stable. If it were not stable, then $D^+(M) - M \neq \emptyset$. Since $N_c(\beta)$ is positively invariant (Lemma 3.1) $D^+(M) \subset \overline{N_c(\beta)}$. Consider a point $x \in D^+(M) - M$; since M is an attractor there exist points $y, z \in M, y \in \Lambda^-(x) \cap M, z \in \Lambda^+(x) \cap M$ (V, 1.26.2). Consider then monotonic sequences $\{t_n\} \subset R^+$ and $\{\tau_n\} \subset R^-$ with $t_n \to +\infty$ and $\tau_n \to -\infty$, $xt_n \to z$ and $x\tau_n \to y$. Consider the corresponding value of the function φ. We have $\varphi(xt_n) < \varphi(x) < \varphi(x\tau_n)$ while $\varphi(x\tau_n) \to \varphi(y) = \beta$ and $\varphi(xt_n) \to \varphi(z) = \alpha$ with $\varphi(y) = \beta > \alpha = \varphi(z)$, which contradicts the fact that $\varphi(x)$ vanishes on M.

Again from Theorems 2.5 and 3.9 we can derive the classical asymptotic stability theorem.

3.10 **Corollary.** Consider the dynamical system 3.9.1. Let $v = \varphi(x)$ and $w = \psi(x)$ be real-valued functions defined in an open neighborhood G of a compact connected invariant set $M \subset R^n$. Assume that

3.10.1 $v = \varphi(x) \in \mathcal{C}^1, \quad x \in G,$

3.10.2 $\varphi(x) = 0,\ x \in M;\ \varphi(x) > 0,\ x \notin M,$

3.10.3 $\psi(x) = \langle \text{grad}\, \varphi(x), f(x) \rangle < 0,\ x \in G - M.$

Then M is asymptotically stable.

3.11 *Remark.* The conditions of Theorem 3.9 imply that the real-valued function $\varphi(x)$ is positive definite in $\overline{N_c(\beta)}$. If this were not true there would exist points $y \in (N_c(\beta) - M)$ in which $\varphi(x)$ has a minimum and therefore grad $\varphi(x) = 0$, which contradicts condition 3.9.5 of Theorem 3.9. This shows that Theorem 3.9 is completely equivalent to the classical theorem on asymptotic stability (see, for instance, Bhatia and Szegö [1], Theorem 3.6.15).

In the proof of Theorem 3.9 the asymptotic stability of the compact set M was deduced from the sign of $\psi(x)$ and from the fact that M is an attractor. The same result can be obtained under different assumptions if we notice that if $\psi(x)$ is sign definite for the set M and M is stable, then M is asymptotically stable. This is done in the following theorem.

3.12 Theorem. Consider a dynamical system defined by the differential equation

3.12.1 $\dot{x} = f(x), \quad x \in R^n, \quad f\colon R^n \to R^n.$

Let $v = \varphi(x)$ and $w = \psi(x)$ be real-valued functions defined in an open neighborhood G of a continuum $M \subset R^n$. Assume that

3.12.2 $v = \varphi(x) \in \mathcal{C}^1,$

3.12.3 $\varphi(x) = 0$ if $x \in M,$

3.12.4 $\psi(x) = \langle \operatorname{grad} \varphi(x), \psi(x)\rangle \neq 0$ if $x \in G - M,$

3.12.5 M is stable.

Then M is asymptotically stable.

Proof. As in the proof of Theorem 3.9 we can show that for all $x \in G - M$, the set $\Lambda^+(x)$ is non-empty, compact and such that $\Lambda^+(x) \subset M$.

Theorem 3.9 gives also an estimate 3.9.6 of the region of attraction $A(M)$. This estimate can be improved in the following corollaries which we give without proof.

3.13 Corollary. Let the conditions of Theorem 3.9 be satisfied. Let, in addition,

3.13.1 $N = \cup \{N_c(\beta) \subset G\colon \beta > 0, \overline{N_c(\beta)} \subset G\}.$

Then

3.13.2 $A(M) \supset N.$

3.14 Corollary. Let the conditions of Theorem 3.9 be satisfied and $G \equiv R^n$. Let

3.14.1 $R^n = \cup \{N_c(\beta), \beta > 0\}.$

Then the compact, connected set $M \subset R^n$ is globally asymptotically stable.

By using the same technique used in the proof of Theorem 2.5 it is possible to give a sufficient condition for a real-valued function $v = \varphi(x)$ to satisfy condition 3.14.1.

3.15 **Theorem.** Let $v = \varphi(x)$ be a continuous real-valued function defined in R^n, such that

3.15.1 $\varphi(x) = 0$, if $x \in M$, $\varphi(x) > 0$ if $x \in \mathscr{C}M$.

Assume that $\lim_{\|x\| \to \infty} \varphi(x)$ exists. Then condition 3.14.1 is satisfied.

3.16 *Remark.* Theorems giving sufficient conditions for negative stability and for negative asymptotic stability (complete instability) can be proved either by inverting the sign conditions 3.5.7 in Corollary 3.5 and 3.9.5 of Theorem 3.9 or by considering both in Theorem 3.3 and 3.9 the sets $O_c(\beta)$ in place of the sets $N_c(\beta)$. Corollaries 3.13 and 3.14 will then apply to the region of negative attraction $A^-(M)$.

By means of continuously differentiable real-valued functions we can also give sufficient conditions for the (not necessarily complete) instability of a compact, connected set $M \subset R^n$. This result applies to the case of a continuum $M \subset R^n$ which is unstable, but neither completely unstable, nor attracting. This theorem gives also an estimate of the region of instability (V, 1.13).

3.17 **Theorem.** Consider a dynamical system represented by the differential equation

3.17.1 $\dot{x} = f(x), \quad x \in R^n, \quad f\colon R^n \to R^n.$

Let $v = \varphi(x)$ and $w = \psi(x)$ be real-valued functions defined on the closure $\bar{B}$ of an open set $B \subset R^n$. Assume that

3.17.2 $v = \varphi(x) \in \mathscr{C}^1$, on $\bar{B}$,

3.17.3 $\psi(x) = \langle \operatorname{grad} \varphi(x), f(x) \rangle$.

Let $M \subset \partial B$ be a continuum and $\eta > 0$ a real number such that

3.17.4 $\varphi(x) = 0, \quad x \in \partial B \cap S(M, \eta)$,

3.17.5 $\psi(x) > 0, \quad x \in B \cap S(M, \eta)$,

3.17.6 there is a sequence $\{x^n\} \subset B$, $x^n \to x \in M$, such that $\varphi(x^n) > 0$.

Then M is unstable and

3.17.7 $D^+(M, B) \cap H(M, \tilde{\eta}) \neq \emptyset, \quad 0 < \tilde{\eta} < \eta.$

Proof. For each real $\beta > 0$, consider the set

3.17.8 $O_B(\beta) = \{x \in B\colon \varphi(x) > \beta\}$.

Since B is an open set also $O_B(\beta)$ is open. In addition from the hypothesis 3.17.4, for each $\beta > 0$, we have:

3.17.9 $\partial O_B(\beta) \cap (\partial B \cap S(M, \eta)) = \emptyset$.

The set $O_B(\beta)$, in general, is not connected. On the other hand from the hypothesis 3.17.5 it follows that if $x \in O_B(\beta) \cap S(M, \eta)$, then either there exists $\tau > 0$ for which $x\tau \in H(M, \eta)$, or $\gamma^+(x) \subseteq \overline{(O_B(\beta) \cap S(M, \eta))}$. In the latter case $\gamma^+(x)$ is compact, hence $\Lambda^+(x)$ is non-empty and compact and in addition $\Lambda^+(x) \cap (\partial B \cap S(M, \eta)) = \emptyset$. Since if $z_1 \in \Lambda^+(x)$, $\psi(z_1) = 0$, we have that $z_1 \in H(M, \eta)$. Thus for each $\tilde{\eta} < \eta$ there exists a real number $\tau_{\tilde{\eta}} > 0$, such that $x\tau_{\tilde{\eta}} \in H(M, \tilde{\eta})$. From the hypothesis 3.17.6, given a point x^n in the sequence $\{x^n\} \subset B$, for a sufficiently small $\beta > 0$, we have that $x^n \in O_B(\beta)$ and therefore for each $\tilde{\eta} < \eta$ there exists a real number $\tau_n > 0$ such that $x^n\tau_n \in H(M, \tilde{\eta})$, which proves the theorem.

From the proof of Theorem 3.17 the following corollaries can be derived.

3.18 **Corollary.** If the hypotheses of Theorem 3.17 are satisfied for an arbitrary real number $\eta > 0$, then M is globally unstable.

3.19 **Corollary.** If the hypotheses of Theorem 3.17 are satisfied, then each compact set $K \subset S(M, \eta) \cap \partial B$ is unstable.

The following stronger theorem is given without proof.

3.20 **Theorem.** Let $M \subset R^n$ be a compact, connected set and let the hypotheses of Theorem 3.17 be satisfied. Let $K \subset R^n$ be a compact, connected set such that $M \subset \partial K$ and $(D^+(M, B) - M) \not\subset K$. Then K is unstable.

3.21 *Remark.* Theorems equivalent to Theorem 3.17, Corollaries 3.18 and 3.19 and Theorem 3.20, respectively, can be proved with different sign conditions on $\varphi(x)$ and $\psi(x)$, namely with conditions 3.17.5 and 3.17.6, respectively replaced by:

3.21.1 $\psi(x) < 0$ if $x \in B \cap S(M, \eta)$

and

3.21.2 there exists a sequence $\{x^n\} \subset B$, $x^n \to x \in M$, with $\varphi(x^n) < 0$.

The proof of those theorems and corollaries is left as exercise to the reader.

3.22 *Exercise.* Prove Theorem 3.9 without assuming that the continuum M is invariant.

4. Extension Theorems

The results derived in the previous section regarding the estimate of the region of attraction $A(M)$ of the compact, connected, asymptotically stable set $M \subset R^n$ (Corollary 3.13 and Corollary 3.14) are deceptively strong. Their application is, usually, not easy since they require a proof of compactness of the set $\bar{N}$ (3.13.1) which is generally difficult. On the other hand, condition 3.9.5 of Theorem 3.9 imposes a sign condition on the real-valued function $\psi(x)$, which, from the hypothesis made on $f(x)$, can immediately be translated into a sign condition on the vector grad $\varphi(x)$, from which, in turn, one should be able to derive an estimate of the set N.

This will be done in this section.

4.1 **Theorem** (Local Extension Theorem). Consider a dynamical system represented by the differential equation

4.1.1 $\dot{x} = f(x), \quad x \in R^n, \quad f\colon R^n \to R.$

Let $v = \varphi(x)$ and $w = \psi(x)$ be real-valued functions defined on R^n. Let $M \subset R^n$ be a compact, connected set. Assume that

4.1.2 $v = \varphi(x) \in \mathscr{C}^1,$

4.1.3 $\varphi(x) = 0, \quad x \in M,$

4.1.4 $\psi(x) = \langle \operatorname{grad} \varphi(x), f(x) \rangle,$

4.1.5 $\beta^0 > 0$ is a real number, such that for each sequence $\{x_n\} \subset N(\beta^0)$, with $x_n \not\to M$, and $\psi(x_n) \to 0$, we have $\varphi(x_n) \to \beta^0$.

4.1.6 M is positively asymptotically stable.

Then

4.1.7 $A(M) \supseteq N(\beta^0).$

If, in addition,

4.1.8 $\psi(x) \neq 0 \quad \text{if} \quad x \in \partial N(\beta^0),$

then

4.1.9 $A(M) \supset \overline{N(\beta^0)}.$

In condition 4.1.5 the statement $x_n \not\to$ stands for that there is a neighborhood U of M and a subsequence $\{x_{n_k}\}$ of $\{x_n\}$ such that $x_{n_k} \notin U$.

Proof. If the theorem were false, then $\partial A(M) \cap N(\beta^0) \neq \emptyset$. Let $x^0 \in \partial A(M) \cap N(\beta^0)$. Since the differential equation 4.1.1 defines a dynamical system, the set $\partial A(M)$ is invariant (V, 1.3.2 and II, 1.6).

Thus $\gamma(x^0) \subset \partial A(M)$. Let next

4.1.10 $\mu = \inf\{|\psi(y)|\colon y \in \gamma(x^0) \cap N(\beta^0)\}.$

Condition 4.1.5 implies that $\psi(x)$ cannot change sign in $N(\beta^0) - M$, then the following four cases may arise:

4.1.11 $\mu = 0, \quad \psi(x) > 0 \quad \text{if} \quad x \in N(\beta^0) - M,$

4.1.12 $\mu = 0, \quad \psi(x) < 0 \quad \text{if} \quad x \in N(\beta^0) - M,$

4.1.13 $\mu > 0, \quad \psi(x) > 0 \quad \text{if} \quad x \in N(\beta^0) - M,$

4.1.14 $\mu > 0, \quad \psi(x) < 0 \quad \text{if} \quad x \in N(\beta^0) - M.$

In the case 4.1.11 there exists a sequence $\{x_n\} \subset \gamma(x^0) \cap N(\beta^0)$ such that $\psi(x_n) \to 0$ and therefore from 4.1.5, $\varphi(x_n) \to \beta^0$. Now, since $\{x_n\} \subset \gamma(x^0) \subset \partial A(M)$ and $\varphi(x)$ is continuous, there exist points $y \in A(M) \cap N(\beta^0)$ such that $\varphi(y) > 0$.

Since $\psi(x) > 0$ for $x \in N(\beta^0) - M$, it follows that

4.1.15 $\varphi(yt) \geqq \varphi(y)$ for all t with $y[0, t] \subset N(\beta^0)$.

Since $\varphi(x) = \beta^0$ for $x \in \partial N(\beta^0)$, from 4.1.15 it then follows that $\Lambda^+(y) \not\subset M$, which contradicts the hypothesis 4.1.6 and proves the theorem in the case 4.1.11.

In the case 4.1.12 we have that the set $N(\beta^0)$ is positively invariant (Lemma 3.1). Thus $\gamma^+(x_0) \subset \partial A(M) \cap N(\beta^0)$. Since $\mu = 0$ we have that there exists a sequence $\{x_n\} \subset \gamma^+(x_0)$, such that $\varphi(x^n) \to \beta^0$. This contradicts the fact that $\varphi(x_n) \leqq \varphi(x^0) < \beta^0$ for all n and proves the theorem in the case 4.1.12.

In the case 4.1.13 we have that

4.1.16 $\varphi(x_0 t) \geqq \varphi(x_0) + \mu t$ for all t with $x_0[0, t] \subset N(\beta^0)$.

Hence there exists $\tau > 0$, such that $\varphi(x_0 \tau) = \beta^0$ and then we proceed as in the case 4.1.11 and prove the theorem in this case.

In the case 4.1.14, since $N(\beta^0)$ is positively invariant (Lemma 3.1), we have

4.1.17 $\varphi(x_0 t) \leqq \varphi(x_0) - \mu t$ for all $t \in R^+$.

Hence there exists $\tau \in R^+$ such that $\varphi(x_0 \tau) < 0$. Since $\varphi(x)$ is continuous there exists then a point $y \in A(M) \cap N(\beta^0)$, such that $\varphi(y) < 0$, which is absurd.

Thus in all cases

4.1.18 $\partial A(M) \cap N(\beta^0) = \emptyset$.

Since $A(M) \cap N(\beta^0) \neq \emptyset$, it follows that $A(M) \supset N(\beta^0)$ (4.1.7), which completes the proof of the first part of the theorem.

Assume next that, in addition, condition 4.1.8 holds; we shall prove that this implies that $\partial A(M) \cap \partial N(\beta^0) = \emptyset$. For this let $x_0 \in \partial A(M) \cap \partial N(\beta^0)$.

Let next $\psi(x_0) > 0$. Since $A(M) \supset N(\beta^0)$, there exists $y \in A(M)$, such that $\varphi(y) > 0$. On the other hand, if $\psi(x_0) > 0$, we have that $\psi(x) > 0$ for all $x \in N(\beta^0) - M$. Following the same arguments as in the case 4.1.11 we reach a contradiction.

Let next $\psi(x_0) < 0$. Then $\psi(x) < 0$ for $x \in N(\beta^0) - M$. Let $\gamma^+(x_0) \subset \partial A(M)$. For $t \in R^+$, sufficiently small, we have $\varphi(x_0 t) < \beta^0$. Thus, either $x^0 t \in N(\beta^0)$, which is absurd, or x^0 belongs to the boundary of a component of $N'(\beta^0)$, say $N_1(\beta^0) \neq N(\beta^0)$. Now for $t \in R^+$, sufficiently small, $\varphi(x_0 t) < \beta^0$. Hence $x_0 t \in N'(\beta^0)$, $x_0 t \in N(\beta^0)$, $x_0 t \in N_1(\beta^0)$. Thus $N(\beta^0)$ and $N_1(\beta^0)$ cannot be two different components of $N'(\beta^0)$ and the proof follows from that of the previous case.

This concludes the proof of the theorem.

4.2 *Remark.* It is possible to prove a theorem, analogous to Theorem 4.1, where instead of the sets $N(\beta^0)$, $\beta^0 > 0$, we consider the sets $O(\tilde{\beta}^0)$, $\tilde{\beta}^0 < 0$. This is left as an exercise to the reader.

We shall next prove a theorem regarding the global stability properties of a compact, connected set $M \subset R^n$. The statement of this theorem is stronger than that of Theorem 4.1.

4.3 **Theorem** (Global Extension Theorem). Consider the dynamical system defined by the differential equation

4.3.1 $\dot{x} = f(x), \quad x \in R^n, \quad f\colon R^n \to R^n.$

Let $M \subset R^n$ be a compact connected set and $v = \varphi(x)$ and $w = \psi(x)$ be real-valued functions on R^n. Assume that

4.3.2 $v = \varphi(x) \in \mathcal{C}^1,$

4.3.3 $\varphi(x) = 0 \quad \text{if} \quad x \in M,$

4.3.4 $\psi(x) = \langle \operatorname{grad} \varphi(x), f(x) \rangle,$

4.3.5 $\{x_n\} \subset R^n, \quad \psi(x_n) \to 0 \quad \text{imply} \quad x_n \to M.$

Then whatever the stability properties of M are, they are global.

Proof. From Theorem 3.12 it follows that, under the hypotheses of Theorem 4.3, positive and negative stability imply positive and negative asymptotic stability, respectively. Hence the local stability properties of M can be only the following: instability (in both directions), positive asymptotic stability and negative asymptotic stability (complete instability). In the first case the theorem is true, since the conditions of Theorem 3.17 (Remark 3.21) are satisfied for $B = R^n$ and any real number $\eta > 0$. Indeed, in this case there exists a sequence $\{x_n\} \subset R^n$, $x_n \to M$ along which $\operatorname{sign} \varphi(x_n) = \operatorname{sign} \psi(x_n)$: if this were not true, then either the hypotheses of Theorem 3.9 or those of the theorem of negative asymptotic stability (Remark 3.16) would be satisfied. M would then be either positively or negatively asymptotically stable which contradicts the hypothesis. Hence M is globally unstable (in the sense of V, 1.13).

Consider next the case that M is asymptotically stable. Then, if the theorem were not true, $\partial A(M) \neq \emptyset$. Consider next a point $x_0 \in \partial A(M)$, since $\partial A(M)$ is invariant, $\gamma^+(x_0) \subset \partial A(M)$. Let next

4.3.6 $\mu = \inf\{|\psi(y)|: y \in \gamma^+(x_0)\}$.

From the hypothesis 4.3.5 it follows that $\mu > 0$.

Assume now that $\psi(x) > 0$ for all $x \in \partial A(M)$. Then

4.3.7 $\varphi(x_0 t) \geqq \varphi(x_0) + \mu t$.

Hence for a sufficiently large $t \in R^+$, we have $\varphi(x_0 t) > 0$. Thus M cannot be asymptotically stable, which contradicts the hypothesis.

If, instead, we assume that $\psi(x) < 0$ for all $x \in \mathscr{C}(M)$, then

4.3.8 $\varphi(x_0 t) \leqq \varphi(x_0) - \mu t$.

Again, for $t \in R^+$ sufficiently large we have $\varphi(x_0 t) < 0$, which contradicts the hypothesis made.

Finally in the case in which M is completely unstable, we shall replace equation 4.3.1 with the equation

4.3.9 $\dot{x} = -f(x)$.

If the hypotheses of the theorem were satisfied by equation 4.3.1 they are still satisfied by equation 4.3.9. Now however, M is asymptotically stable. The proof of this case is therefore reduced to the previous one.

This completes the proof of the theorem.

4.4 *Remark*. Theorems 4.1 and 4.3 still leave some open problems and some unanswered questions, some of which will be clarified in the next section. The most important of these problems is concerned with the identification of the set $N(\beta^0)$, where β^0 is defined in 4.1.5, for the case in which $\psi(x) \neq 0$ if $x \in R^n - M$. This is therefore a case in which only for some sequences $\{x_n\} \subset N(\beta^0)$, $x_n \to \infty$, we have $\psi(x_n) \to 0$. Clearly only two situations may now arise: the case in which $N(\beta^0) = \bigcup_{0 \leq \beta < \beta^0} N(\beta) = R^n$ and the case in which $N(\beta^0)$ is not compact, but $N(\beta^0) \neq R^n$. The distinction between these two cases is an open problem. The finite version of this problem, which has already been mentioned in Theorems 2.3 and 2.4, is the following. Assume that there exists a point $x_0 \in R^n$, such that $\psi(x_0) = 0$, but for all diverging sequences $\{x_n\} \subset N(\beta^0)$, $x_n \to \infty$, $\psi(x_n) \not\to 0$. Then $N(\beta^0)$ may or may not have a compact closure (see Theorem 2.4) however, for all $\varepsilon > 0$ and all $\tilde{\beta} < \beta^0 - \varepsilon$, the set $N_c(\tilde{\beta})$ is non-empty. On the other hand, if there does exist a sequence $\{x_n\} \subset N(\beta^0)$, $x_n \to \infty$, then $\psi(x_n) \to 0$ may not hold. In this case, if for some $\beta > 0$, $N_c(\beta) \neq \emptyset$, it is not trivial to find the largest real number $\tilde{\beta}$ such that $N_c(\tilde{\beta}) \neq \emptyset$. This problem is not related to that of the estimate of $A(M)$ (since $A(M)$ is estimated from 4.1.7 and 4.1.8 anyhow), but with the problem of the geometrical properties of $A(M)$. Clearly when the Liapunov functions $v = \varphi(x)$ are polynomials (which do not have singular point at infinity) these problems do not arise.

We close this section by illustrating with an example the use of Theorem 4.1.

4.5 *Example.* Consider the differential equation V, 2.4.1 and the real-valued function V, 2.4.4. Let $M = \{(0, 0)\}$. The $(0, 0)$ is locally asymptotically stable. The total time derivative of the function V, 2.2.4 with respect to the equation V, 2.3.1 has the form

4.5.1 $$\psi(x) = \begin{cases} -2y^2 + \dfrac{2x^2}{(1+x^2)^2} & \text{for } x^2y^2 \geqq 1, \\ -2y^2 + \dfrac{2x^2(x^2y^2-1)}{(1+x^2)^2} & \text{for } x^2y^2 < 1. \end{cases}$$

We see that the only points in which $\psi(x)$ vanishes are $(0, 0)$ and $y = 0$, $|x| \to +\infty$. Condition 4.1.5 holds if $\beta^0 = 1$, but not if $\beta^0 > 1$, and indeed $A(M) \supseteq \left\{(x, y) : y^2 + \dfrac{x^2}{1+x^2} < 1\right\}$.

5. The Structure of Liapunov Functions

The extension theorems proved in the previous section are still too restrictive in the sense that (see next section: Theorems 6.3 and 6.4), it is possible to characterize the property of positive or negative asymptotic stability of a continuum $M \subset R^n$ by means of Liapunov functions whose total time derivative $\psi(x)$ is only semidefinite. In order to prove these theorems some additional information on the structure of the sets $N'(\beta)$, $N(\beta)$ and $N_c(\beta)$, must be obtained. For that we shall apply the stability theorems proved in sections 3 and 4 to the differential system $\dot{x} = \operatorname{grad} \varphi(x)$. Thus the real-valued function $\varphi(x)$ is now assumed to be twice continuously differentiable.

We shall begin our investigation by asking ourselves what properties $\varphi(x)$ must have in order that it is a Liapunov function which characterizes the positive (or negative) asymptotic stability properties of a continuum M and has a sign-definite total time derivative $\psi(x)$.

The first obvious conditions, 5.1 and 5.2, are of geometric type; their proof is an immediate consequence of the theorems of section 3 and it is left as an exercise to the reader.

5.1 Proposition. A necessary condition for a real-valued function $\varphi(x)$ to be a Liapunov function with negative definite $\psi(x)$, suitable to characterize the properties of asymptotic stability of a compact, connected set $M \subset R^n$, is that

5.1.1 $\quad N(\beta) = N_c(\beta)$ for all $\beta < \beta^0$

and

5.1.2 $$N(\beta^0) = \bigcup_{0 \le \beta < \beta^0} N_c(\beta),$$

where the sets $N(\beta)$ and $N_c(\beta)$ are defined in 2.1.7 and 2.1.8 and β^0 is defined in 4.1.5.

In the case of global properties, we have

5.2 Proposition. A necessary condition for a real-valued function $\varphi(x)$ to be a Liapunov function with negative definite $\psi(x)$, suitable to characterize the properties of global asymptotic stability of a compact, connected set $M \subset R^n$, is that

5.2.1 $\quad N(\beta) = N_c(\beta)$ for all β

and

5.2.2 $$R^n = \bigcup_{\beta \ge 0} N_c(\beta).$$

Conditions 5.1.1, 5.1.2, 5.2.1 and 5.2.2 clarify the structure of Liapunov functions used in the characterization of the positive or negative asymptotic stability properties of compact, connected sets. From the practical point of view, however, such conditions are not easy to satisfy and they are usually replaced by analytical conditions of the following type.

5.3 Proposition. Let $v = \varphi(x)$, $\varphi\colon R^n \to R$. Let $M \subset R^n$ be a compact and connected set. Let

5.3.1 $\quad \varphi(x) = 0 \quad \text{for} \quad x \in M.$

If there exist two strictly increasing functions $\alpha(\mu)$ and $\beta(\mu)$ with $\alpha(0) = \beta(0) = 0$ such that

5.3.2 $$\alpha(\varrho(M, x)) \le |\varphi(x)| \le \beta(\varrho(M, x)),$$

then such a real-valued function can be used for the characterization of positive (or negative) asymptotic stability properties of the set M. If in addition

5.3.3 $\quad \alpha(\mu) \to +\infty$ as $\mu \to +\infty$,

then the real-valued function $\varphi(x)$ can be used for the characterization of positive (or negative) global asymptotic stability properties of the set M.

Notice that condition 5.3.2 is a necessary and sufficient condition for a continuous real-valued function $v = \varphi(x)$ to be definite.

5.4 *Remark.* Clearly (see Theorem 2.5) condition 5.3.2 implies condition 5.1.1 and together with condition 5.3.3 implies condition 5.2.2 while the converse is not true, for instance the function 2.6.1 satisfies condition 5.1.1, but not condition 5.3.2. On the other hand condition 5.2.2 clearly does not imply condition 5.3.3; this can be easily seen by taking a function $\varphi(x)$ which satisfies in the whole space R^n conditions 5.3.2 and 5.3.3 and therefore 5.2.2, then considering the function $\frac{\varphi(x)}{1+\varphi(x)}$ which satisfies 5.2.2 but not 5.3.3.

These remarks may have somewhat clarified the relationship between the classical conditions (5.3.2 and 5.3.3) and the geometrical conditions 5.1.1 and 5.1.2.

Still we have to discuss a problem of great practical importance, that of the relationships between the topological properties of $N'(\beta)$ and the existence of critical points of a continuously differentiable, real-valued function $\varphi(x)$ (the points $y \in R^n$, such that $\operatorname{grad} \varphi(y) = 0$).

The following theorems are an application of Theorems 4.1 and 4.3 to a differential equation uniquely defined by the continuously differentiable real-valued function $\varphi(x)\colon \varphi\colon R^n \to R^n$, namely, the differential equation

5.5 $$\dot{x} = \operatorname{grad} \varphi(x),$$

and others of a similar kind.

5.6 **Theorem.** Let $v = \varphi(x)$, $\varphi\colon R^n \to R$. Let $M \subset R^n$ be a compact, connected set. Assume that the following conditions are satisfied:

5.6.1 $\varphi(x) \in \mathscr{C}^2$,

5.6.2 $\varphi(x) = 0$ if $x \in M$,

5.6.3 there exists a real number $\beta > 0$ such that $N_c(\beta) \neq \emptyset$,

5.6.4 for each sequence $\{x_n\} \subset R^n$, $\operatorname{grad} \varphi(x_n) \to 0$ implies $x_n \to M$.

Then for each $\beta > 0$

5.6.5 $$N_c(\beta) \neq \emptyset$$

and in addition

5.6.6 $$\bigcup_{\beta \geq 0} N_c(\beta) = R^n.$$

Proof. Consider the differential equation

5.6.7 $$\dot{x} = -\frac{\operatorname{grad} \varphi(x)}{1 + \|\operatorname{grad} \varphi(x)\|} = g(x).$$

This equation defines a dynamical system since all its solutions are unique and defined on R (see section I, 2.1). We shall investigate its stability properties by means of the Liapunov function $v = \varphi(x)$. Consider then the real-valued function

5.6.8 $$\psi(x) = \langle \text{grad}\, \varphi(x), g(x)\rangle = -\frac{\|\text{grad}\, \varphi(x)\|^2}{1 + \|\text{grad}\, \varphi(x)\|} \leqq 0.$$

From the condition 5.6.4 it follows that for each sequence $\{x_n\} \subset R^n$, $\psi(x_n) \to 0$ implies $x_n \to M$. We are now in the position of applying Theorem 4.3, and conclude that the compact set $M \subset R^n$, is globally asymptotically stable for the differential equation 5.6.7.

We shall next prove that for each $\beta > 0$ the set $N'(\beta)$ is connected. Assume by contradiction that there exist two components of $N'(\beta)$, say $N_1(\beta)$ and $N_2(\beta)$, such that $N_1(\beta) \cap N_2(\beta) = \emptyset$ and let $N_1(\beta) \supset M$. Let now $x_0 \in N_2(\beta)$, since M is globally asymptotically stable, we have $\gamma^+(x_0) \cap \partial N_1(\beta) \neq \emptyset$. Let $y \in \gamma^+(x_0) \cap \partial N_1(\beta)$, we must have $\varphi(y) = \beta$. On the other hand $\varphi(y) < \varphi(x_0) < \beta$ which is absurd. We shall prove next that for each $\beta > 0$ the set $\overline{N(\beta)}$ is compact. Let $\tilde{\beta}$ be such that $\overline{N(\tilde{\beta})}$ is not compact.

From Theorem 4.3, M is globally asymptotically stable, hence for all $x \in \partial N(\tilde{\beta})$, $\varrho(xt, M) \to 0$ as $t \to +\infty$.

Let next

5.6.9 $$-\mu = \sup \psi(x) \text{ for } x \in R^n - N_c(\beta),$$

where $\beta > 0$ is defined by 5.6.3. From the hypothesis 5.6.4, we have that $\mu > 0$. Let next

5.6.10 $$T = \frac{\tilde{\beta} - \beta}{\mu}.$$

Then for all $\tau > T$ and all $x \in \partial N(\tilde{\beta})$ we have $x\tau \in N_c(\beta)$. In particular we have $x_n\tau = y_n \in N_c(\beta)$ for a sequence $\{x_n\} \subset \partial N(\tilde{\beta})$, $\|x_n\| \to +\infty$. Let now $y_n \to y^*$, hence $y_n(-\tau) \to y^*(-\tau)$. On the other hand $y_n(-\tau) = x_n$, which contradicts the assumption that $\|x_n\| \to +\infty$, and proves the theorem.

In a similar way, by applying Theorem 4.1 to the differential equation 5.6.7, we can prove the following local result.

5.7 Theorem. Let $v = \varphi(x)$, $\varphi\colon R^n \to R$. Let $M \subset R^n$ be a compact, connected set. Assume that conditions 5.6.1, 5.6.2 and 5.6.3 of Theorem 5.6 are satisfied. In addition assume that

5.7.1 $\beta^0 > \beta$ is such that if $\{x_n\} \subset N(\beta^0)$ and $x_n \not\to M$, then $\text{grad}\, \varphi(x_n) \to 0$ implies $\varphi(x_n) \to \beta^0$.

Then

5.7.2 $N(\beta) = N_c(\beta)$ for all $\beta < \beta^0$

and

5.7.3 $N(\beta^0) = \bigcup_{0 \le \beta \le \beta^0} N(\beta)$.

5.8 *Remark.* Clearly, if the differential equation

5.8.1 $\dot{x} = f(x), \quad f\colon R^n \to R^n$

defines a dynamical system and

5.8.2 $\psi(x) = \langle \operatorname{grad} \varphi(x), f(x) \rangle$,

then the conditions of Theorems 4.1 and 4.3 on $\psi(x)$ imply the conditions of Theorems 5.6 and 5.7, respectively. Thus in Theorems 4.1 and 4.3 the sets $N(\beta)$ do satisfy the conditions 5.6.5, 5.6.6, 5.7.2 and 5.7.3 respectively.

One can prove a theorem similar to Theorem 5.6 also for the analytical conditions 5.3.2 and 5.3.3. Thus we have

5.9 **Theorem.** Let $v = \varphi(x)$, be a real-valued function defined on R^n. Let $M \subset R^n$ be a compact, connected set. Assume that

5.9.1 $v = \varphi(x) \in \mathscr{C}^2$,

5.9.2 $\varphi(x) = 0$ for $x \in M$,

5.9.3 $\varphi(x) > 0$ for $x \in H(M, \delta)$, $\delta > 0$, ,

5.9.4 for $\{x_n\} \subset R^n$, $\operatorname{grad} \varphi(x_n) \to 0$ implies $x_n \to M$.

Then there exist two strictly increasing functions $\alpha(\mu)$ and $\beta(\mu)$, such that

5.9.5 $\alpha(\varrho(x, M)) \leqq \varphi(x) \leqq \beta(\varrho(x, M))$, $\alpha(0) = \beta(0) = 0$

and, furthermore,

5.9.6 $\lim_{\mu \to \infty} \alpha(\mu) = +\infty$.

In addition to this, if

5.9.7 $\nu = \min\{\varphi(x) \colon x \in H(M, \delta)\}$,

then

5.9.8 $\varphi(x) > \nu$ for $x \in \mathscr{C}(S[M, \delta])$.

Proof. Consider again the dynamical system 5.6.7 and the real-valued function 5.6.8.

The functions $v = \varphi(x)$ and $w = \psi(x)$ satisfy the conditions of Theorem 4.3. Thus the set M is globally asymptotically stable for the system 5.6.7. Notice now that for any $x_0 \in \mathcal{C}(S(M, \delta))$, there is a $\tau > 0$ such that $x_0\tau \in H(M, \delta)$. Thus $\varphi(x_0\tau) \geqq \nu$. Since $\varphi(x_0t)$, for any $x_0 \in \mathcal{C}(M)$, is a strictly decreasing function of t, we conclude that $\varphi(x_0) > \varphi(x_0\tau) \geqq \nu$. Lastly as $\varphi(x) = 0$, for $x \in M$, $\varphi(x) > 0$ for $x \notin M$, and $\varphi(x)$ is continuous, we can define two continuous increasing functions $\alpha(\eta)$ and $\beta(\eta)$ by

$$\alpha(\eta) = \min\{\varphi(x) : x \in H(M, \eta)\}$$

and

$$\beta(\eta) = \max\{\varphi(x) : x \in H(M, \eta)\}.$$

Notice that $\alpha(0) = \beta(0) = 0$ and $\beta(\eta) \geqq \alpha(\eta) > 0$ for $\eta > 0$. There exist strictly increasing continuous functions $\bar{\alpha}(\eta)$ and $\bar{\beta}(\eta)$ such that

$$\bar{\alpha}(\eta) \leqq \alpha(\eta) \leqq \beta(\eta) \leqq \bar{\beta}(\eta).$$

With these $\bar{\alpha}(\eta)$ and $\bar{\beta}(\eta)$ we have

$$\bar{\alpha}(\varrho(x, M)) \leqq \varphi(x) \leqq \bar{\beta}(\varrho(x, M)).$$

It remains to be proved that there exists one function $\alpha(\eta)$ which satisfies condition 5.9.6. We shall prove first that for all $x_0 \in \mathcal{C}(M)$, $\varphi(x_0t) \to +\infty$ as $t \to -\infty$. In fact, notice that for any $x_0 \in \mathcal{C}(M)$

$$\varphi(x_0t) = \varphi(x_0) + \int_0^t \psi(x_0\tau)\, d\tau.$$

For $\tau \leqq 0$ we have $\varrho(x_0\tau\, M) > \eta > 0$. Thus

$$\psi(x_0\tau) \leqq \max\{\psi(x) : x \notin S(M, \delta)\} = -\chi < 0$$

where $\delta > 0$ is such that $x_0\tau \notin S(M, \delta)$ for $\tau \leqq 0$. Hence for $t \leqq 0$

$$\varphi(x_0t) \geqq \varphi(x_0) + \int_0^t \chi\, d\tau = \varphi(x_0) - \chi t.$$

Thus $\varphi(x_0t) \to +\infty$ as $t \to -\infty$.

Now the existence of our $\alpha(\eta)$ with the property that $\alpha(\eta) \to +\infty$ as $\eta \to +\infty$ is equivalent to the property that $\varphi(x) \to +\infty$ as $\varrho(x, M) \to +\infty$. Assume that the last assertion is not true. Then there is a $h > 0$ such that the surfaces $\varphi(x) = k$ are compact for $k < h$ and non-compact for $k \geqq h$. Consider the open set P defined by

$$P = \{x : 0 < k < \varphi(x) < h\}.$$

This open set is bounded away from the set M and is bounded by the surfaces $\varphi(x) = k$ and $\varphi(x) = h$. We claim that there is a point x_0, $\varphi(x_0) = k$, such that $x_0t \in P$ for $t < 0$. For, otherwise, for every x_0, with $\varphi(x_0) = k$ we will have a unique $t < 0$ such that $x_0t \in \{x : \varphi(x) = h\}$, so that, following the argument in the proof of Theorem V, 2.9, there will be continuous map of the compact set $\varphi(x) = k$ on to the non-compact set $\varphi(x) = h$ which is impossible. Notice that for such an x_0 we have $\varphi(x_0 t) < h$ for $t \leqq 0$ which contradicts the fact that $\varphi(x_0 t) \to +\infty$ as $t \to -\infty$ for all $x_0 \notin M$. The theorem is completely proved.

6. Theorems Requiring Semidefinite Derivatives

In this section we shall prove theorems which allow us to deduce properties of attraction and asymptotic stability of a compact set $M \subset R^n$ using conditions different from those of Theorem 3.9.

In particular we allow $\psi(x)$ to be only semidefinite. After the proof of the main theorem (6.1) we shall prove the corresponding extension theorems relaxing some conditions on $\varphi(x)$ in Theorem 6.1. The proofs of these extension theorems depend upon the results proved in the previous section.

6.1 Theorem. Consider the dynamical system represented by the differential equation

6.1.1 $\quad \dot{x} = f(x), \quad x \in R^n, \quad f\colon R^n \to R^n.$

Let Ω be a compact, positively invariant set. Let $v = \varphi(x)$ and $w = \psi(x)$ be real-valued functions defined on Ω. Assume that

(i) $v = \varphi(x) \in \mathcal{C}^1$, $x \in \Omega$,

(ii) $v = \varphi(x) \geqq 0$, $x \in \Omega$,

(iii) $\psi(x) = \langle \operatorname{grad} \varphi(x), f(x) \rangle$,

(iv) $w = \psi(x) \leqq 0$, $x \in \Omega$.

Consider the following sets:

(I) S is the largest invariant set in Ω,

(II) $M = \{x \in \Omega\colon \varphi(x) = 0\}$,

(III) $P = \{x \in \Omega\colon \psi(x) = 0\}$,

(IV) Q is the largest invariant set in P ($Q \subseteq S$),

(V) U is the largest invariant set contained in M.

Then

(a) The sets S, P and Q are closed,

(b) $M \subseteq P$ and M is stable relative to Ω,

(c) Q is attracting relative to Ω,

(d) S is asymptotically stable relative to Ω,

(e) $\partial Q \cap \partial S \neq \emptyset$,

(f) $S = D^+(Q, \Omega) = \{y$: there is a sequence $\{x_n\} \subset \Omega$ and a sequence $\{t_n\} \subset R^+$ such that $x_n \to Q$ and $x_n t_n \to y\}$ where $D^+(Q, \Omega)$ is the first positive prolongation of Q relative to Ω,

(g) $S \supset M$, and is the largest asymptotically stable set relative to Ω,

(h) if for all $x \in \partial Q$, $\varphi(x) = \text{const}$, Q is asymptotically stable relative to Ω and $Q = S$,

(i) Q minimal implies $Q = S = U$,

(j) $M = Q$ implies Q is asymptotically stable relative to Ω and $M = S$,

(k) if either Ω, S or Q are homeomorphic to the unit ball, then M contains a rest point,

(l) if either $Q \subset \mathscr{I}\Omega$ or $S \subset \mathscr{I}\Omega$, then the words "relative to Ω" may be deleted from the above statements.

Proof.

(a) This is clear, for if a set is invariant, so is its closure.

(b) For all $x \in M$, from (ii) and (iv) we have $\psi(x) = 0$ which implies $M \subseteq P$. Stability then follows from Theorems V, 4.2, 4.5, 4.6.

(c) Since for each $y \in \Omega$, $\Lambda^+(y) \subset \Omega$ and $\psi(x) = 0$ for $x \in \Lambda^+(y)$, we get $\Lambda^+(y) \subset Q$ as $\Lambda^+(y)$ is invariant.

(d) Since $S \supset Q$, S is attracting relative to Ω. Stability follows as S is the largest invariant set in Ω, $D^+(S, \Omega) \subset \Omega$, so that $D^+(S, \Omega) = S$.

(e) For if $\partial Q \cap \partial S = \emptyset$, then $\partial S \subset \Omega$ being invariant, we have $x \in \partial S$, $\Lambda^+(x) \neq \emptyset$ and $\Lambda^+(x) \cap Q = \emptyset$, since $Q \subset S$, we conclude that Q is not attracting relative to Ω, contradicting (c).

(f) Follows from proof of (d).

(g) Obvious (V, 5.12).

(h) We need only show that Q is stable relative to Ω. If not, then there is a sequence $\{x_n\}$ in Ω, $x_n \to x \in \partial Q$, and a sequence $\{t_n\}$, $t_n \geqq 0$, such that $x_n t_n \to y \notin Q$. Indeed $y \in \Omega$ and $x_n [0, t_n] \subset \Omega$ as Ω is positively invariant and compact. Since $\varphi(x_n) \geqq \varphi(x_n t_n)$, we get $\varphi(x) \geqq \varphi(y)$ by continuity. However, if $z \in \Lambda^+(y)$, we get $\varphi(z) \leqq \varphi(y)$, and since $z \in \partial Q$, we have $\varphi(x) \geq \varphi(y) \geq \varphi(z) = \varphi(x)$, showing that $\varphi(y) = \varphi(x)$. This shows, however, that $\gamma(y) \subset \Omega$ and $\varphi(x)$ is constant on $\gamma(y)$. Consequently $\psi(x) = 0$ on $\gamma(y)$, showing that $\gamma(y) \subset Q$, a contradiction.

(i) If Q is minimal, then $\varphi(x)$ is constant on Q and the result follows from (h) and the fact that $U \subset S$.

(j) Follows from (h) and (c).

(k) Follows from V, 3,8.

(l) Follows from V, 5.2.

In view of Lemma 3.1 from Theorem 6.1 it immediately follows:

6.2 **Corollary.** Consider the dynamical system defined by the ordinary differential equation

6.2.1 $\dot{x} = f(x), \quad x \in R^n, \quad f\colon R^n \to R^n.$

Let $v = \varphi(x)$ and $w = \psi(x)$ be real-valued functions defined on R^n, which satisfy the hypotheses of Theorem 3.3. Then for each real number $\beta > 0$ such that the set $N_c(\beta) \neq \emptyset$, Theorem 6.1 holds with

6.2.2 $\Omega = \bar{N}_c(\beta).$

In view of the Theorems 5.6 and 5.7 which allow an estimate of largest $N_c(\beta)$ from the properties of grad $\varphi(x)$, we are now in the position of proving the following theorems:

6.3 **Theorem.** Consider the dynamical system, defined by the ordinary differential equation

6.3.1 $\dot{x} = f(x), \quad x \in R^n, \quad f\colon R^n \to R^n.$

Let $v = \varphi(x)$ and $w = \psi(x)$ be real-valued functions, which satisfy the assumptions of Theorem 6.1 in the whole space R^n. In addition assume that

6.3.2 $v = \varphi(x) \in \mathcal{C}^2$, $x \in R^n$.

6.3.3 The sets S, Q, P and M defined as in Theorem 6.1 with $\Omega = R^n$ are compact.

6.3.4 There exist two real numbers β and β', $\beta < \beta'$ such that

(i) The set $H(\beta)$ has a compact component which will be denoted $H_c(\beta)$. If we denote with $N_c(\tilde{\beta})$ the component with compact closure of $N(\beta')$ which contains $H_c(\beta)$, $N_c(\beta') \neq \emptyset$.

(ii) For all sequences $\{x^n\} \subset R^n$, grad $\varphi(x^n) \to 0$ implies $x^n \to H_c(\beta)$.

Then Theorem 6.1 holds with

6.3.5 $\Omega = R^n.$

Proof. From Theorem 5.6 we have that $R^n = \bigcup_{k \geq \beta} N_c(k)$. Then since the sets S, M, P, Q are compact there exists a real number $\bar{\beta}$ such that all the above mentioned sets are contained in $N_c(\beta)$ for all $\beta \geq \bar{\beta}$.

6.4 Theorem. Consider the dynamical system defined by the ordinary differential equation

6.4.1 $\dot{x} = f(x), \quad x \in R^n, \quad f\colon R^n \to R^n.$

Let $v = \varphi(x)$ and $w = \psi(x)$ be real-valued functions which satisfy the assumptions of Theorem 6.1 in the whole space R^n. In addition assume that

6.4.2 $v = \varphi(x) \in \mathscr{C}^2, \quad x \in R^n.$

6.4.3 There exist two real numbers β and β' with $\beta < \beta'$ such that condition 6.3.4 of Theorem 6.3 is satisfied, and such that there exists a real number $\beta^0 > \beta'$ such that if $\{x_n\} \subset N(\beta^0)$, $x_n \nrightarrow H_c(\beta)$, then $\operatorname{grad} \varphi(x_n) \to 0$ implies $\varphi(x_n) \to \beta^0$.

Then Theorem 6.1 holds with

$$\Omega = \overline{N(\tilde{\beta})} \text{ for all } \tilde{\beta} \text{ with } \beta < \tilde{\beta} \leq \beta'.$$

Proof. From Theorem 5.7 it follows that $\overline{N(\tilde{\beta})}$ is compact for all $\beta < \tilde{\beta} \leq \beta'$" and that $N_c(\beta) = N(\beta)$.

Clearly the sets S, M, P, Q must be referred to the $\overline{N(\tilde{\beta})}$. For the application of the results the following corollary is useful.

6.5 Corollary. Consider the dynamical system defined by the ordinary differential equation

6.5.1 $\dot{x} = f(x), \quad x \in R^n, \quad f\colon R^n \to R^n.$

Let $v = \varphi(x)$ and $w = \psi(x)$ be real-valued functions defined in R^n. Let $M \subset R^n$ be a compact, connected set. Assume that

6.5.2 $v = \varphi(x) \in \mathscr{C}^2,$

6.5.3 $\varphi(x) = 0$ if $x \in M,$

6.5.4 $\psi(x) = \langle \operatorname{grad} \varphi(x), f(x) \rangle,$

6.5.5 for all $\{x_n\} \subset R^n$, $\operatorname{grad} \varphi(x_n) \to 0$ implies $x_n \to M,$

6.5.6 $\psi(x) = 0$ if $x \in E \subset R^n$, $M \subset E,$

6.5.7 M is the only invariant set contained in the set E.

Then if M is asymptotically stable, it is globally asymptotically stable, if M is negatively asymptotically stable, it is globally negatively asymptotically stable and if M is unstable, it is globally unstable.

7. On the Use of Higher Derivatives of a Liapunov Function

In the previous chapter the stability properties of sets with respect to the flow defined by the solutions of ordinary differential equations

7.1 $$\dot{x} = f(x), \quad f\colon R^n \to R^n,$$

have been characterized by the properties of a real-valued function $v = \varphi(x)$ and its total time derivative along the solutions of the differential equation 7.1.

7.2 $$\psi_1(x) = \langle \operatorname{grad} \varphi(x), f(x) \rangle.$$

In this section we shall briefly summarize some recent results obtained by various authors on the use of the total time derivative of order n of the real-valued function $v = \varphi(x)$ along the solutions of 7.1, which is defined as follows

7.3 $$\psi_2(x) = \langle \operatorname{grad} \psi_1(x), f(x) \rangle, \ldots, \psi_n(x) = \langle \operatorname{grad} \psi_{n-1}(x),\ f(x) \rangle,$$

where $f(x) \in \mathcal{C}^{n-1}$ and $\varphi(x) \in \mathcal{C}^n$.

Most of the results obtained are not strictly stability results, but they lead to a more complete analysis of the qualitative behavior of the differential equation 7.1. This analysis is in accordance with the classification due to Nemytskii [13] of trajectories in the neighborhood of an isolated singular point into hyperbolic, parabolic and elliptic sectors.

The first use of $\psi_2(t)$ for the characterization of such qualitative properties seems to be due to N. P. Papush [1]. The aim of his work is to identify the type of the Nemytskii classification of the solutions of 7.1 in a neighborhood of an equilibrium point by means of suitable sign combinations of φ, ψ_1 and ψ_2.

More recently M. B. Kudaev [1] has derived additional results on the behavior of the trajectories of the differential equation 7.1 in a neighborhood of an equilibrium point by suitable sign combinations of φ, ψ_1, ψ_2 and ψ_3.

Most of the results by Kudaev have been recently sharpened by J. Yorke [2] whose results are stated next. Notice that these results by Yorke have the extremely important and unique feature of having local conditions.

7.4 Theorem. Let $v = \varphi(x)$ be a real-valued function defined in R^n. Let $N(\beta)$ be a bounded component of the set $\{x \in R^n\colon \varphi(x) < \beta\}$. Assume that

7.4.1 $v = \varphi(x) \in \mathscr{C}^2$,

7.4.2 for all $z \in \partial N(\beta)$, $\psi_1(z) = 0$ implies $\psi_2(z) > 0$,

7.4.3 there exists $y \in \partial N(\beta)$ such that $\psi_1(y) \leqq 0$,

7.4.4 $N(\beta)$ contains a compact invariant subset.

Then there exists a point $z \in \partial N(\beta)$ such that for the differential equation 7.1

7.4.5 $\gamma^+(z) \subset N(\beta)$.

7.5 Theorem. If in Theorem 7.4 conditions 7.4.1 and 7.4.2 are satisfied and instead of 7.4.3 and 7.4.4 we assume that

7.5.1 the set $\{y \in \partial N(\beta) \colon \psi_1(y) > 0\}$ is non-empty and non-connected.

Then the set $\{z \in R^n \colon \gamma^+(z) \subset N(\beta)\}$ has dimension at least $n - 1$.

7.6 Theorem. Let $M \subset R^n$ be a compact invariant set and let $v = \varphi(x) \in \mathscr{C}^2$ be such that

7.6.1 $\varphi(x) = 0$ for all $x \in M$,

7.6.2 $\varphi(x) \geqq 0$ for all $x \in R^n$,

7.6.3 $\varphi(x) \to \infty$ as $\|x\| \to \infty$,

7.6.4 $\psi_1(x) = 0$ implies $\psi_2(x) > 0$ for all $x \in R^n - M$,

7.6.5 $\psi_1(x) = 0$ for some $y \in R^n - M$.

Then either there exist points z_1 and z_2 in $R^n - M$, such that

7.6.6 $\Lambda^+(z_1) \subset M$, $|z_1 t| \to +\infty$ as $t \to -\infty$,

7.6.7 $\Lambda^-(z_2) \subset M$, $|z_2 t| \to +\infty$ as $t \to +\infty$,

or for all $z \in R^n - M$

7.6.8 $|zt| \to \infty$ as $|t| \to \infty$.

Recently J. A. Yorke [7] proved the following theorem which provides a characterization of a certain type of compact invariant unstable attractors relative to an open set $U \subset R^n$.

7.7 Theorem. Let $v = \varphi(x)$ be a real-valued function defined in an open set $U \subset R^n$. Let $M \subset U$ be compact and such that $U - M$ is simply connected. Assume that

7.7.1 $v = \varphi(x) \in \mathscr{C}^2$,

7.7.2 for each $x \in U - M$, either $\psi_1(x) \neq 0$ or $\psi_2(x) \neq 0$,

where $\psi_1(x)$ and $\psi_2(x)$ are defined by 7.2 and 7.3, respectively. Then for each $x \in U$,

7.7.3 $\Lambda^+(x) \subset M$ and $\Lambda^-(x) \subset M$.

Sketch of Proof. Let

7.7.4 $E = \{x \in U - M: \psi_1(x) = 0\}$.

Then E is closed relatively to $U - M$. Condition 7.7.2 implies that grad $\psi_1(x) \neq 0$ for all $x \in E$. It can be proved that E is the disjoint union of connected manifolds E_i, and that there exist at most a countable number of such manifolds E_i. Since $U - M$ is simply connected, homology theory implies that each M_i separates $U - M$ into components one of which, P_i, is positively invariant and the other, N_i, is negatively invariant. If there exists a point $y \in (U - M) \cap \Lambda^+(x)$, then the image point xt must eventually be in the same connected component of $U - M$ as y, for $y \notin M$. Thus $\psi_1(xt) \neq 0$ for large values of t. Then $\varphi(xt)$ is monotonic and $\varphi(x)$ must be constant on the invariant set $\Lambda^+(x)$. Since $\varphi(yt)$ is constant $\psi_1(y) = \psi_2(y) = 0$ which contradicts 7.7.2. Similarly for $\Lambda^-(x)$.

Notes and References

The idea of characterizing the stability properties of differential equations by means of the sign properties of a real-valued function is due to LIAPUNOV [1]. A similar idea in a much more geometrical context, quite near to our point of view is to be found in the work of POINCARÉ [1, vol. 1, p. 73 ff.]. Here POINCARÉ developes in R^2 a method which enables to prove the absence of limit cycles in certain domains of the plane. This information is derived by analyzing the properties of the set $\psi(x) = \langle \text{grad}\, \varphi(x), f(x) \rangle = 0$ (contact curve) where $v = \varphi(x)$ is a "topographical system", i.e., a real-valued function such that the curves $\varphi(x) = \text{const}$ are non-intersecting, closed and differentiable.

Methods quite close to LIAPUNOV's have also been suggested by M. HADAMARD [1] and D. C. LEWIS [2].

For the investigation of the qualitative properties of ordinary differential equations methods other than that of LIAPUNOV have been suggested among which we recall the topological method of T. WAŻEWSKI [1] (see also F. ALBRECHT [1], N. ONUCHIC [1], A. PLIŚ [1, 3] and Z. SZMYDT).

In the literature various qualitative properties of differential equations which are not considered in this book are labeled as "stability" properties. These are essentially properties which have meaning only in the framework of differential equations or of differentiable flows, therefore these are beyond the aim of this book. Perhaps the most important of these "stability" properties is the so-called structural stability, i.e., the invariance of the global qualitative structure of a differential equation $\dot{x} = f(x)$ for "small" changes in the vector $f(x)$. These problems are investigated in the works of L. MARKUS [7], M. PEIXOTO, S. SMALE and J. MOSER, where additional references can be found.

Another "stability" property for differential equations is that of "total stability" or stability under constantly acting perturbations which deals with the problem of the invariance of the local qualitative structure of a differential equation $\dot{x} = f(x, t)$, when a continuously acting vector-valued function $g(x, t)$ is added to the vector $f(x, t)$. Works on this problem were done by I. MALKIN [8], P. SEIBERT [2, 3], STRAUSS and YORKE [3], and others.

A rather complete survey of the problems in the qualitative theory of differential equations is given by S. LEFSCHETZ [3].

We want to emphasize that LIAPUNOV was originally interested only in the stability properties of a given motion. He formulated this problem as follows. Let $\dot{y} = g(y, t)$. Let $y^1 = y^1(t)$ be a solution of such equation. In order to investigate the stability of $y^1(t)$, for all $\varepsilon > 0$, we shall consider solutions $y = y(t)$ such that $\|y^1(t_0) - y(t_0)\| < \eta$ and see if this implies that $\|y^1(t) - y(t)\| < \varepsilon$ for all $t \geq t_0$. This can be easily done by defining the new variable $x = y - y^1(t)$. Then from the differential equation $\dot{y} = g(y, t)$ one can obtain a new differential equation $\dot{x} = f(x, t) = g(x + y^1(t), t) - g(y^1(t), t)$. This equation is called *equation of the perturbed motion*. Notice that the stability problem for the motion $y^1 = y^1(t)$ for the equation $\dot{y} = g(y, t)$ is now reduced to the stability problem for the equilibrium point $x = 0$ of the equation $\dot{x} = f(x, t)$.

From the theorems shown in this chapter, it can be seen that the central point of the analysis of the stability properties of a compact set is the construction of continuously differentiable, real-valued function $v = \varphi(x)$ such that the corresponding total time derivative (1.4) along the solutions of the differential equation 1.1 is such that $\psi(x) = 0$ if $x \in M$ and $\psi(x) \neq 0$ if $x \notin M$. Only in this case the theorems of section 3 provide a complete classification, since, if $\varphi(x)$ is definite for M and sign $\varphi(x) \neq$ sign $\psi(x)$, M is asymptotically stable, if $\varphi(x)$ is definite and sign $\varphi(x) =$ sign $\psi(x)$, M is negatively asymptotically stable, if $\varphi(x)$ is indefinite then M is (not completely) unstable; $\varphi(x)$ cannot be semidefinite for M. In addition if $\psi(x)$ is definite then the extension theorems of section 4 can be applied. For the construction of such a pair of functions $\varphi(x)$ and $\psi(x)$, equation 1.4 can be regarded in different ways according to which are the unknowns and which the given function in the relationship 1.4. For instance if $\varphi(x)$ and $f(x)$ are given, then clearly $\psi(x) = d\varphi(xt)/dt$, on the other hand if $\psi(x)$ and $f(x)$ are given, then they give rise first to an equation of the type $\psi(x) = \langle a(x), f(x)\rangle$ where $a(x)$ is an unknown vector. Then if $\psi(x)$ and $f(x)$ are such that $a(x)$ satisfies the integrability condition $\partial a_j/\partial x_i = \partial a_i/\partial x_j$, then there exists a real-valued function $\varphi(x)$, such that the relation 1.4 can be satisfied for the given $\psi(x)$ and $f(x)$. One can also require $\psi(x)$ to have a special form. For instance we can assume that $\psi(x) = \theta(x)\,\beta(\varphi(x))$, then find $v = \varphi(x)$ which satisfies the equation $\theta(x)\,\beta(\varphi(x)) = \langle \operatorname{grad} \varphi(x), f(x)\rangle$ under the assumption that the integral $\int_0^\infty \frac{d\varphi}{\beta(\varphi)}$ exists. The above given equation was introduced by G. P. SZEGÖ [3] and it is a generalization of the one introduced by V. I. ZUBOV [1], given in V, 2.16.8.

These two approaches: that of choosing $\varphi(x)$ (or at least its form) and of fixing $\psi(x)$ first, represent two different ways of constructing Liapunov functions.

For the problem of constructing Liapunov functions the reader is referred to the books by W. HAHN [1, 2] where additional references are given.

Section 2. Most of the geometric ideas used in section 2 were introduced by A. STRAUSS [5], who developed some earlier results of J. P. LASALLE [4].

Property 2.3 can also be proved by the same technique used in the proof of Theorem 5.6.

Section 3. All results presented in section 3 are natural extensions of the results of LIAPUNOV to the particular case of compact sets. The theory presented here follows in spirit the one of A. STRAUSS, even if some of the results are different. Theorem 3.9 on asymptotic stability is not formulated in the traditional way, but it is equivalent to the classical theorem. However, the proof of the classical theorem on asymptotic stability (see, for instance, BHATIA and SZEGÖ [1], Theorem 3.6.15) is much more involved than that of Theorem 3.9. This is due to the fact that in the classical theorem it was essentially shown that M is a uniform attractor and the number T

appearing in V, 1.29.2 was explicitly computed. Thus in that proof one was able to prove more properties about the system than the ones essentially needed.

Notice that in the local theorems 3.9, and Corollary 3.13 estimation of the region of attraction $A(M)$ requires that one has to make sure that $\overline{N(\beta)}$ is compact. Thus 3.13.2 essentially expresses the classical condition "∂N is the level surface of the Liapunov function $v = \varphi(x)$ which is bounded and $\partial N \cap (E - M) \neq \emptyset$, while $N - M \cap E = \emptyset$, where $E = \{x \in R^n : \psi(x) = 0\}$." The condition of Theorem 3.15 on global asymptotic stability is due to E. A. Barbashin and N. N. Krasovskii [1].

The instability theorem 3.17 is an extension of the well known theorem due to N. Chetaev [1, 3]. Recently J. A. Yorke [4] proved the following theorem which is another extension of Chetaev's result, which we state for the case of a dynamical system defined by the differential equation $\dot{x} = f(x)$, $f\colon U \to R^n$ and $U \subset R^n$ open.

Theorem. Let $G \subset U$ be closed. Let $G(\nu) = \{x \in G : \varphi(x) = \nu$ and $\psi(x) = 0\}$ and $G^+ = \{x \in G : \varphi(x) > 0\}$. Assume that (i) G is positively invariant; (ii) $\{0\} \in \overline{G}^+$, $\varphi(0) = 0$, $\psi(x) \geq 0$ for $x \in G$; (iii) for $\nu > 0$, $G(\nu)$ has no (non-empty) compact invariant subsets. Then $\{0\}$ is unstable.

A similar result was also proved by N. P. Bhatia [9].

Almost all results proved in sections 1–3 are proved in the more general framework of the theory of flows without uniqueness in the monograph by G. P. Szegö and G. Treccani [1].

Section 4. Preliminary ideas leading to the extension theorems can be found in the works of D. R. Ingwerson [1] and W. Leighton [1]. A complete preliminary statement was given by G. P. Szegö [4]. Its complete proof for the case of global asymptotic stability is due to N. P. Bhatia and G. P. Szegö [2].

The complete global extension theorem 4.3 is due to G. P. Szegö [5].

There is a basic difference between the application of Theorem 3.9 and Corollary 3.13 on the one hand and that of Theorem 4.1 on the other. Namely in order to use Theorem 3.9 for an estimate of $A(M)$ one had to prove that for a certain β, $\overline{N(\beta)}$ is compact, while in Theorem 4.1 this operation does not have to be performed at all and in addition $\overline{N(\beta^0)}$ may well be not compact. Clearly however, in this case, one has to make sure (see condition 4.1.5) that in the open set $N(\beta^0) - M$ there do not exist points in which $\psi(x) = 0$ nor diverging sequences along which $\psi(x) \to 0$.

The use of the extension theorem as compared with the classical asymptotic stability theorem has been shown in various papers (see, for instance, G. P. Szegö [7], G. P. Szegö, C. Olech and A. Cellina [1], G. P. Szegö, G. Arienti and C. Sutti [1]). These theorems are useful for instance when one is interested in the stability properties of the rest point $x = 0$ of an ordinary differential equation $\dot{x} = g(x)$, $g(0) = 0$, $g\colon R^n \to R^n$, which admits a non-singular linear approximation in the neighborhood of the point $x = 0$, i.e., such that the Jacobian matrix $J = \{\partial g_i/\partial x_j\}$ evaluated at the point $x = 0$ (and therefore with constant elements) is nonsingular. In this case the local stability properties of the rest point $x = 0$ coincide with those of the rest point $x = 0$ of the linear differential equation with constant coefficients: $\dot{x} = Jx$. The stability properties of equation 4.4.3 are easily analyzed by algebraic criteria (Routh, Hurwitz, etc., see e.g., Gantmacher [1]). In this situation the extension theorems allow one to solve problems which are not practically solvable within the classical theory, for instance when one is able to construct a continuously differentiable Liapunov function vanishing at the rest point $x = 0$, which is locally asymptotically stable, and such that $\psi(x)$ satisfies conditions 4.1.5 or 4.3.5. On the other hand $\varphi(x)$ may have a structure making it very difficult to analyze its sign properties (for instance it is a polynomial of order higher than two), as required in the classical theory. The extension theorems are proved for the case of differential equations without uniqueness in the monograph by G. P. SZEGÖ and G. TRECCANI [1].

Section 5. By applying the extension theorems for the case of differential equations without uniqueness (G. P. SZEGÖ and G. TRECCANI [1]) to the study of the geometrical problems presented in section 5, all those results can be proved in a stronger form, namely under the assumption that $\varphi(x) \in \mathscr{C}^1$, but not necessarily $\in \mathscr{C}^2$. Similar results for functions of the class $\mathscr{C}^1$, derived from Theorem 5.9 are proved for such functions and can be found in the monograph by N. P. BHATIA and G. P. SZEGÖ [1].

The technique used in the proof of Theorem 5.9 can also be applied in the proof of relationships between the properties of the set $N'(\beta)$ and those of the vector grad $\varphi(x)$. Most of the relationships are essentially generalization to R^n of the well known Rolle's theorem. A first result along the line was proved by G. P. SZEGÖ [8]. Additional results are proved by G. P. SZEGÖ and G. TRECCANI [1].

Section 6. All results of Theorem 6.1 can be found in the monograph of N. P. BHATIA and G. P. SZEGÖ [1] and contain as special cases previous results by E. A. BARBASHIN and N. N. KRASOVSKII [1] and by J. P. LASALLE [4, 6, 7]. A weaker version of Theorem 6.1 was proved for flows without uniqueness by G. P. SZEGÖ and G. TRECCANI [1] where also theorems 6.3, 6.4 and 6.5 are proved for the case of functions $\varphi(x) \in \mathscr{C}^1$.

Finally, we would like to point out that the problem of the existence of Liapunov functions for differential equations (converse problem) has been discussed by many authors, notably J. J. MASSERA [5, 6], N. N. KRASOVSKII [3, 6, 7, 8, 9], K. P. PERSIDSKII [2], I. VRKOC [1] and J. KURZWEIL [1, 2] and J. KURZWEIL and I. VRKOC [1].

Some results on the existence of $\mathscr{C}^\infty$-Liapunov functions for asymptotically stable rest points were then proved by ZUBOV [1].

Results on the existence of $\mathscr{C}^\infty$-Liapunov functions for compact asymptotically stable sets are given by F. W. WILSON [1]. J. L. MASSERA [2] and J. KURZWEIL simultaneously proved the existence of a $\mathscr{C}^\infty$-Liapunov function for the uniformly asymptotically stable rest point of the differential equation $\dot{x} = f(x, t)$. The method used in the proof is quite different from the method of ZUBOV. T. YOSHIZAWA [1] gives simpler converse theorems, but is able to prove only the existence of Lipschitz continuous Liapunov functions.

Section 7. Most of the results of this section are due to J. A. YORKE [1, 7]. The reader may consult the original papers for the complete proofs.

Chapter IX

Non-continuous Liapunov Functions for Ordinary Differential Equations

1. Introduction

Throughout this chapter we shall assume that the ordinary differential equation

1.1 $\dot{x} = f(x), \quad x \in R^n,$

where $f\colon R^n \to R^n$ is continuous, has unique solutions in the space R^n.

This chapter is devoted to the characterization of the (weak) attraction properties of an isolated rest point $x = 0$ of the differential equation 1.1.

As it must be clear from the results of the previous section, there need not exist any continuous Liapunov $v = \varphi(x)$ function with a continuous total time derivative along the solutions of the differential equation 1.1 in the case in which the rest point $x = 0$ of equation 1.1 is an unstable (weak) attractor. The characterization of weak attractors can however be done by means of lower-semicontinuous real-valued functions and their lower-right-hand-side Dini derivates along the solutions of equation 1.1. By means of these functions we are able to give a characterization of (weak) attraction which follows exactly the theory (extension theorems) presented in the previous section for the case of asymptotically stable sets.

Throughout this section, for a mapping $\varphi\colon R^n \to R$ we shall write

1.2 $$\liminf_{x \to x^0} \varphi(x) = \lim_{\varepsilon \to 0} \left[\inf\{\varphi(x) : x \in S(x^0, \varepsilon)\}\right].$$

In this section we shall often use lower and upper-semicontinuous real-valued functions on a set $U \subset R^n$. If $U \subset R^n$, a function $\varphi\colon U \to R$ is lower-semicontinuous if and only if

1.3 $$\liminf_{y \to x} \varphi(y) \geqq \varphi(x) \quad \text{for all } x \in U,$$

and upper-semicontinuous if and only if

1.4 $\limsup_{y \to x} \varphi(y) \leq \varphi(x)$ for all $x \in U$.

If $\varphi: R^n \to R$ is a lower-semicontinuous real-valued function and $f: R^n \to R^n$ is a continuous function, we shall consider the following extended real-valued function

1.5 $$\underline{D}^+(xt)|_{t=0} = \liminf_{\tau \to 0^+} \tau^{-1}[\varphi(x\tau) - \varphi(x)],$$

which is the lower-right-hand-side Dini derivative of the function $\varphi(x)$ along the solutions of the differential equation 1.1.

The stability results of the previous chapter were all essentially based upon the relationships VIII, 1.3 and VIII, 1.5 which connect the real-valued function $\psi(x)$ with the function $\varphi'(x) = \left.\frac{d\varphi(xt)}{dt}\right|_{t=0}$. Notice that while for the estimation of $\psi(x)$ no knowledge about the trajectories of the system is required, the estimation of $\varphi'(x)$ requires the knowledge of the trajectories.

Similarly for the case of lower-semicontinuous real-valued functions one needs a connection between $\underline{D}^+(xt)|_{t=0}$ and the real-valued function, defined below in 1.6.1, for the evaluation of which no knowledge of the trajectories is required. This connection is provided by the following theorem due to YORKE [3].

1.6 Theorem. Let $v = \varphi(x)$, $\varphi: U \to R$ be a lower-semicontinuous real-valued function defined on an open set $U \subset R^n$. Assume that there exists a real-valued function $w = \xi(x)$, $\xi: U \to R^n$, which is defined and continuous in U and such that the extended real-valued function

1.6.1 $$\psi(x) = \varphi^*(x) = \liminf_{\substack{\tau \to 0^+ \\ z \to f(x)}} \tau^{-1}[\varphi(x + \tau z) - \varphi(x)]$$
$$= \liminf_{\substack{\tau \to 0^+ \\ \|y\| \to 0}} \tau^{-1}[\varphi(x + \tau f(x) + \tau y) - \varphi(x)]$$

satisfies the condition

1.6.2 $\varphi^*(x) \leq \xi(x)$ for all $x \in U$.

Then

1.6.3 $$\varphi(xt) - \varphi(x) \leq \int_0^t \xi(x\tau)\, d\tau$$

for all $x \in U$ and $t \in R$, such that $x[0, t] \subset U$ and

1.6.4 $$\varphi^*(x) \leq \liminf_{\tau \to 0^+} \tau^{-1}[\varphi(x\tau) - \varphi(x)] = \underline{D}^+(xt)|_{t=0}.$$

Sketch of Proof. Let $\varepsilon > 0$. Then there is an integer N such that for $n \geqq N$, the inequality

$$1.6.5 \qquad \varepsilon + \int_0^t \xi\,(xs)\,ds \geqq \sum_{k=0}^{n-1} \xi\left(x\frac{kt}{n}\right)\frac{t}{n}$$

holds.

From 1.6.2, for each integer $n \geqq N$ and k with $0 \leqq k \leqq n-1$, there exist $h_{(n,k)}$, $y_{(n,k)}$, satisfying $0 < h_{(n,k)} \leqq t/n$, $|y_{(n,k)}| \leqq t/n$, such that

$$1.6.6 \qquad \xi\left(x\frac{kt}{n}\right) \geqq h_{(n,k)}^{-1}\left[\varphi\left(x\frac{kt}{n} + h_{(n,k)}\,f\left(x\frac{kt}{n}\right) + h_{(n,k)}\,y_{(n,k)}\right) - \varphi\left(x\frac{kt}{n}\right)\right] - \frac{\varepsilon}{tn}.$$

Since φ is lower-semicontinuous, and f is uniformly bounded on the compact trajectory segment $x\,[0, t]$, we may further choose $h_{(n,k)}$ and $y_{(n,k)}$ such that the first term on the right hand side is either non-negative (this is possible at points $x\,(kt/n)$ where φ is not a maximum), or (where φ has a maximum and is hence continuous) the inequality

$$\varphi\left(x\frac{kt}{n} + h_{(n,k)}\,f\left(x\frac{kt}{n}\right) + h_{(n,k)}\,y_{(n,k)}\right) - \varphi\left(x\frac{kt}{n}\right) \geqq -h_{(n,k)}\frac{\varepsilon}{n}$$

holds. With such a choice, 1.6.5 yields on using 1.6.6, the inequality

$$\begin{aligned} 1.6.7 \qquad \varepsilon + \int_0^t \xi\,(xs)\,ds &\geqq -\frac{\varepsilon}{n} - \varepsilon \\ &\quad + \sum_{k=0}^{n-1}\left[\varphi\left(x\frac{kt}{n} + h_{(n,k)}\,f\left(x\frac{kt}{n}\right) + h_{(n,k)}\,y_{(n,k)}\right) \right. \\ &\quad \left. - \varphi\left(x\frac{kt}{n}\right)\right] \\ &\geqq -\frac{\varepsilon}{n} - \varepsilon + \varphi\,(xt) - \varphi\,(x) \\ &\quad + \sum_{k=1}^{n}\left[\varphi\left(x\frac{k-1}{n}t + h_{(n,k-1)}\,f\left(x\frac{k-1}{n}t\right.\right.\right. \\ &\quad \left.\left. + h_{(n,k-t)}\,y_{(n,k-1)}\right) - \varphi\left(x\frac{kt}{n}\right)\right]. \end{aligned}$$

In the last sum, the terms are either non-negative, or no less than $h_{(n,k-1)}\,\varepsilon/n$. Therefore we have

$$\varepsilon + \int_0^t \xi\,(xs)\,ds \geqq -3\varepsilon + \varphi\,(xt) - \varphi\,(x).$$

Since $\varepsilon > 0$ was arbitrary, the inequality 1.6.3 holds. This proves the theorem.

2. A Characterization of Weak Attractors

In this section we shall apply Theorem 1.6 for providing a complete characterization of (unstable) (weak) attractors. The first theorem (2.1) provides a sufficient condition for a non-empty compact set $M \subset R^n$ to be an attractor. Under an additional hypothesis on the systems this condition is also necessary (Remark 2.2.3). Theorems 2.4 and 2.5 on the other hand provide necessary and sufficient conditions for a non-empty compact set $M \subset R^n$ to be a global weak attractor and an attractor, respectively.

2.1 **Theorem.** Consider the dynamical system represented by the differential equation

2.1.1 $\dot{x} = f(x), \quad x \in R^n, \quad f\colon R^n \to R^n.$

Let $M \subset R^n$ be a compact set and let $v = \varphi(x), \varphi\colon \mathscr{C}(M) \to R$ be a lower-semicontinuous real-valued function. Assume that

2.1.2 there exists a continuous real-valued function $w = \xi(x)$, ξ: $\mathscr{C}(M) \to R^-$ such that the extended real-valued function $\varphi^*(x)$ (1.6.1) satisfies the condition $\varphi^*(x) \leqq \xi(x) < 0$ for all $x \in \mathscr{C}(M)$.

2.1.3 For each $\nu \in R$, let $K'(\nu) = \{x\colon \varphi(x) \leqq \nu\}$, and assume that either $K'(\nu) \cup M$ is compact or is all of R^n.

Then for each $x \in R^n$, either

2.1.4 $x\tau_n \in M$ for some sequence $\tau_n \to +\infty$,

or

2.1.5 $\varrho(xt, M) \to 0$ as $t \to +\infty$.

Proof. Let 2.1.4 not hold for an $x \in R^n$; that is, for some α_0, $xt \in \mathscr{C}(M)$ for $t \in [\alpha_0, +\infty)$. Let $S^1 = K'(\varphi(x\alpha_1))$. Then, if $\alpha_1 > \alpha_0$, for $t > \alpha_1$, $xt \in S^1$ and $S^1 \cup M$ is compact. It suffices to prove that $\Lambda^+(xt)$ (which is non-empty since $S^1 \cup M$ is compact) is a subset of M. Choose $y \in \Lambda^+(xt)$. Suppose $y \notin M$. Then there exist γ and δ such that $\varrho(y, M) > 2\gamma > 0$ and on the set $B = S[y, 2\gamma]$, $\varphi^* \leqq \xi < -\delta < 0$. Let $\mu = \max\limits_B |f|$. Then for any t

2.1.6 $\varrho(xt - y) < \gamma \Rightarrow x(t + \tau) \in B$ for $|\tau| \leqq \gamma/\mu$.

Choose $t_i \to +\infty$ such that $\alpha_0 < t_1$ and $t_{i+1} > t_i + \gamma/\mu$ and $|xt_i - y| < \gamma$. Since φ is lower-semicontinuous and $\varphi(xt)$ is monotonically decreasing for all i,

2.1.7 $\varphi(y) \leqq \liminf\limits_{i\to\infty} \varphi(xt_i) < \varphi(xt_i).$

But

$$\varphi(xt_{i+1}) - \varphi(xt_1) \leqq \int_{t_1}^{t_{i+1}} \xi(x\tau)\, d\tau \leqq \sum_{j=1}^{i} \int_{t_j}^{t_j+\gamma/\mu} \xi(x\tau)\, d\tau \leqq -\gamma\, \delta i/\mu \to -\infty$$

as $i \to \infty$. Hence, $\liminf\limits_{i\to\infty} \varphi(xt_i) = -\infty$, contradicting 2.1.7. Therefore $y \in M$, and $\Lambda^+(xt) \subset M$, proving 2.1.5.

2.2 *Remarks.*

2.2.1 Notice that if M is positively invariant, the conclusion is that M is a global attractor.

2.2.2 In many examples of attractors, such a φ exists. For instance, if we make the additional hypothesis that for each $y \notin M$ the arc length of the half-trajectory $\gamma^+(y)$ is finite, then a function φ as described below exists, letting $\varphi(y) = \int_0^\infty \left|\frac{dy\tau}{d\tau}\right| d\tau$, the arc length. Then φ is a function as described below with

$$\varphi^*(y) \leqq \frac{d}{dt}\varphi(yt)\,|_{t=0} = -|f(y)| = \xi(y).$$

2.2.3 Notice that 2.1.1 implies that $K'(\nu)$ is closed relative to $\mathcal{C}(M)$, i.e., the limit points of $K'(\nu)$ in $\mathcal{C}(M)$ are in $K'(\nu)$; therefore, 2.1.3 may be guaranteed, for example, by assuming $\varphi(x) \to \infty$ as $|x| \to \infty$.

2.2.4 It can be seen from the proof that φ satisfies $\varphi(x) \geqq \liminf_{y \to M} \varphi(y)$ for all $x \in \mathcal{C}(M)$. If φ was assumed to be defined and continuous on R^n and constant on M, and if M was invariant, then M would be stable.

2.2.5 The assumption 2.1.2 is similar to saying φ^* is negative definite. The assumption "$\varphi^*(y) < 0$", however, does not imply that $\varphi(xt)$ is not constant since φ^* is not necessarily a continuous function. (Examples can be given.)

2.3 *Example of an Attractor.* Let dimension $n = 2$, and let x be in the usual polar coordinates (ϱ, θ). Let M be $\{(0, 0), (1, 0)\}$. Then M is invariant for

2.3.1 $$\begin{cases} \dot\theta = (2\pi - \theta)\,\theta + \max\{0, 1 - \varrho\}, \\ \dot\varrho = -|1 - \varrho|. \end{cases}$$

The phase portrait is given in Fig. 2.3.2.

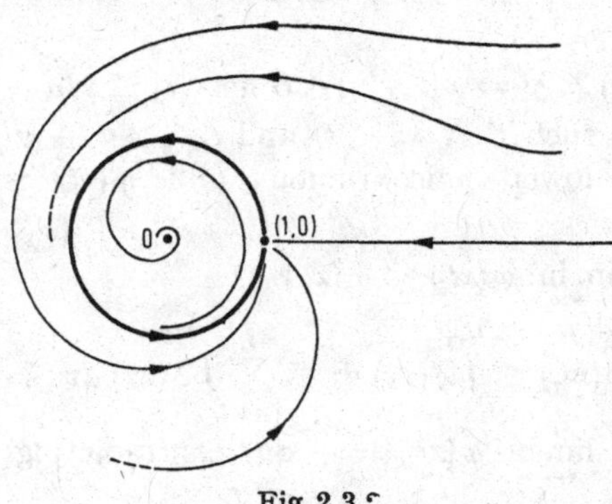

Fig. 2.3.2

Define φ on $\mathscr{C}(M)$:

2.3.3 $$\varphi(\varrho, \theta) = \begin{cases} -\log(1-\varrho) & \text{for } 0 < \varrho < 1, \\ -\theta + \varrho & \text{for } \varrho \geqq 1. \end{cases}$$

Then

2.3.4 $$\varphi^*(\varrho, \theta) = \begin{cases} -1 & \text{for } \varrho < 1, \\ -\dot{\theta} - |1-\varrho| & \text{for } \varrho \geqq 1. \end{cases}$$

Then if we assume $\xi(\theta, \varrho) = \max\{-1, -\dot{\theta} - |1-\varrho|\}$, we have $\varphi^* \leqq \xi < 0$ on $\mathscr{C}(M)$. For $\varrho \geqq 1$, $\varphi(\varrho, \theta) \geqq \varrho - 2\pi$ and $\varphi(\varrho, \theta) \to \infty$ as $|x| = \varrho \to \infty$. Checking φ for $\varrho = 1$ in particular, we see φ is lower-semicontinuous. Thus M is an attractor.

2.4 Theorem. Consider the differential equation 2.1.1. Let $M \subset R^n$ be a compact set. The set M is a global weak attractor if and only if the following conditions are satisfied.

2.4.1 For each $\delta > 0$, there exists a compact set M_δ, with $M \subset M_\delta \subset S[M, \delta]$, a lower-semicontinuous real-valued function $v = \varphi_\delta(x)$, $\varphi_\delta\colon \mathscr{C}(M_\delta) \to R$ and a continuous real-valued function $w = \xi_\delta(x)$, $\xi_\delta\colon \mathscr{C}(M_\delta) \to (-\infty, 0)$ such that

2.4.2 $\varphi_\delta^*(x) \leqq \xi_\delta(x) < 0$ for all $x \in \mathscr{C}(M_\delta)$.

2.4.3 Let $K'_\delta(\nu) = \{x\colon \varphi_\delta(x) \leqq \nu\}$. For each real ν and δ $K'_\delta(\nu) \cup M_\delta$ is compact or is all of R^n.

Proof. Assume 2.4.1 holds. Let M in Theorem 2.1 be M_δ in Theorem 2.4; for each y there exists a sequence $\tau_n \to \infty$ such that $y\tau_n \to M_\delta$ (whether 2.1.4 or 2.1.5 is satisfied). It follows that letting $\delta = 1/2n$ there exists $t_n > n$ such that $\varrho(yt_n, M_\delta) < 1/2n$. Hence, $yt_n \to M$ as $n \to \infty$. Therefore M is a weak attractor.

To show the converse, assume M is a weak attractor. Let $M_\delta = S[M, \delta]$ for all $\delta > 0$. For $y \in \mathscr{C}(M_\delta)$ let $\varphi_\delta(y) = \inf\{t > 0\colon yt \in M_\delta\}$. To see that φ_δ is lower-semicontinuous, fix y and let $\nu \in [0, \varphi_\delta(y))$. Then $\inf_{t\in[0,\nu]} \varrho(yt, M) > \delta$ and for some neighborhood N of y, for all $z \in N$, $zt \notin S[M, \delta]$ for $t \in [0, \nu]$. Hence, $\varphi_\delta(z) > \nu$ for $z \in N$, and φ_δ is lower-semicontinuous. Also note that

2.4.4 $$\left.\frac{d\varphi(yt)}{dt}\right|_{t=0} = -1 \quad \text{for} \quad y \in \mathscr{C}(M_\delta).$$

So 1.6.3 is satisfied letting $\xi = -1 = \xi_\delta$. Since 1.6.3 is equivalent to 1.6.2, $\varphi^* \leqq -1 = \xi$ on $\mathscr{C}(M_\delta)$.

2.5 Theorem. Consider the differential equation 2.1.1. Let $M \subset R^n$ be a compact invariant set. The set M is a global attractor if and only if 2.4.1 holds requiring in addition that, for each $\delta > 0$, the sets M_δ can be chosen to be invariant.

Proof. From Theorem 2.1, if each M_δ is invariant and 2.4.1 holds, each M_δ is an attractor. It follows that M is an attractor. To see the converse, let M_δ be the largest invariant set in $S[M, \delta]$. Since the union of invariant sets is invariant, M_δ exists (perhaps $= M$), and since the closure of an invariant set is invariant, M_δ is closed. Since M is an attractor, we may define

$$\varphi_\delta(y) = \sup\{t: \varrho(yt, M) > \delta\} \text{ for } y \in \mathcal{C}(M_\delta).$$

Note that if "$\varrho(yt, M) \leqq \delta$ for all t", then $\gamma(y) \cup M_\delta$ would be an invariant set in $S[M, \delta]$, so $\varphi_\delta(y)$ is defined for all $y \in \mathcal{C}(M_\delta)$.

To see φ_δ is lower-semicontinuous, note that if $\varrho(yt, M) > \delta$ for some t and y, then for all z in some neighborhood N of y, $\varrho(zt, M) > \delta$, and $\varphi_\delta(z) > t$, so $\liminf_{z \to y} \varphi_\delta(z) \geqq \varphi_\delta(y)$. As in the previous proof, 2.4.4 holds and $\varphi^* \leqq -1 = \xi_\delta$ on $\mathcal{C}(M_\delta)$.

2.6 *Remarks.*

2.6.1 For all $\delta > 0$, we may have $M_\delta = M$. If in 2.4.1 the largest invariant subset of $S[M, \delta]$ is M, then $M_\delta = M$. If $M_{\delta_0} = M$ for some $\delta_0 > 0$, then we may let $M_\delta = M_{\delta_0}$ and $\varphi_\delta = \varphi_{\delta_0}$ for all δ.

2.6.2 Both 2.4.1 and 2.4.3 use families of Liapunov functions; if M is invariant in 2.4.1 and is not a global attractor, then it can be seen from Theorem 2.1 that $M_\delta \neq M$; if $M_\delta = M$ for some δ, then 2.2.1 implies $xt \to M$ if $t \to +\infty$ for all x and $x \in R^n$, and M is an attractor.

3. Piecewise Differentiable Liapunov Functions

In this section we shall assume that the ordinary differential equation

3.1 $$\dot{x} = f(x), \quad x \in R^n, \quad f(0) = 0,$$

where $f: R^n \to R^n$ is continuous, defines a dynamical system (R^n, R, π) and that the rest point $x = 0$ is isolated.

In this section we shall use a rather restrictive class of lower-semicontinuous real-valued functions $v = \varphi(x)$, $\varphi: R^n \to R^n$, namely, piecewise differentiable functions. On these functions and on their corresponding functions $\psi(x)$ (1.6.1) we shall impose additional conditions. These conditions will allow us to prove the extension theorems and to construct a special type of Liapunov function on the region of instability (V, 1.13) of an unstable attracting rest point (section 6). In addition to this application the introduction of this special kind of lower-semicontinuous function has been suggested by problems arising in the numerical construction of Liapunov functions where piecewise differentiable Liapunov functions are simpler to construct than higher order polynomials.

We shall begin this presentation by defining a few properties of the real-valued function φ.

3.2 **Property.** Let $v = \varphi(x)$, $\varphi\colon U \to R^n$, where $U \subset R^n$ is an open set with $\{0\} \subset U$, be lower-semicontinuous. We shall say that $\varphi(x)$ has the property 3.2 with respect to the dynamical system 3.1 if and only if the function $\psi(x)$ (1.6.1) satisfies $\psi(0) = 0$ and for each $\varepsilon > 0$ there exists a real-valued function $\alpha_\varepsilon(\mu)$, $\alpha_\varepsilon(0) = 0$, $\alpha_\varepsilon(\mu) > 0$ for $\mu > 0$, which is non-decreasing and continuous on $[\varepsilon, +\infty)$ and for each $x \in U$

3.2.1 $$\psi(x) \leqq -\alpha_\varepsilon(\|x\|).$$

3.3 **Property.** Let $v = \varphi(x)$, $\varphi\colon U \to R^n$, where $U \subset R^n$ is an open set, be lower-semicontinuous. Let $D \subset U$ be the set of all points in which $\varphi(x)$ is not continuously differentiable. We shall say that $\varphi(x)$ has the property 3.3 if and only if for each $x \in D^* \equiv \overline{D} - \{0\}$, there exists a real number $\delta > 0$ and a real-valued function $\omega_{x,\delta}(y)$, which is lower-semicontinuous in the sphere $S(x, \delta)$, and such that for all $y \in S(x, \delta)$

3.3.1 $$\omega^*_{x,\delta}(y) = \liminf_{\substack{\tau \to 0^+ \\ z \to f(y)}} \tau^{-1}[\omega_{x,\delta}(y + \tau z) - \omega_{x,\delta}(y)] \leqq -\varepsilon, \quad \varepsilon > 0$$

and moreover

3.3.2 $$z \in D^* \cap S(x, \delta),\ z \neq x \text{ implies } \varphi(z) = \varphi(x).$$

The following example shows that there exist lower-semicontinuous functions and differential equations for which the property 3.3 is satisfied.

3.4 *Example.* Consider the following planar system of ordinary differential equations, described in polar coordinates:

3.4.1 $$\begin{cases} \dot r = -r \sin\theta, \\ \dot\theta = r, \end{cases}$$

where r and θ are the radial and angular coordinates respectively.

Consider for each point $(r_0, 0)$ $r_0 \neq 0$, the neighborhood

3.4.2 $$S = \{|r - r_0| < \sigma;\quad 2\pi - \varepsilon < \theta < 2\pi;\ 0 \leqq \theta < \varepsilon\}.$$

On each set S define the function

3.4.3 $$\omega(r, \theta) = \begin{cases} -\theta & \text{for } 0 \leqq \theta < \varepsilon, \\ e^{-\theta} & \text{for } 2\pi - \varepsilon < \theta < 2\pi. \end{cases}$$

The real-valued function 3.4.3 is lower-semicontinuous and has the value 0 for $\theta = 0$, $|r - r_0| < \sigma$. For each point of the set S (3.4.2), in which $\varepsilon > \theta > 0$, the function $\omega(r, \theta)$ is continuously differentiable with $\dot\omega = \langle \operatorname{grad}\omega, f\rangle = -r$, while for each point of the set S in which

$\theta < 2\pi$, we have $\dot{\omega} = -re^{-\theta}$. Consider then a point in the set S (3.4.2) in which $\theta = 0$, i.e. the point $(r, 0) \in S$. Then

3.4.4 $$\omega^*(r, 0) = \liminf_{\substack{\tau \to 0^+ \\ \tilde{r} \to 0}} \tau^{-1} [\omega(r + \tau\tilde{r}, \tau r + \tau\tilde{\theta})]$$
$$= \lim_{\alpha \to 0^+} \left[\inf_{0 \leq \tau < \alpha} \tau^{-1} \omega(r, \tau r + \tau\tilde{\theta}) \right].$$

Now for a given $\varepsilon > 0$, if $\alpha < \dfrac{\varepsilon}{r + 2\pi}$, we have $0 < \tau(r + \tilde{\theta}) < \varepsilon$ and hence $\tau^{-1}\omega(r, \tau r + \tau\tilde{\theta}) = -(r + \tilde{\theta})$. Thus

3.4.5 $$\omega^*(r, 0) = -(r + 2\pi).$$

In summary for each point $(r, \theta) \in S$

3.4.6 $$\omega^*(r, \theta) \leqq -e^{-2\pi} r = -k_1, \quad k_1 > 0,$$

while for each point of the axis $\theta = 0$

3.4.7 $$\omega(r, 0) = 0.$$

Thus each lower-semicontinuous real-valued function which is continuously differentiable on the whole plane (r, θ), except on the axis $\theta = 0$, has the property 3.3 for the system 3.4.1.

Notice that in this example the condition on ω^* holds indeed only locally (for each $\alpha < \varepsilon/r + 2\pi$) and not globally on the axis $\theta = 0$.

From the definitions given the following proposition can be easily derived.

3.5 Proposition. If $v = \varphi(x)$ is a lower-semicontinuous real-valued function which has the property 3.3 in the set $U \subset R^n$, then for each $x \in U$ the set $\gamma(x) \cap D^*$ is a set of isolated points.

We shall prove next that if a real-valued function has the property 3.3, then some properties of the function which hold in the set $U - D^*$, hold for the whole set U.

The following theorem is a special case of Theorem 1.6. We shall however give a complete proof, since under the stronger assumptions made on the function $\varphi(x)$ its proof is quite simple.

3.6 Theorem. Let $v = \varphi(x)$, $\varphi\colon U \to R^n$, where $U \subset R^n$ is an open set with $\{0\} \subset U$, be lower-semicontinuous and have the properties 3.2 and 3.3 with respect to the dynamical system 3.1, while for each $x \in D^*$, $\psi(x)$ (1.6.1) is finite. Then

3.6.1 $$\varphi(xt) - \varphi(x) \leqq -\int_0^t \alpha_\varepsilon(\|x\tau\|)\, d\tau$$

or all $x \in U$ and $t \in R^+$ such that $x[0, t] \subset U - S(0, \varepsilon)$.

Proof. Clearly in the interval $[0, t]$ there exists (Proposition 3.5) at the most a finite number of points τ_k such that $x\tau_k \in D^*$. For each $x\tau_k$ there exists then a real number $\sigma_k(\tau_k)$ such that

$$\{x(\tau_k - \sigma_k, \tau_k) \cup x(\tau_k, \tau_k + \sigma_k)\} \cap D^* = \emptyset.$$

Let $\theta = \min_{[0,t]} \sigma_k$ and $K = \max_{[0,t]} \psi(x\tau_k)$. Then from the inequality 3.2.1

3.6.2 $$\varphi(xt) - \varphi(x) \leqq \sum_{k=1}^{n} \left\{ \int_{\tau_k-\theta}^{\tau_k+\theta} K \, d\tau \right\} + \sum_{k=1}^{n-1} \left\{ \int_{\tau_k+\theta}^{\tau_{k+1}-\theta} - \alpha_\varepsilon(\|x\|) \, d\tau \right\} + \int_0^{\tau_1-\theta} - \alpha_\varepsilon(\|x\|) \, d\tau + \int_{\tau_n+\theta}^{t} - \alpha_\varepsilon(\|x\|) \, d\tau.$$

Now by letting $\theta \to 0^+$ from 3.6.2 we obtain

$$\varphi(xt) - \varphi(x) \leqq - \int_0^t \alpha_\varepsilon(\|x\tau\|) \, d\tau$$

for all $x[0, t] \subset U - S(0, \varepsilon)$, which proves the theorem.

All the results above suggest the following definition:

3.7 **Definition.** Let $U \subset R^n$ be an open set such that $0 \in U$. Consider the dynamical system defined by the differential equation 3.1. Let $v = \varphi(x)$: $\varphi = U \to R$ be a lower-semicontinuous real-valued function. Let $D \subset U$ be the set on which $\varphi(x)$ is not continuously differentiable. Let $D^* = \bar{D} - \{0\}$. We shall say that $\varphi(x)$ *has the property* 3.7, if and only if $\varphi(x)$ has the properties 3.2 and 3.3 and in addition, $\psi(x)$ is finite for each $x \in D^*$.

We shall say that the extended real-valued function $w = \psi(x)$ has the property 3.7 whenever it has the properties stated above. A similar definition can be given for upper-semicontinuous real-valued functions.

3.8 *Remark.* It is well known that for the case of a continuous real-valued function $v = \varphi(x)$, φ: $U \to R$, $U \subset R^n$, $0 \in U$, the condition:

3.8.1 there exists a continuous, nondecreasing real-valued function $\alpha(\mu)$, α: $R \to R$, $\alpha(0) = 0$ and $\alpha(\mu) > 0$ for $\mu > 0$, and such that

$$\varphi(x) \leqq -\alpha(\|x\|) \text{ for each } x \in U,$$

is equivalent to the condition:

3.8.2 $\varphi(x) < 0$ for $x \in U - \{0\}$ and if $\{x^n\} \subset U$, $\varphi(x^n) \to 0$ implies $x^n \to \{0\}$.

This equivalence is not generally true for semicontinuous functions.

4. Local Results

We shall next apply the results of the previous section to the study of some local properties of the flow in the neighborhood of the rest point $x = 0$ of the dynamical system 3.1. Here we shall give some results on properties which cannot be characterized by continuously differentiable Liapunov functions.

4.1 **Theorem.** Let $U \subset R^n$ be an open neighborhood of the rest point $x = 0$, and $\bar{U}$ be compact. Let $v = \varphi(x)$ have the property 3.7. Then the set $U - \{0\}$ does not contain any recurrent trajectory.

Proof. Let $\gamma(x) \subset U - \{0\}$ be a recurrent trajectory. Then $\overline{\gamma(x)}$ is a compact minimal set. Thus $\varrho[\overline{\gamma(x)}, \{0\}] = \sigma > 0$. We can choose $\alpha_\sigma(\mu)$ such that 3.2.1 holds. For Theorem 3.6 for each $t > 0$:

4.1.1 $$\varphi(xt) - \varphi(x) \leqq -\int_0^t \alpha_\sigma(\|x\tau\|)\, d\tau \leqq -\int_0^t \alpha_\sigma(\sigma)\, d\tau = -\alpha_\sigma(\sigma)\, t.$$

On the other hand if $\gamma(x)$ is recurrent then $x \in \Lambda^+(x)$, hence there exists a sequence $\{t^n\} \subset R^+$, $t^n \to +\infty$, such that $xt^n \to x$. Thus from 4.1.1

4.1.2 $$\varphi(xt^n) \leqq \varphi(x) - \alpha_\sigma(\sigma)\, t^n$$

and $\varphi(xt^n) \to -\infty$ for $n \to -\infty$ which implies that $v = \varphi(x)$ is not lower-semicontinuous in the point $x \in U$ and violates the hypothesis.

4.2 **Theorem.** Let $v = \varphi(x)$, $\varphi\colon U \to R$ be a lower-semicontinuous real-valued function defined on an open set $U \subset R^n$, with $0 \in U$, and such that

4.2.1 the extended real-valued function $\psi(x)$ (1.6.1) has the property 3.7 on the set U,

4.2.2 $\varphi(0) = 0$, $\varphi(x) > 0$ for $x \in U - \{0\}$,

4.2.3 the component containing $\{0\}$ of the set $\{x \in U : \varphi(x) \leqq \beta;\ \beta > 0\}$ is either compact or is U.

Then $\{0\}$ is an attractor and $A(\{0\}) \supset U$.

Proof. Let $y \in U - \{0\}$ and assume $\Lambda^+(y) = \emptyset$. It must be that $\varphi(y) < \sup\{\varphi(x) : x \in U\}$: as U is open, there is $\tau < 0$ such that $y\tau \in U$. By applying the inequality 3.6.1 to the point $y\tau$, we have that $\varphi(y\tau) > \varphi(y)$. It follows that $\{x \in U\colon \varphi(x) \leqq \varphi(y)\} \neq U$ and so it is compact. Then $\Lambda^+(y) \neq 0$ and compact. We shall prove that $\Lambda^+(y) = \{0\}$. By Theorem 4.1, $\{0\} \subset \Lambda^+(y)$.

Assume $z \in \Lambda^+(y)$, $z \neq \{0\}$. There is then a sequence $\{t^n\} \subset R^+$, $t^n \to +\infty$ such that $yt^n \to z$. Clearly we can construct two sequences, $\{\tau^n\}$

and $\{s^n\}$ such that $y\tau^n \in H(z, \frac{1}{3}\,||z||)$ and $ys^n \in H(z, \frac{2}{3}\,||z||)$ and for each n, $s^n < \tau^n < t^n$.

We may also assume that for $s^n < t < \tau^n$, $yt \in S(z, \frac{2}{3}\,||z||)$ and that $\tau^n - s^n \geqq \sigma > 0$ for each n. From the inequality 3.6.1 it follows that

$$\varphi(yt^n) - \varphi(y) < \sum_{j=1}^{n} (\varphi(y\tau^j) - \varphi(ys^j))$$

$$\leqq - \sum_{j=1}^{n} \int_{s^j}^{\tau^j} \alpha_{\frac{2}{3}||z||}(\tfrac{2}{3}\,||z||)\, d\tau = - \alpha_{\frac{2}{3}||z||}(\tfrac{2}{3}\,||z||) \sum_{j=1}^{n} (\tau^j - s^j)$$

$$\leqq - n\alpha_{\frac{2}{3}||z||}(\tfrac{2}{3}\,||z||)\,\sigma.$$

Then for sufficiently large n, $\varphi(yt^n) < \frac{1}{2}\varphi(z)$ and so $\varphi(x)$ cannot be lower-semicontinuous in $z \in U$. We have proved that for each $y \in U$, $\Lambda^+(y) = \{0\}$ and so $\{0\}$ is an attractor and its region of attraction contains U.

5. Extension Theorems

We shall prove first the global extension theorem, i.e., a theorem which allows us to extend to the whole space the local stability properties of the rest point $x = 0$, which are assumed to be known.

5.1 Theorem. Let $v = \varphi(x)$, $\varphi\colon R^n \to R$ be a real-valued function which satisfies the following conditions:

5.1.1 $v = \varphi(x)$ is lower-semicontinuous, i.e., if $x^n \to x$, then

$$\liminf_{x^n \to x} \varphi(x^n) \geqq \varphi(x),$$

5.1.2 $\varphi(0) = 0$,

5.1.3 the extended real-valued function $\psi(x)$ (1.6.1) has the property 3.7 on R^n.

Then if $\{0\}$ is a positive attractor, it is a positive global attractor.

Proof. Assume that the rest point $x = 0$ is a positive attractor and A is its region of attraction. Let $x_0 \in \partial A$, since ∂A is invariant, $\gamma(x_0) \subset \partial A$. Now

5.1.4 $\inf\{||y||\colon y \in \gamma(x_0)\} = k > 0.$

As $\psi(x)$ has the property 3.7, we can choose $\alpha_k(\mu)$ such that

5.1.5 $\varphi(x_0t) - \varphi(x_0) \leqq - \int_0^t \alpha_k(||x_0\tau||)\, d\tau \leqq -\alpha_k(k)\, t.$

Thus there exists $\tau > 0$, such that $x_0\tau \in \partial A$ and $\varphi(x_0\tau) < 0$. Since $\varphi(x)$ has the property 3.3, for each $\tau' > \tau$, $|\tau' - \tau| < \varepsilon$, $\varepsilon > 0$ sufficiently small, $\varphi(x)$ is continuous in $x\tau'$ and $\varphi(x^0\tau') \leqq \varphi(x^0\tau) < 0$. Since $x\tau' \in \partial A$, there exists $z \in A$ such that $\varphi(z) < 0$. Hence

$$\varphi(zt) < \varphi(z) \text{ for all } t > 0.$$

Consider now a sequence $\{t_n\}$, $t_n \to +\infty$, such that $zt_n \to \{0\}$. We have $\varphi(zt_n) < \varphi(z) < 0$ and therefore

$$\liminf_{n\to+\infty} \varphi(zt_n) < \varphi(0),$$

which contradicts the assumption 5.1.1 and proves that $\partial A = \emptyset$. Thus $A = R^n$.

5.2 *Remark.* The most interesting case in which Theorem 5.1 can be applied is the case of an unstable attractor. This is a case in which it is not possible to construct a continuous Liapunov function with sign-definite total time derivative. On the other hand it is possible to construct a lower-semicontinuous real-valued function $v = \varphi(x)$ which satisfies the hypotheses of Theorem 5.1. It is to be noticed that such a real-valued function must be discontinuous at the point $x = 0$.

A theorem analogous to Theorem 5.1 can be proved also for upper-semicontinuous real-valued function $v = \varphi(x)$. Its proof is left as an exercise to the reader.

5.3 **Theorem.** Let $v = \varphi(x)$, $\varphi: R^n \to R$ be a real-valued function, which satisfies the following conditions:

5.3.1 $v = \varphi(x)$ is upper-semicontinuous,

5.3.2 $\varphi(0) = 0$,

5.3.3 the extended real-valued function $\psi(x)$ defined in 1.6.1 has the property 3.7 on R^n.

Then if $\{0\}$ is a negative attractor, it is a negative global attractor.

We shall proceed next with the proof of the local extension theorem. Its proof requires the sets introduced in Definition VIII, 2.1. We therefore apply Definition VIII, 2.1 to the case of a lower-semicontinuous real-valued function $v = \varphi(x)$, and to the case in which $M = \{0\}$. In this case, in general, the properties VIII, 2.2 do not hold, instead we have

5.4 **Properties.** Let $v = \varphi(x)$ be a lower-semicontinuous real-valued function defined on R^n. Consider the sets $N(\beta)$ and $O(\beta)$ defined as in VIII, 2.1.7.

5.4.1 If $v = \varphi(x)$ is bounded in a neighborhood of $\{0\}$, there exists $\beta > 0$, such that $N(\beta)$ contains an open neighborhood of $\{0\}$.

5.4.2 If $v = \varphi(x)$ is discontinuous at the point $x = 0$, then there exists a real number $\beta > 0$ such that $\{0\} \subset \partial N(\beta)$.

5.4.3 $\varphi(x) \leqq \beta$ for $x \in \partial N(\beta)$.

5.4.4 $O(\beta)$ is open, while $N(\beta)$ may not be open.

We are now in the position of proving the local extension theorem for attracting rest points.

5.5 **Theorem.** Consider the differential equation 3.1 which defines a dynamical system. Let $x = 0$ be an isolated rest point which is an attractor. Let $v = \varphi(x)$ be a lower-semicontinuous real-valued function such that

5.5.1 $\varphi(0) = 0$,

5.5.2 there exists a real number $\beta^0 > 0$ such that the extended real-valued function, $w = \psi(x)$, defined in 1.6.1 has the property 3.7 in $N(\beta^0)$.

Then $A(\{0\}) \supset N(\beta^0)$.

Proof. Assume if possible that $x_0 \in (\partial A(\{0\}) \cap N(\beta^0))$. From the hypotheses made, the set $N(\beta^0) \cap \partial A(\{0\})$ is positively invariant, hence $\gamma^+(x^0) \subset [N(\beta^0) \cap \partial A(\{0\})]$. Now by proceeding as in the proof of Theorem 5.1, we reach a contradiction which proves $\partial A(\{0\}) \cap N(\beta^0) = \emptyset$ and hence $N(\beta^0) \subset A(\{0\})$.

6. Non-continuous Liapunov Functions on the Region of Weak Attraction

The functions used in the previous section in the case of an attractor were constructed in a whole neighborhood of the attractor, even in the case of an unstable attractor. In this latter case one can obtain deeper insight in the structure of the flow by constructing a Liapunov function on the region of instability (V, 1.13).

The proof of the main theorems rests upon the following Lemma whose proof is similar to that of Theorem V, 2.16 and it is left as an exercise to the reader.

6.1 **Lemma.** Let the rest point $x = 0$ of the dynamical system defined by the ordinary differential equation 3.1 be asymptotically stable relative to a closed, positively invariant set $U \subset R^n$ with $\{0\} \subset U$. Then there exist in U two continuous, real-valued functions $\varphi(x)$ and $\theta(x)$

with the following properties:

6.1.1 $\quad -1 < \varphi(x) < 0$ for $||x|| > 0$,

6.1.2 $\quad \theta(x) > 0$ for $||x|| > 0$,

6.1.3 $\quad \varphi(0) = \theta(0) = 0$,

6.1.4 $$\dot{\varphi} = \frac{d\varphi(xt)}{dt}\bigg|_{t=0} = \theta(x)\,(1 + \varphi(x)).$$

6.2 **Corollary.** Let all the conditions of Lemma 6.1 be satisfied, but instead of asymptotically stable assume that $\{0\}$ is negatively asymptotically stable relative to U. Then there exist on U two continuous, real-valued functions $\varphi(x)$ and $\theta(x)$ which satisfy conditions 6.1.1, 6.1.2, 6.1.3 and

6.2.1 $$\dot{\varphi} = \frac{d\varphi(xt)}{dt}\bigg|_{t=0} = -\theta(x)\,(1 + \varphi(x)).$$

The results 6.1 and 6.2 will be applied next to the construction of a Liapunov function $v = \varphi(x)$ on the region of instability $I(\{0\}) = D^+(\{0\}) - \{0\}$ of the rest point $x = 0$ of the ordinary differential equation 3.1. This rest point will be assumed to be an unstable weak attractor. We recall that in this case the prolongation of $\{0\}$, $D^+(\{0\})$ is compact.

It is desired that such a Liapunov function has the following

6.3 **Properties.**

6.3.1 $\quad v = \varphi(x)$, $\varphi\colon I(\{0\}) \to R$ is lower-semicontinuous

on the set

6.3.2 $\quad I = I(\{0\}) = D^+(\{0\}) - \{0\}$.

6.3.3 For each $\delta > 0$ there exists $\varepsilon > 0$ such that

$$\varphi^*(x) = \liminf_{\substack{\tau \to 0 \\ y \to f(x)}} [\varphi(x + \tau y) - \varphi(x)]\,\tau^{-1} < -\varepsilon$$

for all $x \in I$ with $|x| < \delta$.

6.3.4 If $D \subset I(\{0\})$ is the set of all points in $I(\{0\})$ in which $\varphi(x)$ is either not continuous or does not have a continuous total time derivative along the trajectories of the system, then the set $\bar{D} - \{0\}$ is intersected by each trajectory on a set of isolated points.

The following theorem gives a sufficient condition for a Liapunov function to have on the set $I(\{0\})$ the properties 6.3.1, 6.3.3 and 6.3.4.

6.4 **Theorem.** Consider a dynamical system defined by the differential equation 3.1. Let its rest point $x = 0$ be an unstable weak attractor. Assume that there exist in I (6.3.2) two compact sets P and N with the following properties:

6.4.1 P is positively invariant.

6.4.2 N is negatively invariant.

6.4.3 $P \cup N = I(\{0\}) \cup \{0\} = D^+(\{0\})$.

6.4.4 For each $x \in I$ there exists one and only one real number $\tau(x) \in R$ such that $x\tau(x) \in P \cap N$.

6.4.5 $J^+(x, P) = \{0\}$ for each $x \in P$.

6.4.6 $J^-(x, N) = \{0\}$ for each $x \in N$.

Then there exists on $D^+(\{0\})$ a Liapunov function with the properties 6.3.1, 6.3.3 and 6.3.4.

Proof. The rest point $x = 0$ is positively (negatively) asymptotically stable relatively to P (to N). By applying Lemma 6.1 and Corollary 6.2 one can construct respectively on P and on N the real-valued functions $\varphi_P(x)$ and $\theta_P(x)$ and $\varphi_N(x)$ and $\theta_N(x)$ which satisfy the properties 6.1.1—6.1.4 and 6.3.1, respectively. Then the real-valued function

6.4.7
$$\varphi(x) = \begin{cases} 0 & \text{for } x = 0, \\ -\varphi_P(x) & \text{for } x \in P, \\ \varphi_N(x) + 3 & \text{for } x \in N - P \end{cases}$$

satisfies the conditions 6.3.1, 6.3.3 and 6.3.4.

6.5 Theorem. Let $\{0\}$ be as in Theorem 6.4 and let S be a section of I (6.3.2), i.e. let $S \subset I$ be such that for all $x \in I$ there exists one and only one real number $\tau(x) \in R$, such that $x\tau(x) \in S$. Let next

6.5.1 $P = \overline{\gamma^+(S)}$ and $N = \overline{\gamma^-(S)}$.

Then the following statements are equivalent:

6.5.2 The sets P and N (6.5.1) have the properties 6.4.1—6.4.6.

6.5.3 $x_n \to x \in P - S$ or $x_n \to x \in N - S$ implies $\operatorname{sign} \tau(x_n) = \operatorname{sign} \tau(x)$.

6.5.4 For each $\varepsilon > 0$, $\tau(x)$ is bounded on the set $I - S(0, \varepsilon)$.

Proof. Notice, first of all, that under the hypotheses made on the system there always exists an infinite number of sections of I, since I contains neither periodic orbits nor rest points (IV, 1.10, 2.4). Let us now proceed with the proof of the theorem.

6.5.3 $\to$ 6.5.2. As already pointed out $D^+(\{0\})$ is a compact set, which from the definition of the section S is such that $P \cup N = D^+(\{0\}) = I \cup \{0\}$. If it were true that $P \cap N \neq S \cup \{0\}$, since $S \subset P \cup N$, there would exist a point $z \notin S$, $z \in P \cap N = \overline{\gamma^+(S)} \cap \overline{\gamma^-(S)}$. Then there exist sequences $\{x_n\}$, $\{y_n\} \subset S$ and sequences $\{\tau_n\} \subset R^+$ and $\{\sigma_n\} \subset R^-$ such that $x_n\tau_n \to z$ and $y_n\sigma_n \to z$.

Since $z \in P \cap N \subset P \cup N$ if $z \neq 0$ then $z \in I$. Hence there exists $\tau(z) \in R$ such that $z\tau(z) \in S$. From 6.5.3 both τ^n and σ^n have a sign which is equal to the sign of $\tau(z)$ and since they have signs which opposite to one another it follows that $\tau(z) = 0$, i.e., $z \in S$. Finally since the set $J^+(x, P)$ is invariant and $J^+(x, P) \subset P$, if $z \neq 0$ and $z \in J^+(x, P)$, then $\gamma^-(z) \subset P$. Then there exists $\tau(z) < 0$ such that $z\tau(z) \in S$ and for $t < \tau(z)$, $zt \in P \cap N$, from which $zt \in S$. This implies that $t = \tau(z)$ which is absurd. The same argument can be applied to $J^+(x, N)$. This concludes this part of the theorem.

6.5.2 → 6.5.3. Let $x \in P - S$ be such that there exists a sequence $\{x_n\} \subset R^n$ with $x_n \to x$, $x_n\tau(x_n) \in S$, $\operatorname{sign}\tau(x_n) \neq \operatorname{sign}\tau(x)$. Since $x \in P - S$, then $\tau(x) < 0$. Then $\tau(x_n) \to +\infty$ or else there exists a real number $\tau'(x) > 0$ such that $x\tau'(x) \in S$.

Then $x_n\tau(x_n) \to 0$. Now $y_n = x_n\tau(x_n) \subset S \subset N$ and since $J^-(\{0\}, N) = \{0\}$ we must have $y_n(-\tau(x_n)) \to 0$. This is absurd if $y_n(-\tau(x_n)) = x_n \to x \neq 0$. This concludes this part of the theorem. The fact that 6.5.2 ↔ 6.5.4 can be analogously proved.

6.6 *Remark.* Notice that if for a certain section $S \subset I$ the real-valued function $\tau(x)$ can be selected to be continuous, then the properties 6.5.3 and 6.5.4 are obviously true, then from Theorem 6.5 the sets P and N have the properties 6.4.1—6.4.6. Since the set $I \subset R^n$ (6.3.2) is invariant, and locally compact, then a necessary and sufficient condition for the dynamical system to be dispersive on I (IV, 2.4) is that there can be selected on I a section with $\tau(x)$ continuous. On the other hand the dynamical system is dispersive on I if and only if $J^+(I, I) = \{0\}$.

Thus

6.7 **Theorem.** Consider a dynamical system defined by the ordinary differential equation 3.1. Assume that the rest point $x = 0$ is a weak attractor and let I (6.3.1) be its region of instability. Assume that

6.7.1 $\quad J^+(I, I) = \{0\}.$

Then there exists in $D^+(\{0\})$ a Liapunov function which satisfies the properties 6.3.1, 6.3.3 and 6.3.4.

Proof. The theorem will be proved by showing that the hypothesis 6.7.1 is equivalent to the conditions 6.4.5 and 6.4.6. Clearly if both 6.4.5 and 6.4.6 are true also 6.6.1 is true. Conversely if $J^+(I, I) = \{0\}$, then from what was pointed out in Remark 6.6 it follows that $J^+(x, P) = \{0\}$

and $J^-(x, N) = \{0\}$ where $P = \overline{\gamma^+(S)}$ and $N = \overline{\gamma^-(S)}$ and S is a section of I with $\tau(x)$ continuous.

6.8 *Remark.* The function which can be constructed on the set I, if the condition 6.7.1 is satisfied, has a property which is stronger than 6.3.4. In this latter case, in fact, the set of discontinuity D is intersected in one and only one point by each trajectory in I. It is easy to see that the existence of this type of Liapunov function is a necessary and sufficient condition for 6.7.1 to be satisfied.

6.9 **Theorem.** Let rest point $x = 0$ of the dynamical system 6.3.1 be an unstable weak attractor which is such that on the set I there exists a Liapunov function $v = \varphi(x)$ which satisfies the conditions 6.3.1, 6.3.2 and 6.3.3.

In addition condition 6.3.4 is satisfied in the strongest form, i.e., the set $D - \{0\}$ is intersected by each positive semi-trajectory in one point at the most. Then the relationship 6.7.1 is true.

Proof. It is clear that D is a section of I. For this it is enough to show that for each $x \in I$ there exists $\tau(x) \in R$, such that $x\tau(x) \in D$. If this were not true we would have $\gamma(x) \subset I - D$ and $\varphi(x)$ would be continuous relatively to I and would have a continuous total time derivative along $\gamma(x)$. From 6.3.3 we would in addition have $\Lambda^+(x) = \Lambda^-(x) = \{0\}$, which is absurd. Then D is a section of I. Let next $P = \gamma^+(D)$ and $N = \gamma^-(D)$, we have $J^+(I, I) = J^+(I, N) \cup J^+(I, P)$. If $J^+(I, I) \neq \{0\}$ we have $J^+(I, P) \neq \{0\}$. Now $J^+(I, P)$ is an invariant set contained in P, hence there would exist in P a trajectory which would not intersect D, which is absurd.

Notes and References

The idea of using non-differentiable Liapunov functions for the characterization of the stability properties of differential equations was originated by H. Okamura [1] and fully developed by T. Yoshizawa in many of his papers (see T. Yoshizawa [1]). For the differential equation $\dot{x} = f(x, t)$ Okamura and Yoshizawa use Lipschitz continuous functions $\varphi(t, x)$ and the derivative $\dot{\varphi}(t, x) = \limsup_{\tau \to 0^+} \tau^{-1}[\varphi(t + \tau, x + \tau f(x, t)) - \varphi(t, x)]$. In Chapter V we have already used continuous Liapunov functions and have been able to prove very directly the converse theorems. In this case however we did not need to prove that φ is Lipschitz because the derivative is taken along trajectories.

Section 1. The theorems presented in this section are due to J. A. Yorke [3], and proved for the case of the differential equation $\dot{x} = f(x, t)$. The proof we give is different from the original. We notice that in the above mentioned paper two different cases are taken into account: the case in which the differential equation has uniqueness of solution and the case in which it has not. In this latter case the statement of the theorem is less general since it is proved that for each point $x \in U$ there exists at least one solution $\varphi(x, t, t_0)$ of the differential equation, with $\varphi(x, t_0, t_0) = x$, which satisfies the inequality 1.6.3. Indeed J. A. YORKE gives an example of a second order differential equation without uniqueness for which there exists one solution which does not satisfy the inequality 1.6.3 when all conditions of the theorem are met.

Section 2. The results of this section are due to N. P. BHATIA, G. P. SZEGÖ and J. A. YORKE [1]. They constitute a motivation for the introduction of non-continuous Liapunov functions and provide a Liapunov-type characterization of attractors and weak attractors.

Section 3–5. The results of these sections are due to G. P. SZEGÖ and G. TRECCANI [3].

The problem which has motivated the research of these sections is that of the numerical construction of piecewise quadratic Liapunov functions, as an alternative to high order polynomials. Piecewise quadratic Liapunov functions have the definite advantage on higher order polynomials to allow the computation of the hypervolume of set bounded by a level surface in a much simpler way. The details on the construction of Liapunov functions can be found in the paper by G. P. SZEGÖ, G. ARIENTI and C. SUTTI [1].

Section 6. The motivation of Theorem 6.7 lies in the need for a finer analysis of the (instability) properties of flows as shown in section VI, 2. The results of this section are due to G. P. SZEGÖ and G. TRECCANI [4].

References

Aeppli, A.; Markus, L.

1. Integral equivalence of vector fields on manifolds and bifurcation of differential systems. Amer. J. Math. **85**, 633–654 (1963).

Albrecht, F.

1. Un théorème de comportement asymptotique des solutions des équations des systèmes d'équations différentielles. Bull. Acad. Polon. Sci., Cl. III, **4**, 737–739 (1956).

Andrea, S. A.

1. On homeomorphisms of the plane, and their embedding in flows. Bull. Amer. Math. Soc. **71**, 381–383 (1965).

Andronov, A.; Chaikin, C. B.

1. Theory of Oscillations (Russian). Moscow 1937. English translation, Princeton: Princeton Univ. Press 1949.

Anosov, D. V.

1. Geodesic flows on closed Riemann manifolds with negative curvature (Russian). Trudy Mat. Inst. Steklov, no. 90 (1967). English translation: Proc. Steklov Inst. Math., no. 90, Amer. Math. Soc., Providence, 1969.
2. Roughness of geodesic flows on compact Riemannian manifolds of negative curvature (Russian). Dokl. Akad. Nauk SSSR **145**, 707–709 (1962). English translation: Soviet Math. (Dokl.) **3**, 1068–1070 (1962).
3. Ergodic properties of geodesic flows on closed Riemannian manifolds of negative curvature (Russian). Dokl. Akad. Nauk SSSR **151**, 1250–1273 (1963). English translation: Soviet Math. (Dokl.) **4**, 1153–1155 (1963).

Antosiewicz, H. A.

1. A survey of Liapunov's second method. In: Contributions to the Theory of Nonlinear Oscillations, vol. 4 (Annals of Math. Studies, no. 41). Princeton: Princeton Univ. Press 1958, pp. 141–166.

Antosiewicz, H. A.; Davis, P.

1. Some implications of Liapunov's conditions of stability. J. Rat. Mech. Anal. **3**, 447–457 (1954).

Antosiewicz, H. A.; Dugundji, J.

1. Parallelizable flows and Liapunov's second method. Ann. Math. **73**, 543–555 (1961).

Arienti, G. (see Szegö, G. P.)

Arnol'd, V. L.; Avez, A.

1. Problèmes ergodiques de la mécanique classique. Paris: Gauthier-Villars 1967.

Arzelà, C.

1. Funzioni di linee. Atti R. Accad. Lincei Rend. **5**, 342–348 (1889).
2. Sulle funzioni di linee. Mem. R. Accad. Bologna **5**, 225–244 (1895).

Ascoli, G.

1. Le curve limiti di una varietà data di curve. Mem. R. Accad. Lincei **18**, 521–586 (1883/84).

Auslander, J.

1. Mean-L-stable systems. Illinois J. Math. **3**, 566–579 (1959).
2. On the proximal relation in topological dynamics. Proc. Amer. Math. Soc. **11**, 890–895 (1960).
3. Generalized recurrence in dynamical systems. In: Contributions to Differential Equations, vol. 3. New York: Wiley **1964**, pp. 55–74.
4. Transformation groups without minimal sets. Math. Systems Theory **2**, 93–95 (1968).

Auslander, J.; Bhatia, N. P.; Seibert, P.

1. Attractors in dynamical systems. Bol. Soc. Mat. Mexicana **9**, 55–66 (1964).

Auslander, J.; Gottschalk, W. H.

1. (Editors) Topological Dynamics. Proceedings of an International Symposium held at Colorado State University, Fort Collins, Colorado, August 1967. New York-Amsterdam: Benjamin **1968**.

Auslander, J.; Hahn, F.

1. Point transitive flows, algebras of functions and the Bebutov system. Fund. Math. **60**, 117–137 (1967).

Auslander, J.; Seibert, P.

1. Prolongations and generalized Liapunov functions. In: Nonlinear Differential Equations and Nonlinear Mechanics, J. P. LaSalle and S. Lefschetz (Editors). New York: Academic Press **1963**, pp. 456–462.
2. Prolongations and stability in dynamical systems. Ann. Inst. Fourier (Grenoble) **14**, 237–267 (1964).

Auslander, L.; Green, L. W.

1. G-induced flows. Amer. J. Math. **88**, 43–60 (1966).

Auslander, L.; Green, L. W.; Hahn, F.

1. Flows on Homogeneous Spaces (Annals of Math. Studies, no. 53). Princeton: Princeton Univ. Press **1963**.

Auslander, L.; Hahn, F.

1. Real functions coming from flows on compact spaces and concepts of almost periodicity. Trans. Amer. Math. Soc. **106**, 415–426 (1963).

Avez, A. (see also Arnol'd, V. L.)

1. Ergodic Theory of Dynamical Systems. Lecture Notes, University of Minnesota, **1967**.
2. Anosov diffeomorphisms. In: Topological Dynamics, J. Auslander and W. H. Gottschalk (Editors). New York-Amsterdam: Benjamin **1968**, pp. 17–51.

Baidosov, V. A. (see also Barbashin, E. A.)

1. Invariant functions of dynamical systems (Russian). Izv. Vyssh. Uchebn. Zaved., Mat. no. 1 (8), pp. 9–15 (1959).
2. On homeomorphisms of dynamical systems (Russian). Izv. Vyssh. Uchebn. Zaved., Mat. no. 3 (16), pp. 21–29 (1960).

Barbashin, E. A.

1. Introduction to the Theory of Stability (Russian). Moscow: Izd. "Nauka" **1967**.
2. Sur certaines singularités qui surviennent dans un système dynamique quand l'unicité est en défaut. C. R. (Dokl.) Acad. Sci. URSS **41**, 139–141 (1943).
3. Les singularités locales des points ordinaires d'un système d'équations différentielles. C. R. (Dokl.) Acad. Sci. URSS **41**, 183–186 (1943).
4. Sur la conduite des points sous les transformations homéomorphes de l'espace. C. R. (Dokl.) Acad. Sci. URSS **51**, 3–5 (1946).

5. On homeomorphisms of dynamical systems (Russian). Dokl. Akad. Nauk SSSR **61**, 429–432 (1948).
6. On the theory of generalized dynamical systems (Russian). Uchen. Zap. Moskov. Univ., no. 135, Mat. **2**, 110–133 (1948).
7. On homeomorphisms of dynamical systems (Russian). Mat. Sbornik **27**, 455–470 (1950).
8. Dispersive dynamical systems (Russian). Uspehi Mat. Nauk **5**, 138–139 (1950).
9. On the theory of systems of multivalued transformations of a topological space (Russian). Uchebn. Zap. Ural. Univ., no. 7, pp. 54–60 (1950).
10. The method of sections in the theory of dynamical systems (Russian). Mat. Sbornik **29**, 233–280 (1951).
11. On homeomorphisms of dynamical systems, II (Russian). Mat. Sbornik **29**, 501–518 (1951).
12. On the behavior of points under homeomorphic transformations of a space. (Generalization of theorems of Birkhoff) (Russian). Trudy Ural. Politehn. Inst. **51**, 4–11 (1954).
13. Conditions for the existence of recurrent trajectories in dynamical systems with a cylindrical phase space. Differencial'nye Uravnenija **3**, 1627–1633 (1967).

Barbashin, E. A.; Baidosov, V. A.

1. On the question of topological definition of integral invariants (Russian). Izv. Vyssh. Uchebn. Zaved., Mat. no. 3 (4), pp. 8–12 (1958).

Barbashin, E. A.; Krasovskii, N. N.

1. On stability of motion in the large. Dokl. Akad. Nauk SSSR **86**, 453–456 (1952).
2. On the existence of Liapunov functions in the case of asymptotic stability in the large. Prikl. Mat. Mekh. **18**, 345–350 (1954).

Barbashin, E. A.; Sholohovich, F. A.

1. The mapping of a dynamical system into a time-analytic dynamical system (Russian). Izv. Vyssh. Uchebn. Zaved., Mat. no. 1 (14), pp. 11–15 (1960).

Barbuti, U.

1. Su alcuni teoremi di stabilità. Ann. Scuola Norm. Sup. Pisa **8**, 81–91 (1954).

Barocio, S.

1. On certain critical points of a differential system in the plane. In: Contributions to the Theory of Nonlinear Oscillations, vol. 3 (Annals of Math. Studies, no. 36). Princeton: Princeton Univ. Press 1956, pp. 127–136.
2. On trajectories in the vicinity of a three-dimensional singularity (Spanish). Bol. Soc. Mat. Mexicana **1**, 57–58 (1956).

Bass, R. W.

1. On the regular solutions at a point of singularity of a system of nonlinear differential equations. Amer. J. Math. **77**, 734–742 (1955).
2. Zubov's stability criterion. Bol. Soc. Mat. Mexicana **4**, 26–29 (1959).

Baum, J. D.

1. An equicontinuity condition for transformation groups. Proc. Amer. Math. Soc. **4**, 656–662 (1953).
2. Asymptoticity in topological dynamics. Trans. Amer. Math. Soc. **77**, 506–519 (1954).
3. *P*-recurrence in topological dynamics. Proc. Amer. Math. Soc. **7**, 1146–1154 (1956).
4. An equicontinuity condition in topological dynamics. Proc. Amer. Math. Soc. **12**, 30–32 (1961).
5. Instability and asymptoticity in topological dynamics. Pacific J. Math. **12**, 25–34 (1962).

6. Some remarks on P-limit point stability. In: Topological Dynamics, J. Auslander and W. H. Gottschalk (Editors). New York-Amsterdam: Benjamin 1968, pp. 53–66.

BEBUTOV, M. V.

1. Sur les systèmes dynamiques stables au sens de Liapounoff. C. R. (Dokl.) Acad. Sci. URSS **18**, 155–158 (1938).
2. Sur la représentation des trajectoires d'un système dynamique sur un système de droites parallèles. Bull. Math. Univ. Moscou, Sér. Internat. **2**, fasc. 3, pp. 1–22 (1939).
3. Sur les systèmes dynamiques dans l'espace des fonctions continues. Bull. Math. Univ. Moscou, Sér. Internat. **2**, fasc. 5 (1939).
4. Sur les systèmes dynamiques dans l'espace des fonctions continues. C. R. (Dokl.) Acad. Sci. URSS **27**, 904–906 (1940).
5. On dynamical systems in the space of continuous functions (Russian). Bull. Moskov. Gos. Univ., Mat. **2**, no. 5, pp. 1–52 (1941).

BEBUTOV, M. V.; STEPANOV, V. V.

1. Sur le changement du temps dans les systèmes dynamiques possédant une mesure invariante. C. R. (Dokl.) Acad. Sci. URSS **24**, 217–219 (1939).
2. Sur la mesure invariante dans les systèmes dynamiques qui ne diffèrent que par le temps. Rec. Math. (Mat. Sbornik) **7**, 143–166 (1940).

BECK, A.

1. On invariant sets. Ann. Math. **67**, 99–103 (1958).
2. Continuous flows with closed orbits. Bull. Amer. Math. Soc. **66**, 305–307 (1960).
3. Plane flows with few stagnation points. Bull. Amer. Math. Soc. **71**, 892–896 (1965).
4. Plane flows with closed orbits. Trans. Amer. Math. Soc. **114**, 539–551 (1965).

BELLMAN, R.

1. Stability Theory of Differential Equations. New York-Toronto-London: McGraw-Hill 1953.

BELLMAN, R.; COOKE, K. L.

1. Differential-Difference Equations. New York: Academic Press 1963.

BENDIXSON, I.

1. Sur les courbes définies par des équations différentielles. Acta Math. **24**, 1–88 (1901).

BESICOVITCH, A. S.

1. Almost Periodic Functions. Cambridge: Cambridge Univ. Press 1932. Reprint, New York: Dover 1954.

BHATIA, N. P. (see also AUSLANDER, J.)

1. Stability and Lyapunov functions in dynamical systems. In: Contributions to Differential Equations, vol. 3. New York: Wiley 1964, pp. 175–188.
2. Weak attractors in dynamical systems. Bol. Soc. Mat. Mexicana **11**, 56–64 (1966).
3. Criteria for dispersive flows. Math. Nachr. **32**, 89–93 (1960).
4. Lectures on Ordinary Differential Equations: Stability Theory with Applications. Department of Mathematics, Western Reserve University, Spring, 1964.
5. On asymptotic stability in dynamical systems. Math. Systems Theory **1**, 113–128 (1967).
6. Semidynamical flow near a compact invariant set. In: Topological Dynamics, J. Auslander and W. H. Gottschalk (Editors). New York-Amsterdam: Benjamin 1968, pp. 81–96.

7. Dynamical systems. In: Mathematical Systems Theory and Economics I, H. W. Kuhn and G. P. Szegö (Editors) (Lecture Notes in Operations Research and Mathematical Economics, vol. 11). Berlin-Heidelberg-New York: Springer **1969**, pp. 1–10.

8. Semidynamical systems. In: Mathematical Systems Theory and Economics II, H. W. Kuhn and G. P. Szegö (Editors) (Lecture Notes in Operations Research and Mathematical Economics, vol. 12). Berlin-Heidelberg-New York: Springer **1969**, pp. 303–318.

9. Attraction and non-saddle sets in dynamical systems. J. Diff. Eqs. **8** (1970).

10. Characteristic Properties of Stable Sets and Attractors in Dynamical Systems. Tech. Note BN-637, IFDAM, Univ. of Maryland, College Park, Maryland, January 1970. Invited Lecture at Conference on Rational Mechanics, Rome Feb. 23–26, 1970.

Bhatia, N. P.; Hajek, O.

1. Local Semi-Dynamical Systems (Lecture Notes in Mathematics, vol. 90). Berlin-Heidelberg-New York: Springer **1969**.

Bhatia, N. P.; Lazer, A.; Leighton, W.

1. Applications of the Poincaré-Bendixson theory. Ann. Mat. Pura Appl., Ser. IV, **73**, 27–32 (1966).

Bhatia, N. P.; Lazer, A. C.; Szegö, G. P.

1. On global weak attractors in dynamical systems. J. Math. Anal. Appl. **16**, 544–552 (1966).

Bhatia, N. P.; Szegö, G. P.

1. Dynamical Systems: Stability Theory and Applications (Lecture Notes in Mathematics, vol. 35). Berlin-Heidelberg-New York: Springer **1967**.

2. An extension theorem for asymptotic stability. In: Differential Equations and Dynamical Systems, J. K. Hale and J. P. LaSalle (Editors). New York: Academic Press **1967**, pp. 147–154.

3. Weak attractors in R^n. Math. Systems Theory **1**, 129–134 (1967).

Bhatia, N. P.; Szegö, G. P.; Yorke, J. A.

1. A Liapunov characterization of attractors. Boll. Un. Mat. Ital., Ser. IV, **1**, 222–228 (1969).

Birkhoff, G. D.

1. Dynamical Systems (Amer. Math. Soc. Colloquium Publications, vol. 9). New York **1927**.

2. Collected Works, vol. 1, 2, 3. Amer. Math. Soc., New York, **1950**. Reprint, New York: Dover **1968**.

3. Über gewisse Zentralbewegungen dynamischer Systeme. Kgl. Ges. d. Wiss. Göttingen, Nachrichten, Math.-phys. Kl., **1926**, Heft 1, pp. 81–92.

4. On the periodic motions of dynamical systems. Acta Math. **50**, 359–379 (1927).

5. Stability and the equations of dynamics. Amer. J. Math. **49**, 1–38 (1927).

6. Proof of a recurrence theorem for strongly transitive systems. Proc. Nat. Acad. Sci. USA **17**, 650–655 (1931).

Bochner, S.

1. Beiträge zur Theorie der fastperiodischen Funktionen, I. Math. Ann. **96**, 119–147 (1926).

Bohr, H.

1. Zur Theorie der fastperiodischen Funktionen, I, II, III. Acta Math. **45**, 29–127 (1924); **46**, 101–214 (1925); **47**, 237–281 (1926).

Bouquet, J. C.; Briot, C. A. A.

1. Recherches sur les fonctions définies par les équations différentielles. J. École Polytech. (Paris) **21**, cah. 36, pp. 133–198 (1856).

BRAUER, F.

1. Liapunov functions and comparison theorems. In: Nonlinear Differential Equations and Nonlinear Mechanics, J. P. LaSalle and S. Lefschetz (Editors). New York: Academic Press 1963, pp. 435–441.
2. The use of comparison theorems for ordinary differential equations. In: Stability Problems of Solutions of Differential Equations, A. Ghizzetti (Editor). Gubbio: Oderisi 1966.

BRAUER, F.; STERNBERG, SH.

1. Local uniqueness, existence in the large, and the convergence of successive approximations. Amer. J. Math. **80**, 797 (1958).

BRIOT, C. A. A. (see BOUQUET, J. C.)

BRONSTEIN, I. U. (see also SIBIRSKII, K. S.)

1. Motions in partially ordered dynamical systems (Russian). Uchen. Zap. Kishinev. Univ., no. 39, pp. 249–251 (1959).
2. On dynamical systems without uniqueness, as semigroups of non-single-valued mappings of a topological space (Russian). Dokl. Akad. Nauk SSSR **144**, 954–957 (1962). English translation: Soviet Math. (Dokl.) **3**, 824–827 (1962).
3. Recurrence, periodicity and transitivity in dynamical systems without uniqueness (Russian). Dokl. Akad. Nauk SSSR **151**, 15–18 (1963). English translation: Soviet Math. (Dokl.) **4**, 889–892 (1963).
4. On dynamical systems without uniqueness, as semigroups of non-singlevalued mappings of a topological space (Russian). Izv. Akad. Nauk Moldav. SSR, Ser. Estestven. Tekh. Nauk, no. 1, pp. 3–18 (1963).
5. Two examples of dynamical systems (Russian). Izv. Akad. Nauk Moldav. SSR, Ser. Estestven. Tekh. Nauk, no. 1, pp. 73–74 (1963).
6. Recurrent points and minimal sets in dynamical systems without uniqueness (Russian). Izv. Akad. Nauk Moldav. SSR, Ser. Estestven. Tekh. Nauk, no. 7, pp. 14–21 (1965).
7. On homogeneous minimal sets (Russian). Papers on Algebra and Analysis, Kishinev, 1965, pp. 115–118.
8. Recurrent points and minimal sets in dynamical systems without uniqueness. Bul. Akad. Shtiince RSS Moldoven., 1965, no. 7, pp. 14–21.
9. A contribution to the theory of distal minimal sets and distal functions (Russian). Dokl. Akad. Nauk SSSR **172**, 255–257 (1967). English translation: Soviet Math. (Dokl.) **8**, 59–61 (1967).

BRONSTEIN, I. U.; SHCHERBAKOV, B. A.

1. Certain properties of Lagrange stable funnels of generalized dynamical systems (Russian). Izv. Akad. Nauk Moldav. SSR, Ser. Estestven. Tekh. Nauk, no. 5, pp. 99–102 (1962).

BROUWER, L. E. J.

1. On continuous vector distributions, I, II, III. Verh. Nederl. Akad. Wetensch., Afd. Natuurk., Sec. I, **11**, 850–858 (1909); **12**, 716–734 (1910); **13**, 171–186 (1910).

BROWDER, F. E.

1. On the iteration of transformations in noncompact minimal dynamical systems. Proc. Amer. Math. Soc. **9**, 773–780 (1958).
2. On a generalization of the Schauder fixed point theorem. Duke Math. J. **26**, 291–303 (1959).
3. On the continuity of fixed points under deformations of continuous mappings. Summa Brasil. Math. **4**, 183–191 (1960).

4. On the fixed points index for continuous mappings of locally connected spaces. Summa Brasil. Math. **4**, 253–293 (1960).

BROWN, M.

1. The monotone union of open n-cells is an open n-cell. Proc. Amer. Math. Soc. **12**, 812–814 (1961).

BUDAK, B. M.

1. Dispersive dynamical systems (Russian). Vestnik Moskov. Univ., no. 8, pp. 135–137 (1947).
2. The concept of motion in a generalized dynamical system (Russian). Uchen. Zap. Moskov. Univ., no. 155, Mat., **5**, 174–194 (1952).

BUSHAW, D.

1. Dynamical polysystems and optimization. In: Contributions to Differential Equations, vol. 2, New York: Wiley 1963.
2. A stability criterion for general systems. Math. Systems Theory **1**, 79–88 (1967).
3. Dynamical polysystems — a survey. In: Proceedings U.S.-Japan Seminar on Differential and Functional Equations, W. A. Harris, Jr. and Y. Sibuya (Editors). New York-Amsterdam: Benjamin 1967, pp. 13–26.
4. Stability of Liapunov and Poisson types. SIAM Rev. **11**, 214–225 (1969).

CAIRNS, S. S.

1. (Editor) Differential and Combinatorial Topology. Princeton: Princeton Univ. Press 1965.

CARATHEODORY, C.

1. Vorlesungen über reelle Funktionen. Leipzig: Teubner 1927. Reprint, New York: Chelsea 1958/60.
2. Calculus of Variations and Partial Differential Equations of the First Order (German). Leipzig-Berlin: Teubner 1935. English translation in two volumes, San Francisco: Holden-Day 1965/67.

CARTWRIGHT, M. L.

1. Some decomposition theorems for certain invariant continua and their minimal sets. Fund. Math. **48**, 229–250 (1960).
2. From non-linear oscillations to topological dynamics. J. London Math. Soc. **39**, 193–201 (1964).
3. Topological problems of nonlinear mechanics. Abh. Deutsch. Akad. Wiss. Berlin, Kl. Math., Phys., Tech. (1965) no. 1. — III. Konferenz über nichtlineare Schwingungen, Berlin 1964, Teil I. Berlin: Akademie-Verlag 1965, pp. 135–142.
4. Equicontinuous mappings of plane minimal sets. Proc. London Math. Soc. **14A**, 51–54 (1965).
5. Almost periodic differential equations and almost periodic flows. J. Diff. Eqs. **5**, 117–135 (1969).

CAUCHY, A. L.

1. Oeuvres complètes, vol. 1. Paris: Gauthier-Villars 1888.

CELLINA, A. (see SZEGÖ, G. P.)

CESARI, L.

1. Asymptotic Behavior and Stability Problems in Ordinary Differential Equations (Ergebnisse der Mathematik und ihrer Grenzgebiete, Neue Folge, Heft 16). Berlin-Göttingen-Heidelberg: Springer 1959.

CHAIKIN, C. B. (see ANDRONOV, A.)

CHEN, K. T.

1. Equivalence and decomposition of vector fields about an elementary critica point. Amer. J. Math. **85**, 693–722 (1963).

CHERRY, T. M.

1. Topological properties of solutions of ordinary differential equations. Amer. J. Math. **59**, 957–982 (1937).
2. Analytic quasi-periodic curves of discontinuous type on a torus. Proc. London Math. Soc. **44**, 175–215 (1938).
3. The pathology of differential equations. J. Austral. Math. Soc. **1**, 1–16 (1959).

CHETAEV, N.

1. The Stability of Motion. Moscow: GITTL 1946; 2nd ed. 1959. English translation, London: Pergamon Press 1961.
2. On stability in the sense of Poisson (Russian). Zap. Kazansk. Mat. Obshch., 1929.
3. Un théorème sur l'instabilité. C. R. (Dokl.) Acad. Sci. URSS (N.S.) **2**, 529–531 (1934).
4. On instability of the equilibrium in certain cases where the force function does not have a maximum (Russian). Uchen. Zap. Kazansk. Univ., 1938.
5. On unstable equilibrium in certain cases when the force function is not maximum (Russian). Prikl. Mat. Mekh. **16**, 89–93 (1952).
6. On certain questions relative to the problem of the stability of unsteady motions (Russian). Prikl. Mat. Mekh. **24**, 6–9 (1960). English translation: J. Appl. Math. Mech. **24**, 5–22 (1960).
7. On the stability of "rough" systems (Russian). Prikl. Mat. Mekh. **24**, 20–22 (1960). English translation: J. Appl. Math. Mech. **24**, 23–26 (1960).

CHU, HSIN

1. On totally minimal sets. Proc. Amer. Math. Soc. **13**, 457–458 (1962).
2. Algebraic topology criteria for minimal sets. Proc. Amer. Math. Soc. **13**, 503–508 (1962).
3. Fixed points in a transformation group. Pacific J. Math. **15**, 1131–1135 (1965).
4. A note on compact transformation groups with a fixed end point. Proc. Amer. Math. Soc. **16**, 581–583 (1965).

CLAY, J. P.

1. Proximity relations in transformation groups. Trans. Amer. Math. Soc. **108**, 88–96 (1963).
2. Invariant attractors in transformation groups. Illinois J. Math. **8**, 473–479 (1964).

CODDINGTON, E. A.; LEVINSON, N.

1. Theory of Ordinary Differential Equations. New York-Toronto-London: McGraw-Hill 1955.

COLEMAN, C.

1. Equivalence of planar dynamical and differential systems. J. Diff. Eqs. **1**, 222–233 (1965).

CONLEY, C. C.

1. Invariant sets in a monkey saddle. In: Proceedings U.S.-Japan Seminar on Differential and Functional Equations, W. A. Harris, Jr. and Y. Sibuya (Editors). New York-Amsterdam: Benjamin 1967, pp. 663–668.
2. Twist mappings, linking, analyticity and periodic solutions which pass close to an unstable periodic solution. In: Topological Dynamics, J. Auslander and W. H. Gottschalk (Editors). New York-Amsterdam: Benjamin 1968, pp. 129–153.
3. On the ultimate behavior of orbits with respect to an unstable critical point, I. Oscillating, asymptotic and capture orbits. J. Diff. Eqs. **5**, 136–158 (1969).

CONLEY, C.; EASTON, R.
1. Isolated invariant sets and isolating blocks (Advances in Differential and Integral Equations). SIAM, Philadelphia, **1969**.

CONTI, R. (see also REISSIG, R. and SANSONE, G.)
1. Sull'equivalenza asintotica dei sistemi di equazioni differenziali. Ann. Mat. Pura Appl. **41**, 95–104 (1965).
2. Limitazioni in ampiezza delle soluzioni di un sistema di equazioni differenziali ed applicazioni. Boll. Un. Mat. Ital. **11**, 344–349 (1956).
3. Sulla prolungabilità delle soluzioni di un sistema di equazioni differenziali ordinarie. Boll. Un. Mat. Ital. **11**, 510–514 (1956).

COOKE, K. L. (see BELLMAN, R.)

COPPEL, W. A.
1. Stability and Asymptotic Behavior of Differential Equations. Boston: Heath & Co. **1965**.

CORDUNEAU, C.
1. Sur la stabilité asymptotique. An. Sti. Univ. "Al. I. Cuza", Iaşi, Sect. I a Mat. **5**, 37–39 (1959).
2. Sur la stabilité asymptotique, II. Rev. Roumaine Math. Pures Appl. **5**, 573–576 (1960).
3. Application des inégalités différentielles à la théorie de la stabilité (Russian). An. Sti. Univ. "Al. I. Cuza", Iaşi, Sect. I a Mat. **6**, 46–58 (1960).
4. Sur certains systèmes différentielles nonlinéaires. An. Sti. Univ. "Al. I. Cuza", Iaşi, Sect. I a Mat. **6**, 257–260 (1960).

CRONIN, J.
1. Fixed Points and Topological Degree in Nonlinear Analysis (Mathematical Surveys, no. 11). Amer. Math. Soc., Providence, **1964**.

D'AMBROSIO, U.; LAKSHMIKANTHAM, V.
1. On ψ-stability and differential inequalities. In: Topological Dynamics, J. Auslander and W. H. Gottschalk (Editors). New York-Amsterdam: Benjamin 1968, pp. 155–164.

DAVIS, P. (see ANTOSIEWICZ, H. A.)

DENJOY, A.
1. Sur les caractéristiques à la surface du tore. C. R. Acad. Sci. Paris **194**, 830–833 (1932).
2. Sur les caractéristiques du tore. C. R. Acad. Sci. Paris **194**, 2014–2016 (1932).
3. Sur les courbes définies par les équations différentielles à la surface du tore. J. Math. Pures Appl. **11**, 333–375 (1932).
4. Les trajectoires à la surface du tore. C. R. Acad. Sci. Paris **223**, 5–8 (1964).
5. Sur les trajectoires du tore. C. R. Acad. Sci. Paris **251**, 175–177 (1960).

DESBROW, D.
1. On connexion, invariance and stability in certain flows. Proc. Cambridge Philos. Soc. **60**, 51–55 (1964).

DEYSACH, L. G.; SELL, G. R.
1. On the existence of almost periodic motions. Michigan Math. J. **12**, 87–95 (1965).

DILIBERTO, S. P.
1. Qualitative behavior for classical dynamical systems (Advances in Differential and Integral Equations). SIAM, Philadelphia, **1969**.

DIRICHLET, G. L.
1. Über die Stabilität des Gleichgewichts. J. reine angew. Math. **32**, 85–88 (1846).

DOWKER, Y. N.
1. On minimal sets in dynamical systems. Quart. J. Math., Oxford Ser., **7**, 5–16 (1956).

DOWKER, Y. N.; FRIEDLANDER, F. G.

1. On limit sets in dynamical systems. Proc. London Math. Soc. **4**, 168–176 (1954).

DUBOSHIN, G. N.

1. Foundations of the Theory of Stability of Motions. Moscow 1957.

DUGUNDJI, J. (see also ANTOSIEWICZ, H. A.)

1. Topology. Boston: Allyn & Bacon 1966.

EASTON, R. (see CONLEY, C.)

EDREI, A.

1. On iteration of mappings of a metric space onto itself. J. London Math. Soc. **26**, 96–103 (1951).

ELLIS, R.

1. Continuity and homeomorphism groups. Proc. Amer. Math. Soc. **4**, 969–973 (1953).
2. A note on the continuity of the inverse. Proc. Amer. Math. Soc. **8**, 372–373 (1957).
3. Locally compact transformation groups. Duke Math. J. **24**, 119–125 (1957).
4. Distal transformation groups. Pacific J. Math. **8**, 401–405 (1958).
5. Universal minimal sets. Proc. Amer. Math. Soc. **11**, 540–543 (1960).
6. A semigroup associated with a transformation group. Trans. Amer. Math. Soc. **94**, 272–281 (1960).
7. Point transitive transformation groups. Trans. Amer. Math. Soc. **101**, 384–395 (1961).
8. Locally coherent minimal sets. Michigan Math. J. **10**, 97–104 (1963).
9. Global sections of transformation groups. Illinois J. Math. **8**, 380–394 (1964).
10. The construction of minimal discrete flows. Amer. J. Math. **87**, 564–574 (1965).
11. Group-like extensions of minimal sets. Trans. Amer. Math. Soc. **127**, 125–135 (1967).
12. The beginning of an algebraic theory of minimal sets. In: Topological Dynamics, J. Auslander and W. H. Gottschalk (Editors). New York-Amsterdam: Benjamin 1968, pp. 165–184.

ELLIS, R.; GOTTSCHALK, W. H.

1. Homeomorphisms of transformation groups. Trans. Amer. Math. Soc. **94**, 258–271 (1960).

EL'SGOL'C, L. E.

1. An estimate for the number of singular points of a dynamical system defined on a manifold (Russian). Mat. Sbornik **26**, 215–223 (1950). English translation: Amer. Math. Soc. Transl. no. 68 (1952).

ENGELKING, R.

1. Quelques remarques concernant les opérations sur les fonctions sémi-continues dans les espaces topologiques. Bull. Acad. Polon. Sci., Ser. Sci. Math. Astronom. Phys. **11**, 719–726 (1963).

ENGLAND, J. W.

1. A characterization of orbits. Proc. Amer. Math. Soc. **17**, 207–209 (1966).

ENGLAND, J. W.; KENT, J. F. III

1. *P*-recurrence and quasi-minimal sets. In: Topological Dynamics, J. Auslander and W. H. Gottschalk (Editors). New York-Amsterdam: Benjamin 1968, pp. 185–204.

ERUGIN, N. P.

1. On certain questions of stability of motion and the qualitative theory of differential equations (Russian). Prikl. Mat. Mekh. **14**, 459–512 (1950).
2. A qualitative investigation of integral curves of a system of differential equations (Russian). Prikl. Mat. Mekh. **14**, 659–664 (1950).

3. Theorems on instability (Russian). Prikl. Mat. Mekh. **16**, 355–361 (1952).
4. The methods of A. M. Liapunov and questions of stability in the large (Russian). Prikl. Mat. Mekh. **17**, 389–400 (1953).
5. Qualitative methods in theory of stability (Russian). Prikl. Mat. Mekh. **19**, 599–616 (1955).

FOLAND, N. E.; UTZ, W. R.
1. The embedding of discrete flows in continuous flows. In: Ergodic Theory. Proceedings of an International Symposium held at Tulane University, New Orleans, Louisiana, October 1961. New York: Academic Press 1963, pp. 121–134.

FOMIN, S. (see also KOLMOGOROV, A. N.)
1. On dynamical systems in a space of functions (Russian). Ukrainsk. Mat. Zh. **2**, no. 2, pp. 25–134 (1950).

FRIEDLANDER, F. G. (see also DOWKER, Y. N.)
1. On the iteration of a continuous mapping of a compact space into itself. Proc. Cambridge Philos. Soc. **46**, 46–56 (1950).

FUKUHARA, M.
1. Sur les systèmes des équations différentielles ordinaires. Japan J. Math. **5**, 345–350 (1929).
2. Sur les systèmes d'équations différentielles ordinaires, II. Japan J. Math. **6**, 269–299 (1930).
3. Sur l'ensemble des courbes intégrales d'un système d'équations différentielles ordinaires. Proc. Imp. Acad. Japan **6**, 360–362 (1930).

FULLER, F. B.
1. Notes on trajectories in a solid torus. Ann. Math. **56**, 438–439 (1952).

FURSTENBERG, H.
1. The structure of distal flows. Amer. J. Math. **85**, 477–515 (1963).
2. Disjointness in ergodic theory, minimal sets and diophantine approximations. Math. Systems Theory **1**, 1–50 (1967).

GANTMACHER, F. M.
1. Theory of Matrices. New York: Chelsea 1960.

GARCIA, M.; HEDLUND, G. A.
1. The structure of minimal sets. Bull. Amer. Math. Soc. **54**, 954–964 (1948).

GARY, J.
1. The topological structure of trajectories. Michigan Math. J. **7**, 225–227 (1960).

GEISS, G. (see SZEGÖ, G. P.)

GHIZZETTI, A.
1. (Editor) Stability Problems of Solutions of Differential Equations. Proceedings of a NATO Advanced Study Institute held in Padova, Italy, August 1965. Gubbio: Oderisi 1966.

GOTTSCHALK, W. H. (see also AUSLANDER, J. and ELLIS, R.)
1. A note on pointwise nonwandering transformations. Bull. Amer. Math. Soc. **52**, 488–489 (1946).
2. Almost periodicity, equi-continuity and total boundedness. Bull. Amer. Math. Soc. **52**, 633–636 (1946).
3. Recursive properties of transformation groups, II. Bull. Amer. Math. Soc. **54**, 381–383 (1948).
4. Transitivity and equicontinuity. Bull. Amer. Math. Soc. **54**, 982–984 (1948).
5. Characterizations of almost periodic transformation groups. Proc. Amer. Math. Soc. **7**, 709–712 (1956).
6. Minimal sets: an introduction to topological dynamics. Bull. Amer. Math. Soc. **64**, 336–351 (1958).

7. The universal curve of Sierpinski is not a minimal set. Notices. Amer. Math. Soc. **6**, 257 (1959).
8. An irreversible minimal set. In: Ergodic Theory. Proceedings of an International Symposium held at Tulane University, New Orleans, Louisiana, October 1961. New York: Academic Press 1963, pp. 135–150.
9. Substitution minimal sets. Trans. Amer. Math. Soc. **109**, 467–491 (1963).
10. Minimal sets occur maximally. Trans. New York Acad. Sci. **26**, 348–353 (1964).
11. A survey of minimal sets. Ann. Inst. Fourier (Grenoble) **14**, 53–60 (1964).
12. Ambits. Invited Lecture given at the Regional Conference on Global Differentiable Dynamics held at Case-Western Reserve University, Cleveland, Ohio, June 1969.

Gottschalk, W. H.; Hedlund, G. A.
1. Recursive properties of transformation groups. Bull. Amer. Math. Soc. **52**, 637–641 (1946).
2. The dynamics of transformation groups. Trans. Amer. Math. Soc. **65**, 348–359 (1949).
3. Asymptotic relations in topological groups. Duke Math. J. **18**, 481–485 (1951).
4. Topological Dynamics (Amer. Math. Soc. Colloquium Publications, vol. 36). Providence 1955.
5. A characterization of the Morse minimal set. Proc. Amer. Math. Soc. **15**, 70–74 (1964).

Grabar, M. I.
1. The representation of dynamical systems as systems of solutions of differential equations. Dokl. Akad. Nauk SSSR **61**, 433–436 (1948).
2. Transformations of dynamical systems into systems of solutions of differential equations (Russian). Vestnik Moskov. Univ., 1952, no. 3, pp. 3–8.
3. On change of time in dynamical systems (Russian). Dokl. Akad. Nauk SSSR **109**, 250–252 (1956).
4. On a sufficient test for isomorphism of dynamical systems (Russian). Dokl. Akad. Nauk SSSR **109**, 431–433 (1956).
5. Isomorphism of dynamical systems differing only in time (Russian). Dokl. Akad. Nauk SSSR **126**, 931–934 (1959).

Green, L. W. (see also Auslander, L.)
1. Transversals to a flow. In: Topological Dynamics, J. Auslander and W. H. Gottschalk (Editors). New York-Amsterdam: Benjamin 1968, pp. 225–242.

Grobman, D. M.
1. Systems of differential equations analogous to linear ones. Dokl. Akad. Nauk SSSR **86**, 19–22 (1952).
2. Homeomorphisms of systems of differential equations. Dokl. Akad. Nauk SSSR **128**, 880–881 (1959).
3. Topological and asymptotic equivalence for systems of differential equations (Russian). Dokl. Akad. Nauk SSSR **140**, 746–747 (1961). English translation: Soviet Math. (Dokl.) **2**, 1201–1253 (1961).
4. Topological classification of the neighborhood of a singular point in n-dimensional space. Mat. Sbornik (N.S.) **56** (**98**), 77–94 (1962).
5. The topological equivalence of dynamical systems (Russian). Dokl. Akad. Nauk SSSR **175**, 1211–1212 (1967). English translation: Soviet Math. (Dokl.) **8**, 985–986 (1967).
6. Global topological equivalence of systems of differential equations. Mat. Sbornik (N.S.) **73**, 600–607 (1967).

7. The neighborhood of a singular point of a system of differential equations with small nonlinearities (Russian). Dokl. Akad. Nauk SSSR **172**, 524–526 (1967). English translation: Soviet Math. (Dokl.) **8**, 124–126 (1967).

HAAS, F.

1. A theorem about characteristics of differential equations on closed manifolds. Proc. Nat. Acad. Sci. USA **38**, 1004–1047 (1952).
2. On the global behavior of differential equations on two-dimensional manifolds. Proc. Amer. Math. Soc. **4**, 630–636 (1953).
3. The global behavior of differential equations on n-dimensional manifolds. Proc. Nat. Acad. Sci. USA **39**, 1258–1260 (1953).
4. Poincaré-Bendixson type theorems for two-dimensional manifolds different from the torus. Ann. of Math. **59**, 292–299 (1954).
5. On the total number of singular points and limit cycles of a differential equation. In: Contributions to the Theory of Nonlinear Oscillations, vol. 3 (Annals of Math. Studies, no. 36). Princeton: Princeton Univ. Press **1956**, pp. 137–172.

HADAMARD, J.

1. Sur les trajectoires en dynamiques. J. de Math., Ser. III, **3**, 331–387 (1897).
2. Sur les intégrales d'un system d'équations différentielles ordinaires, considérées comme fonctions des données initiales. Bull. Soc. Math. France **28**, 64–66 (1900).
3. Sur l'itération et les solutions asymptotiques des équations différentielles. Bull. Soc. Math. France **29**, 224–228 (1901).

HAHN, F. J. (see also AUSLANDER, J. and AUSLANDER, L.)

1. Recursion of set trajectories in a transformation group. Proc. Amer. Math. Soc. **11**, 527–532 (1960).
2. Nets and recurrence in transformation groups. Trans. Amer. Math. Soc. **99**, 193–200 (1961).
3. On affine transformations of compact Abelian groups. Amer. J. Math. **85**, 428–446 (1963); errata **86**, 463–464 (1964).

HAHN, F.; PARRY, W.

1. Minimal dynamical systems with quasi-discrete spectrum. J. London Math. Soc. **40**, 309–324 (1965).
2. Some characteristic properties of dynamical systems with quasi-discrete spectra. Math. Systems Theory **2**, 179–190 (1968).

HAHN, W.

1. Theorie und Anwendung der direkten Methode von Ljapunov (Ergebnisse der Mathematik und ihrer Grenzgebiete, Neue Folge, Heft 22). Berlin-Göttingen-Heidelberg: Springer 1959. English translation: Theory and Application of Ljapunov's Direct Method. Englewood Cliffs, N.J.: Prentice-Hall **1963**.
2. Stability of Motion. Berlin-Heidelberg-New York: Springer **1967**.
3. Über die Anwendung der Methode von Ljapunov auf Differenzengleichungen. Math. Ann. **136**, 430–441 (1958).

HAJEK, O. (see also BHATIA, N. P.)

1. Dynamical Systems in the Plane. New York: Academic Press **1968**.
2. Critical points of abstract dynamical systems. Comm. Math. Univ. Carolinae **5**, 121–124 (1964).
3. Betti numbers of regions of attraction. Comm. Math. Univ. Carolinae **5**, 129–132 (1964).
4. Structure of dynamical systems. Comm. Math. Univ. Carolinae **6**, 53–72 (1965); correction **6**, 211–212 (1965).
5. Flows and periodic motions. Comm. Math. Univ. Carolinae **6**, 165–178 (1965).

6. Sections of dynamical systems in E^2. Czechoslovak Math. J. **15**, 205–211 (1965).
7. Prolongations of sections in local dynamical systems. Czechoslovak Math. J. **16**, 41–45 (1966).
8. Differentiable representation of flows. Comm. Math. Univ. Carolinae **7**, 219–225 (1966).
9. Theory of processes, I. Czechoslovak Math. J. **17**, 159–199 (1967).
10. Theory of processes, II. Czechoslovak Math. J. **17**, 372–398 (1967).
11. Local characterization of local semi-dynamical systems. Math. Systems Theory **2**, 17–26 (1968).
12. Categorical concepts in dynamical systems theory. In: Topological Dynamics, J. Auslander and W. H. Gottschalk (Editors). New York-Amsterdam: Benjamin 1968, pp. 243–258.

HALANAY, A.

1. Differential Equations: Stability Theory, Oscillations, Time-Lags. New York: Academic Press 1965.

HALE, J. K.

1. Integral manifolds of perturbed differential systems. Ann. of Math. **73**, 496–531 (1961).
2. Dynamical systems and stability. J. Math. Anal. Appl. **26**, 39–59 (1969).

HALE, J. K.; LASALLE, J. P.

1. (Editors) Differential Equations and Dynamical Systems. Proceedings of an International Symposium held at the University of Puerto Rico, Mayaguez, Puerto Rico, Dec. 27–30, 1965. New York: Academic Press 1967.

HALE, J. K.; STOKES, A. P.

1. Behavior of solutions near integral manifolds. Arch. Rat. Mech. Anal. **6**, 133–170 (1960).

HALKIN, H.

1. Topological aspects of optimal control of dynamical polysystems. In: Contributions to Differential Equations, vol. 3. New York: Wiley 1964, pp. 377–385.
2. Finitely convex sets of nonlinear differential equations. Math. Systems Theory **1**, 51–54 (1967).

HALMOS, P. R.

1. Lectures on Ergodic Theory. Mathematical Society of Japan, Tokyo 1956. Reprint, New York: Chelsea 1960.

HARRIS, W. A., Jr.; SIBUYA, Y.

1. (Editors) Proceedings U.S.-Japan Seminar on Differential and Functional Equations, held at the University of Minnesota, Minneapolis, Minn. on June 26–30, 1967. New York-Amsterdam: Benjamin 1967.

HARTMAN, P.

1. Ordinary Differential Equations. New York: Wiley 1964.
2. On local homeomorphisms of Euclidean spaces. Bol. Soc. Mat. Mexicana **5**, 220–241 (1960).
3. On stability in the large for systems of ordinary differential equations. Canad. J. Math. **13**, 480–492 (1961).
4. On the local linearization of differential equations. Amer. Math. Monthly **70**, 219–232 (1963).
5. The existence and stability of stationary points. Duke Math. J. **33**, 2, 281–290 (1966).

HARTMAN, P.; OLECH, Cz.

1. On global asymptotic stability of solutions of ordinary differential equations. Trans. Amer. Math. Soc. **104, 154–178** (1962).

HARTMAN, P.; WINTNER, A.

1. Integrability in the large and dynamical stability. Amer. J. Math. **65, 273–278** (1943).
2. On the asymptotic behavior of the solutions of a nonlinear differential equation. Amer. J. Math. **68, 301–308** (1946).

HEDLUND, G. A. (see also GARCIA, M. and GOTTSCHALK, W. H.)

1. Sturmian minimal sets. Amer. J. Math. **66, 605–620** (1944).
2. A class of transformations of the plane. Proc. Cambridge Philos. Soc. **51, 554–564** (1955).
3. Mappings on sequence spaces (Part I). Communications Res. Div. Tech. Rep. no. 1, von Neumann Hall, Princeton, N.J., February 1961.
4. Transformations commuting with the shift. In: Topological Dynamics, J. Auslander and W. H. Gottschalk (Editors). New York-Amsterdam: Benjamin 1968, pp. 259–290.

HILMY, H.

1. Sur la structure d'ensemble des mouvements stables au sens de Poisson. Ann. of Math. **37, 43–45** (1936).
2. Sur les ensembles quasi-minimaux dans les systèmes dynamiques. Ann. of Math. **37, 899–907** (1936).
3. Sur les centres d'attraction minimaux des systèmes dynamiques. Compositio Math. **3, 227–238** (1936).
4. Sur les mouvements des systèmes dynamiques qui admettent l'incompressibilité des domaines. Amer. J. Math. **59, 803–808** (1937).
5. Sur une propriété des ensembles minimaux. C. R. (Dokl.) Acad. Sci. URSS **14, 261–262** (1937).
6. Sur la théorie des ensembles quasi-minimaux. C. R. (Dokl.) Acad. Sci. URSS **15, 113–116** (1937).
7. Sur les théorèmes de récurrence dans la dynamique générale. Amer. J. Math. **61, 149–160** (1939).

HIRASAWA, Y. (see URA, T.)

HOCKING, J.; YOUNG, G.

1. Topology. Reading, Mass.: Addison-Wesley 1961.

HOPF, E.

1. Ergodentheorie (Ergebnisse der Mathematik, vol. 5, no. 8). Berlin: Springer 1937. Reprint, New York: Chelsea 1948.

HORELICK, B.

1. An algebraic approach to the study of minimal sets in topological dynamics. Thesis, Wesleyan University, 1967.

HU, SZE-TSEN

1. Theory of Retracts. Detroit: Wayne State Univ. Press 1955.

INGWERSON, D. R.

1. A modified Liapunov method for nonlinear stability analysis. IRE Trans. Automatic Control **6, 199–210** (1961); discussion **7, 85–88** (1962).

JARNIK, J.; KURZWEIL, J.

1. On invariant sets and invariant manifolds of differential systems. J. Diff. Eqs. **6, 247–263** (1969).

JONES, G. S.

1. (Editor) Seminar on Differential Equations and Dynamical Systems (Lecture Notes in Mathematics, vol. 60). Berlin-Heidelberg-New York: Springer 1968.

2. Asymptotic fixed point theorems and periodic systems of functional differential equations. In: Contributions to Differential Equations, vol. 2. New York: Wiley 1963, pp. 385–405.
3. Stability and asymptotic fixed point theory. Proc. Nat. Acad. Sci. USA **53**, 6, 1262–1264 (1965).
4. The existence of critical points in generalized dynamical systems. In: Seminar on Differential Equations and Dynamical Systems, G. S. Jones (Editor) (Lecture Notes in Mathematics, vol. 60). Berlin-Heidelberg-New York: Springer 1968.

JONES, G. S.; YORKE, J. A.
1. The existence and nonexistence of critical points in bounded flows. J. Diff. Eqs. **6**, 238–246 (1969).

KAHN, P. J.; KNAPP, A. W.
1. Equivariant maps onto minimal flows. Math. Systems Theory **2**, 319–324 (1968).

KAKUTANI, SH.
1. A proof of Beboutov's theorem. J. Diff. Eqs. **4**, 194–201 (1968).

KALMAN, R. E.
1. Algebraic aspects of the theory of dynamical systems. In: Differential Equations and Dynamical Systems. J. K. Hale and J. P. LaSalle (Editors). New York: Academic Press 1968, pp. 133–146.
2. Introduction to the algebraic theory of linear dynamical systems. In: Mathematical Systems Theory and Economics I, H. W. Kuhn and G. P. Szegö (Editors) (Lecture Notes in Operations Research and Mathematical Economics, vol. 11). Berlin-Heidelberg-New York: Springer 1969, pp. 41–66.

KAMKE, E.
1. Differentialgleichungen reeller Funktionen. Leipzig: Akad. Verlagsges. 1930. Reprint, New York: Chelsea 1947.
2. Differentialgleichungen, Lösungsmethoden und Lösungen, I. Gewöhnliche Differentialgleichungen, 7th ed. Leipzig: Akad. Verlagsges. 1961.
3. Differentialgleichungen, Lösungsmethoden und Lösungen, II. Partielle Differentialgleichungen erster Ordnung für eine gesuchte Funktion, 4th ed. Leipzig: Akad. Verlagsges. 1959.
4. Zur Theorie der Systeme gewöhnlicher Differentialgleichungen, II. Acta Math. **58**, 57–85 (1932).

VAN KAMPEN, E. R.
1. The topological transformations of a simple closed curve into itself. Amer. J. Math. **57**, 142–152 (1935).
2. Remarks on systems of ordinary differential equations. Amer. J. Math. **59**, 144–152 (1937)

KAPLAN, W.
1. Regular curve-families filling the plane, I. Duke Math. J. **7**, 154–185 (1940).
2. Regular curve-families filling the plane, II. Duke Math. J. **8**, 11–46 (1941).
3. Differentiability of regular curve families on the sphere. In: Lectures in Topology. Ann Arbor: Univ. of Michigan Press 1941, pp. 299–301.
4. The structure of a curve-family on a surface in the neighborhood of an isolated singularity. Amer. J. Math. **64**, 1–35 (1942).
5. Dynamical systems with indeterminacy. Amer. J. Math. **72**, 573–594 (1950).
6. Analytic ordinary differential equations in the large. In: Proceedings U.S.-Japan Seminar on Differential and Functional Equations, W. A. Harris, Jr. and Y. Sibuya (Editors). New York-Amsterdam: Benjamin 1967, pp. 133–151.

Kato, J.
1. The asymptotic behaviour of the solutions of differential equations or the product space. Arch. Rat. Mech. Anal. **6**, 133–170 (1960).
2. The asymptotic relation of two systems of ordinary differential equations. In: Contributions to Differential Equations, vol. 3. New York: Wiley **1964**, pp. 141–161.
3. Asymptotic equivalences between systems of differential equations and their perturbed systems. Funkcial. Ekvac. **8**, 45–78 (1966).
4. Asymptotic equivalence. In: Seminar on Differential Equations and Dynamical Systems, G. S. Jones (Editor) (Lecture Notes in Mathematics, vol. 60). Berlin-Heidelberg-New York: Springer **1968**, pp. 27–32.
5. A remark on the result of Strauss. In: Seminar on Differential Equations and Dynamical Systems, G. S. Jones (Editor) (Lecture Notes in Mathematics, vol. 60). Berlin-Heidelberg-New York: Springer **1968**, pp. 89–98.

Kayande, A. A.; Lakshmikantham, V.
1. Conditionally invariant sets and vector Liapunov functions. J. Math. Anal. Appl. **14**, 285–293 (1966).

Kelley, J. L.
1. General Topology. New York: Van Nostrand **1955**.

Kent, J. F. III (see England, J. W.)

Keynes, H.
1. The proximal relation in a class of substitution minimal sets. Math. Systems Theory **1**, 165–176 (1967).

Kimura, I. (see Ura, T.)

Knapp, A. W. (see Kahn, P. J.)

Kneser, A.
1. Studien über die Bewegungsvorgänge in der Umgebung instabiler Gleichgewichtslagen, I, II. J. reine angew. Math. **115**, 308–327 (1895); **118**, 186–223 (1897).

Kneser, H.
1. Über die Lösungen eines Systemes gewöhnlicher Differentialgleichungen, das der Lipschitzschen Bedingung nicht genügt. Sitzungsber. Preuss. Akad. Wiss., Phys.-Math. Kl., **1923**, pp. 171–174.
2. Reguläre Kurvenscharen auf den Ringflächen. Math. Ann. **91**, 135–154 (1924).

Kolmogorov, A. N.; Fomin, S. V.
1. Elements of the Theory of Functions and Functional Analysis. Rochester, N.Y.: Graylock **1957**.

Krasovskii, N. N. (see also Barbashin, E. A.)
1. Certain Problems of the Theory of Stability of Motion (Russian). Moscow: Gos. Izd. Fiz.-mat. Lit. **1959**. English translation, Stability of Motions. Stanford: Stanford Univ. Press **1963**.
2. On a problem of stability of motion in the large (Russian). Dokl. Akad. Nauk SSSR **88**, 401–404 (1953).
3. On stability of motion in the large for constantly acting disturbances (Russian). Prikl. Mat. Mekh. **18**, 95–102 (1954).
4. On the inversion of theorems of A. M. Liapunov and N. G. Chetaev on instability for stationary systems of differential equations (Russian). Prikl. Mat. Mekh. **18**, 513–532 (1954).
5. On stability in the large of the solutions of a nonlinear system of differential equations (Russian). Prikl. Mat. Mekh. **18**, 735–737 (1954).
6. Sufficient conditions for stability of solutions of a system of nonlinear differential equations (Russian). Dokl. Akad. Nauk SSSR **98**, 901–904 (1954).

7. On the converse of K. P. Persidskii's theorem on uniform stability (Russian). Prikl. Mat. Mekh. **19**, 273—278 (1955).
8. On conditions of inversion of A. M. Liapunov's theorems on instability for stationary systems of differential equations (Russian). Dokl. Akad. Nauk SSSR **101**, 17—20 (1955).
9. Converse of theorems on Liapunov's second method and questions of stability of motion in the first approximation (Russian). Prikl. Mat. Mekh. **20**, 255—265 (1956).
10. On the converse of theorems of the second method of A. M. Liapunov for investigation of stability of motion (Russian). Uspehi Mat. Nauk **9**, no. 3, pp. 159—164 (1956).
11. On the theory of the second method of A. M. Liapunov for the investigation of stability (Russian). Mat. Sbornik **40**, 57—64 (1956).
12. On stability with large initial perturbations (Russian). Prikl. Mat. Mekh. **21**, 309—319 (1957).

Krecu, V. I. (see Sibirskii, K. S.)

Kreil, K. A.

1. Das qualitative Verhalten der Integralkurven einer gewöhnlichen Differentialgleichung erster Ordnung in der Umgebung eines singulären Punktes. Jber. Deutsch. Math. Verein **57**, 111 (1955).

Krein, M. G.

1. On some questions related to the ideas of Liapunov in the theory of stability. Uspehi Mat. Nauk (N.S.) **3**, no. 3, pp. 166—169 (1948).

Kudaev, M. B.

1. The use of Liapunov functions for investigating the behavior of trajectories of systems of differential equations (Russian). Dokl. Akad. Nauk SSSR **147**, 1285—1287 (1962). English translation: Soviet Math. (Dokl.) pp. 1802—1804.
2. Classification of higher-dimensional systems of ordinary differential equations by the method of Liapunov functions. Differencial'nye Uravnenija **1**, 346—356 (1965). English translation: Diff. Eqs. pp. 263—269.
3. Liapunov function for the region of influence of a single singular point of higher order. Vestnik Moskov. Univ., 1965, no. 1, pp. 3—13.

Kuhn, H. W.; Szegö, G. P.

1. (Editors) Mathematical Systems Theory and Economics, I and II. Proceedings of an International Summer School held in Varenna, Italy, June 1—12, 1967 (Lecture Notes in Operations Research and Mathematical Economics, vol. 11 and 12). Berlin-Heidelberg-New York: Springer 1969.

Kurzweil, J. (see also Jarnik, J.)

1. On the reversibility of the first theorem of Liapunov concerning the stability of motion (Czech). Czechoslovak Math. J. **5**, 382—398 (1955).
2. The converse second Liapunov's theorem concerning the stability of motion (Czech). Czechoslovak Math. J. **6**, 217—259, 455—473 (1956). English translation: Amer. Math. Soc. Transl., Ser. II, **24**, 19—77 (1963).
3. Invariant manifolds for flows. In: Differential Equations and Dynamical Systems, J. K. Hale and J. P. LaSalle (Editors). New York: Academic Press 1967, pp. 431—468.
4. Invariant manifolds of differential systems. Z. angew. Math. Mech. **49**, 11—14 (1969).

Kurzweil, J.; Vrkoc, I.

1. The converse theorems of Liapunov and Persidskii concerning the stability of motion (Czech). Czechoslovak Math. J. **7**, 254—274 (1957). English translation: Amer. Math. Soc. Transl., Ser. II, **29**, 271—288 (1963).

LAGRANGE, J. L.
1. Mécanique analytique. Paris: Desaint 1788.
2. Oeuvres, vol. I (1867) and IV (1869). Paris: Gauthier-Villars.

LAKSHMIKANTHAM, V. (see also D'AMBROSIO, U. and KAYANDE, A. A.)
1. On the boundedness of solutions of nonlinear differential equations. Proc. Amer. Math. Soc. **8**, 1044–1048 (1957).
2. Upper and lower bounds of the norm of solutions of differential equations. Proc. Amer. Math. Soc. **13**, 615–616 (1962); **14**, 509–513 (1963).
3. Notes on a variety of problems of differential systems. Arch. Rat. Mech. Anal. **10**, 305–310 (1962).
4. Vector Liapunov functions and conditional stability. J. Math. Anal. Appl. **10**, 368–377 (1965).

LAKSHMIKANTHAM, V.; LEELA, S.
1. Differential and Integral Inequalities, I, II. New York-London: Academic Press 1969.

LASALLE, J. P. (see also HALE, J. K.)
1. A study of synchronous asymptotic stability. Ann. of Math. **65**, 571–581 (1957).
2. Asymptotic stability criteria. Proc. Symp. Appl. Math. **13**, 299–307 (1962).
3. Some extensions of Lyapunov's second method. IRE Trans. Circuit Theory **7**, 520–527 (1960).
4. The extent of asymptotical stability. Proc. Nat. Acad. Sci. USA **46**, 363–365 (1960).
5. Recent advances in Liapunov stability theory. SIAM Rev. **6**, 1–11 (1964).
6. Liapunov's second method. In: Stability Problems of Solutions of Differential Equations, A. Ghizzetti (Editor). Gubbio: Oderisi 1966.
7. An invariance principle in the theory of stability, differential equations and dynamical systems. In: Differential Equations and Dynamical Systems, J. K. Hale and J. P. LaSalle (Editors). New York: Academic Press 1967, pp. 277–286.
8. Stability theory for ordinary differential equations. J. Diff. Eqs. **4**, 57–65 (1968).

LASALLE, J. P.; LEFSCHETZ, S.
1. Stability by Liapunov's Direct Method with Applications. New York: Academic Press 1961.
2. (Editors) Nonlinear Differential Equations and Nonlinear Mechanics. Proceedings of an International Symposium held at the U.S. Air Force Academy, Colorado Springs, August 1961. New York: Academic Press 1963.

LAZER, A. C. (see BHATIA, N. P.)

LEBEDEV, A. A.
1. The problem of stability in a finite interval of time. Prikl. Mat. Mekh. **18**, 75–94 (1954).
2. On stability of motion during a given interval of time. Prikl. Mat. Mekh. **18**, 139–148 (1954).

LEE, E. B.; MARKUS, L.
1. Foundations of Optimal Control Theory. New York: Wiley 1967.

LEFSCHETZ, S. (see also LASALLE, J. P.)
1. Differential Equations: Geometric Theory. New York: Interscience-Wiley, 1st ed. 1957, 2nd ed. 1962.
2. Liapunov and stability in dynamical systems. Bol. Soc. Mat. Mexicana 3, 25–39 (1958).
3. Geometric differential equations: recent past and proximate future. In: Differential Equations and Dynamical Systems, J. K. Hale and J. P. LaSalle (Editors). New York: Academic Press 1967, pp. 1–14.

Leighton, W. (see also Bhatia, N. P.)
1. Morse theory and Liapunov functions. Rend. Circ. Mat. Palermo, Ser. II, **13**, 1–10 (1966).

Levin, J. J.
1. On the global asymptotic behavior of nonlinear systems of differential equations. Arch. Rat. Mech. Anal. **6**, 65–74 (1960).

Levinson, N. (see Coddington, E. A.)

Lewis, D. C., Jr.
1. Metric properties of differential equations. Amer. J. Math. **71**, 294–312 (1949).
2. Differential equations referred to a variable metric. Amer. J. Math. **73**, 48–58 (1951).
3. Reversible transformations. Pacific J. Math. **11**, 1077–1087 (1961).

Liapunov, A. M.
1. Problème général de la stabilité du mouvement (Annals of Math. Studies, no. 17). Princeton: Princeton Univ. Press 1947 [reproduction of the French translation, Ann. de la Faculté des Sciences de Toulouse **9**, 203–474 (1907); of a Russian memoir, Obshchaya Zadacha Ustoichivosti Dvizheniya, Kharkov, 1892, and of a note, Comm. Soc. Math. Kharkov 3, 265–272 (1893)]. English translation, Stability of Motion. New York: Academic Press 1966.

Lipschitz, R.
1. Sur la possibilité d'intégrer complètement un système donné d'équations différentielles. Bull. Sci. Math. Astron. **10**, 149–159 (1876).

Lojasiewicz, S.
1. Sur l'allure asymptotique des intégrales du système d'équations différentielles au voisinage du point singulier. Ann. Polon. Math. **1**, 34–72 (1954).

Lovingood, J. A.
1. A special class of dynamical polysystems. J. Diff. Eqs. **6**, 326–336 (1969).

Malkin, I. G.
1. Theorie der Stabilität einer Bewegung. München: Oldenbourg 1959 (German translation of the Russian original published in 1952). (A very poor English translation of this book is: Theory of Stability of Motion. Atomic Energy Commission, Transl. no. 3352, Dept. of Commerce, Washington, D.C., 1958.)
2. Das Existenzproblem von Ljapunovschen Funktionen. Izv. fiz.-mat. Obshch. Kazan III, **4**, 51–62 (1930); **5**, 63–84 (1931).
3. Certain questions in the theory of stability of motion in the sense of Liapunov. Amer. Math. Soc. Transl. no. 20, 1950.
4. Verallgemeinerung des Fundamentalsatzes von Liapunoff über die Stabilität der Bewegungen. C. R. (Dokl.) Acad. Sci. URSS **18**, 162–164 (1938).
5. On the stability of motion in the sense of Liapounov (Russian). Mat. Sbornik **3**, 47–100 (1938).
6. Sur un théorème d'existence de Poincaré-Liapounoff. C. R. (Dokl.) Acad. Sci. URSS **27**, 307–310 (1940).
7. Basic theorems of the theory of stability of motion (Russian). Prikl. Mat. Mekh. **6**, 411–448 (1942).
8. Stability in the case of constantly acting disturbances (Russian). Prikl. Mat. Mekh. **8**, 241–245 (1944).

Marchaud, M. A.
1. Sur les champs des demi-droites et les équations différentielles du premier ordre. Bull. Soc. Math. France **63**, 1–38 (1934).
2. Sur les champs continus de demi-cônes convexes et leur intégrales. Compositio Math. **3**, 89–127 (1936).

MARKOV, A. A.

1. Sur une propriété générale des ensembles minimaux de M. Birkhoff. C. R. Acad. Sci. Paris **193**, 823–825 (1931).
2. On a general property of minimal sets (Russian). Rusk. Astron. Zh. 1932.
3. Stabilität im Liapunoffschen Sinne und Fastperiodizität. Math. Z. **36**, 708–738 (1933).
4. Almost periodicity and harmonizability (Russian). L. Trudy Vtorogo Vsesoyuzn. Mat. S'ezda **2**, 227–231 (1936).

MARKUS, L. (see also AEPPLI, A. and LEE, E. B.)

1. Escape times for ordinary differential equations. Rend. Sem. Mat. Politec. Torino **11**, 271–277 (1952).
2. On completeness of invariant measures defined by differential equations. J. Math. Pure Appl. **31**, 341–353 (1952).
3. Invariant measures defined by differential equations. Proc. Ann. Math. Soc. **4**, 89–91 (1953).
4. A topological theory for ordinary differential equations in the plane. Colloque de Topol. et Geom. Diff., Strasbourg 1952.
5. Global structure of ordinary differential equations in the plane. Trans. Amer. Math. Soc. **76**, 127–148 (1954).
6. Asymptotically autonomous differential systems. In: Contributions to the Theory of Nonlinear Oscillations, vol. 3 (Annals of Math. Studies, no. 36). Princeton: Princeton Univ. Press 1956, pp. 17–30.
7. Structurally stable differential systems. Ann. of Math. **73**, 1–19 (1961).
8. Periodic solutions and invariant sets of structurally stable differential systems. In: Symposium Internacional de Ecuaciones Diferenciales Ordinarias, Mexico, 1961, pp. 190–194.
9. The global theory of ordinary differential equations. Lecture Notes, University of Minnesota, 1964/65.
10. Dynamical systems on group manifolds. In: Differential Equations and Dynamical Systems, J. K. Hale and J. P. LaSalle (Editors). New York: Academic Press 1967.
11. Parallel dynamical systems. Topology **8**, 47–57 (1969).

MARKUS, L.; YAMABE, H.

1. Global stability criteria for differential systems. Osaka Math. J. **12**, 305–317 (1960).

MARTIN, M.

1. A problem in arrangements. Bull. Amer. Math. Soc. **40**, 859–864 (1934).

MASSERA, J. L.

1. On Liapunoff's condition of stability. Ann. of Math. **50**, 705–721 (1949).
2. Contributions to stability theory. Ann. of Math. **64**, 182–206 (1956); correction **68**, 202 (1958).
3. On the existence of Liapunov functions. Publ. Inst. Mat. Estad. Montevideo **3**, no. 4, pp. 111–124 (1960).
4. Converse theorems of Liapunov's second method. In: Symposium Internacional de Ecuaciones Diferenciales Ordinarias, Mexico, 1961, pp. 158–163.

MASSERA, J. L.; SCHÄFFER, J. J.

1. Linear Differential Equations and Function Spaces. New York: Academic Press 1966.

MATROSOV, V. M.

1. On the stability of motion. Prikl. Mat. Mekh. **26**, 885–895 (1962). English translation: J. Appl. Math. Mech. **26**, 1337–1353 (1962).

2. On the theory of stability of motion. Prikl. Mat. Mekh. **26**, 992–1002 (1962). English translation: J. Appl. Math. Mech. **26**, 1560–1522 (1962).

MENDELSON, P.

1. On Lagrange stable motions in the neighborhood of critical points. In: Contributions to the Theory of Nonlinear Oscillations, vol. 5 (Annals of Math. Studies, no. 45). Princeton: Princeton Univ. Press **1960**, pp. 219–224.
2. On unstable attractors. Bol. Soc. Mat. Mexicana **5**, 270–276 (1960).

MICHAEL, E.

1. Topologies on spaces of subsets. Trans. Amer. Math. Soc. **71**, 152–182 (1951).

MILLER, R. K.

1. On almost periodic differential equations. Bull. Amer. Math. Soc. **70**, 792–795 (1964).
2. Asymptotic behavior of solutions of nonlinear differential equations. Trans. Amer. Math. Soc. **115**, 400–416 (1965).
3. Almost periodic differential equations as dynamical systems with applications to the existence of a.p. solutions. J. Diff. Eqs. **1**, 337–345 (1965).
4. The topological dynamics of Volterra integral equations (Advances in Differential and Integral Equations). SIAM, Philadelphia, **1969**.

MILLER, R. K.; SELL, G. R.

1. A note on Volterra integral equations and topological dynamics. Bull. Amer. Math. Soc. **74**, 804–809 (1968).

MILNOR, J.

1. Morse Theory (Annals of Math. Studies, no. 51). Princeton: Princeton Univ. Press **1963**.
2. Sommes de variétés différentiables et structures différentiables des sphères. Bull. Soc. Math. France **87**, 439–444 (1959).
3. Differential Topology (Lectures on Modern Mathematics, vol. II), T. L. Saaty (Editor). New York: Wiley **1964**.

MINKEVICH, M. I.

1. The theory of integral funnels in generalized dynamical systems without a hypothesis of uniqueness (Russian). Dokl. Akad. Nauk SSSR **59**, 1049–1052 (1948).
2. Closed integral funnels in generalized dynamical systems without a hypothesis of uniqueness (Russian). Dokl. Akad. Nauk SSSR **60**, 341–343 (1948).
3. Theory of integral funnels in dynamical systems without uniqueness (Russian). Uchen. Zap. Moskov. Univ., no. 135, Mat. **2**, 134–151 (1948).
4. Closed integral funnels in generalized dynamical systems without a hypothesis of uniqueness (Russian). Uchen. Zap. Moskov. Univ., no. 6, Mat. **163**, 73–88 (1952).

MINORSKY, N.

1. Nonlinear Oscillations. Princeton: Van Nostrand **1962**.

MIRANDA, C.

1. Un'osservazione su un teorema di Brouwer. Boll. Un. Mat. Ital. **3**, 5–7 (1940).

MOISEEV, N. D.

1. Summary of the History of Stability (Russian). Moscow **1949**.

MONTEL, P.

1. Sur les suites infinies de fonctions. Ann. Sci. École Norm. Sup., Ser. III, **24**, 233–234 (1907).
2. Sur l'intégral supérieure et l'intégrale inférieure d'une équation différentielle. Bull. Soc. Math. France **50**, 205–217 (1926).

MONTGOMERY, D.; ZIPPIN, L.

1. Topological Transformation Groups. New York: Interscience-Wiley **1955**.

MORSE, M.
1. A one-to-one representation of geodesics on a surface of negative curvature. Amer. J. Math. **43**, 35–51 (1921).
2. Recurrent geodesics on a surface of negative curvature. Trans. Amer. Math. Soc. **22**, 84–100 (1921).
3. Relations between the critical points of a real function of n independent variables. Trans. Amer. Math. Soc. **27**, 345–396 (1925).
4. Symbolic Dynamics. Lecture Notes, Dept. of Mathematics, Princeton University, 1966.

MOSER, J.
1. On a theorem of Anosov. J. Diff. Eqs. **5**, 411–440 (1969).

MOSTERT, P. S.
1. One-parameter transformation groups in the plane. Proc. Amer. Math. Soc. **9**, 462–463 (1958).

MÜLLER, M.
1. Über das Fundamentaltheorem in der Theorie der gewöhnlichen Differentialgleichungen. Math. Z. **26**, 619–645 (1927).
2. Beweis eines Satzes des Herrn H. Kneser über die Gesamtheit der Lösungen, die ein System gewöhnlicher Differentialgleichungen durch einen Punkt schickt. Math. Z. **28**, 349–355 (1928).
3. Neuere Untersuchung über den Fundamentalsatz in der Theorie der gewöhnlichen Differentialgleichungen. Jber. Deutsch. Math. Verein. **37**, 33–48 (1928).

MYSHKIS, A. D.
1. Generalizations of the theorem on a fixed point of a dynamical system inside of a closed trajectory (Russian). Mat. Sbornik **34**, 525–540 (1954).

NAGUMO, M.
1. Eine hinreichende Bedingung für die Unität der Lösung von Differentialgleichungen erster Ordnung. Japan J. Math. **3**, 107–112 (1926).
2. Un théorème relatif à l'ensemble des courbes intégrales d'un système d'équations différentielles ordinaires. Proc. Phys. Math. Soc. Japan, III, **12**, 233–239 (1930).

NAGY, J.
1. Lyapunov's direct method in abstract local semi-flows. Comm. Math. Univ. Carolinae **8**, 257–266 (1957).
2. Stability in continuous local semi-flows. Casopis Pest. Mat. **93**, 8–21 (1968).
3. Stability of sets with respect to abstract processes. In: Mathematical Systems Theory and Economics II, H. W. Kuhn and G. P. Szegö (Editors) (Lecture Notes in Operations Research and Mathematical Economics, vol. 12). Berlin-Heidelberg-New York: Springer 1969, pp. 355–378.

NEMYTSKII, V. V.
1. Sur les systèmes dynamiques instables. C. R. Acad. Sci. Paris **199**, 19–20 (1934).
2. Über vollständig unstabile dynamische Systeme. Ann. Mat. Pura Appl. **14**, 275–286 (1936).
3. Sur les systèmes de courbes remplissant un espace métrique. C. R. (Dokl.) Acad. Sci. URSS **21**, 99–102 (1938).
4. Sur les systèmes de courbes remplissant un espace métrique (Généralisation des théorèmes de Birkhoff). Rec. Math. (Mat. Sbornik) **6**, 283–292 (1939).
5. Systèmes dynamiques sur une multiplicité intégrale limité. C. R. (Dokl.) Acad. Sci. URSS **36**, 535–538 (1945).

6. Sur les familles de courbes du type de Bendixson. C. R. (Dokl.) Acad. Sci. URSS **21**, 103–105 (1938).
7. Les systèmes dynamiques généraux. C. R. (Dokl.) Acad. Sci. URSS **53**, 491–494 (1946).
8. On the theory of orbits of general dynamical systems (Russian). Mat. Sbornik **23**, 161–186 (1948).
9. The structure of one-dimensional limiting integral manifolds in the plane and three-dimensional space (Russian). Vestnik Moskov. Univ., no. 10, 1948, pp. 49–61.
10. Topological problems of the theory of dynamical systems (Russian). Uspehi Mat. Nauk **4**, 91–153 (1949). English translation: Amer. Math. Soc. Transl. no. 103, p. 85 (1954).
11. Generalizations of the theory of dynamical systems (Russian). Uspehi Mat. Nauk **5**, 47–49 (1950).
12. Some general theorems on the distribution of integral curves in the plane. Vestnik Moskov. Univ., Ser. I Mat. Mekh. no. 6, pp. 3–10 (1960).
13. Sur une classe importante des systèmes caractéristiques sur le plan. Ann. Mat. Pura Appl. **49**, 11–24 (1960).
14. Some modern problems in the qualitative theory of ordinary differential equations. Russian Math. Survey, 1965, pp. 1–34.
15. Topological classification of singular points and generalized Liapunov functions. Differencial'nye Uravnenija **3**, 359–370 (1967).

Nemytskii, V. V.; Stepanov, V. V.

1. Qualitative Theory of Differential Equations (Russian). Moscow-Leningrad, 1st ed. 1947, 2nd ed. 1949. English translation, Princeton: Princeton Univ. Press 1960.

Okamura, H.

1. Condition nécessaire et suffisante remplie par les équations différentielles ordinaires sans points de Peano. Mem. Coll. Sci. Kyoto Imp. Univ., Ser. A, **24**, 21–28 (1942).

Olech, Cz. (see also Hartman, P. and Szegö, G. P.)

1. On the asymptotic behavior of the solutions of a system of ordinary nonlinear differential equations. Bull. Acad. Polon. Sci., Cl. III, **4**, 555–561 (1956).
2. Remarks concerning criteria for uniqueness of solutions of ordinary differential equations, Bull. Acad. Polon. Sci., Cl. III, **8**, 661–666 (1960).
3. On the global stability of autonomous systems in the plane. In: Contributions to Differential Equations, vol. 1. New York: Wiley 1963, pp. 389–400.

Onuchic, N. (see also Hartman, P.)

1. Applications of the topological method of Ważewski to certain problems of asymptotic behavior in ordinary differential equations. Pacific J. Math. **11**, 1511–1527 (1961).
2. Relationships among the solutions of two systems of ordinary differential equations. Michigan J. Math. **10**, 129–139 (1963).

Opial, Z.

1. Sur l'allure asymptotique des solutions de certaines équations différentielles de la mécanique nonlinéaire. Ann. Polon. Math. **8**, 105–124 (1960).
2. Sur la dépendence des solutions d'un système d'équations différentielles de leurs seconds membres. Ann. Polon. Math. **8**, 75–89 (1960).

Osgood, W.

1. Beweis der Existenz einer Lösung der Differentialgleichung $dy/dx = f(x, y)$ ohne Hinzunahme der Cauchy-Lipschitzschen Bedingung. Monatsh. Math. Phys. **9**, 331–345 (1898).

OXTOBY, J. C.

1. Stepanoff flows on the torus. Proc. Amer. Math. Soc. **4**, 982–987 (1953).

PAPUSH, N. P.

1. A study of the distribution of integral curves occupying a domain containing one singular point. Mat. Sbornik **38**, 337 (1956).

PARRY, W. (see HAHN, F.)

PEANO, G.

1. Sull'integrabilità delle equazioni differenziali di primo ordine. Att. R. Accad. Torino **21**, 677–685 (1885/86).
2. Démonstration de l'intégrabilité des équations différentielles ordinaires. Math. Ann. **37**, 182–228 (1890).

PEIXOTO, M. C.; PEIXOTO, M. M.

1. Structural stability in the plane with enlarged boundary conditions. An. Acad. Brasil Ci. **31**, 135–160 (1959).

PEIXOTO, M. M. (see also PEIXOTO, M. C.)

1. On structural stability. Math. **69**, 199–222 (1959).
2. Some examples on n-dimensional structural stability. Proc. Nat. Acad. Sci. USA **45**, 633–636 (1959).
3. Structural stability on two-dimensional manifolds. In: Symposium Internacional de Ecuaciones Diferenciales Ordinarias, Mexico, 1961, pp. 188–189.
4. Structural stability on two-dimensional manifolds. Topology **1**, 101–120 (1962).
5. Qualitative theory of differential equations and structure stability. In: Differential Equations and Dynamical Systems, J. K. Hale and J. P. LaSalle (Editors). New York: Academic Press 1967, pp. 469–480.

PEROV, A. I.

1. Investigation of the neighborhood of a critical point of a multi-dimensional differential equation in the analytic case (Russian). Dokl. Akad. Nauk SSSR **166**, 544–547 (1966). English translation: Soviet Math. (Dokl.) **7**, 122–125 (1966).
2. On the structure of a limit set (Russian). Dokl. Akad. Nauk SSSR **176**, 526–529 (1967). English translation: Soviet Math. (Dokl.) **8**, 1142–1145 (1967).

PERRON, O.

1. Über Ein- und Mehrdeutigkeit des Integrales eines Systems von Differentialgleichungen. Math. Ann. **95**, 98–101 (1926).
2. Über Existenz und Nichtexistenz von Integralen partieller Differentialgleichungssysteme im reellen Gebiet. Math. Z. **27**, 549–564 (1928).
3. Eine hinreichende Bedingung für die Unität der Lösung von Differentialgleichungen erster Ordnung. Math. Z. **28**, 216–219 (1928).
4. Über Stabilität und asymptotisches Verhalten der Integrale von Differentialgleichungssystemen. Math. Z. **29**, 129–160 (1928).
5. Über Stabilität und asymptotisches Verhalten der Lösungen eines Systems endlicher Differenzengleichungen. J. reine angew. Math. **161**, 41–61 (1929).
6. Die Stabilitätsfrage bei Differentialgleichungen. Math. Z. **32**, 703–728 (1930).

PERSIDSKII, K. P.

1. Au sujet du problème de stabilité. Bull. Soc. phys.-math. Kazan III, **5**, no. 3, pp. 56–62 (1931).
2. Un théorème sur la stabilité du mouvement. Bull. Soc. phys.-math. Kazan III, **6**, pp. 76–79 (1934).
3. On the stability theory of the solutions of systems of differential equations (Russian). Bull. Soc. phys.-math. Kazan III, **8** (1936).
4. On a theorem of Liapunov (Russian). C. R. (Dokl.) Acad. Sci. URSS **14**, 541–544 (1937).

5. On the theory of stability of solutions of differential equations (Russian). Thesis, Moscow **1946**. Summary: Uspehi Mat. Nauk **1**, no. 1, 5–6, pp. 250–255 (1946).
6. On the stability of the solutions of an infinite system of equations (Russian). Prikl. Mat. Mekh. **12**, 597–612 (1948).
7. On stability of solutions of a system of countably many differential equations (Russian). Izv. Akad. Nauk Kazach. SSR **56**, Ser. Mat. Mekh., no. 2, pp. 3–35 (1948).
8. Countable systems of differential equations and the stability of their solutions (Russian). Uchen. Zap. Kazach. Univ. Mat. Fiz. no. 2 (1949).
9. On stability of solutions of differential equations (Russian). Izv. Akad. Nauk Kazach. SSR **60**, Ser. Mat. Mekh., no. 4, pp. 3–18 (1950).
10. On Liapunov's second method in linear normed spaces (Russian). Vestnik Akad. Nauk Kazach. SSR, no. 7, pp. 89–97 (1958).
11. Inversion of Liapunov's second theorem on instability in linear normed spaces (Russian). Vestnik Akad. Nauk Kazach. SSR, no. 10, pp. 31–35 (1959).

Persidskii, S. K.
1. On the second method of Liapunov (Russian). Izv. Akad. Nauk Kazach. SSR, no. 4, pp. 43–47 (1956).
2. On stability in a finite interval (Russian). Vestnik Akad. Nauk Kazach. SSR, no. 9, pp. 75–80 (1959).
3. Some theorems on the second method of Liapunov (Russian). Vestnik Akad. Nauk Kazach. SSR, no. 2, pp. 70–76 (1960).
4. On Liapunov's second method (Russian). Prikl. Mat. Mekh. **25**, 17–23 (1961). English translation: J. Appl. Math. Mech. **25**, 20–28 (1961).

Petrovskii, I. G.
1. Lectures on the theory of ordinary differential equations (Russian). 5th augmented ed. Moscow: Izd. "Nauka" 1964, p. 272. English translation, Englewood Cliffs, N.J.: Prentice Hall 1966.

Picard, E.
1. Leçons sur quelques équations fonctionnelles. Paris: Gauthier-Villars 1928.

Pliś, A.
1. On a topological method for studying the behavior of the integrals of ordinary differential equations. Bull. Acad. Polon. Sci., Cl. III, **2**, 415–418 (1954).
2. Characteristics of nonlinear partial differential equations. Bull. Acad. Polon. Sci., Cl. III, **2**, 419–422 (1954).
3. Sets filled by asymptotic integrals of ordinary differential equations. Bull. Acad. Polon. Sci., Cl. III, **4**, 749–752 (1956).
4. On sets filled by asymptotic solutions of differential equations. Ann. Inst. Fourier (Grenoble) **14**, 191–194 (1966).

Pliss, V. A.
1. Certain Problems of the Theory of Stability of Motion in the Whole. Leningrad: Izd. Leningradsk. Univ. 1958.

Poincaré, H.
1. Oeuvres. Paris: Gauthier-Villars 1929.
2. Les méthodes nouvelles de la mécanique céleste. Paris: Gauthier-Villars 1892/99. Reprint, New York: Dover 1960.
3. Mémoire sur les courbes définies par une équation différentielle. J. de Math. **7**, 375–422 (1881); **8**, 251–296 (1882); **11**, 187–244 (1885); **12**, 151–217 (1886).

Pugh, C. C.
1. Cross-sections of solution funnels. Bull. Amer. Math. Soc. **70**, 580–583 (1964).
2. The closing lemma and structural stability. Bull. Amer. Math. Soc. **70**, 584–587 (1964).

PUTNAM, C. R.

1. Unilateral stability and almost periodicity. J. Math. Mech. **9**, 915–917 (1960).

REEB, G.

1. Sur certaines propriétés topologiques des trajectoires des systèmes dynamiques. Acad. Roy. Belg., Cl. Sci. Mem. **27**, no. 9 (1952).
2. Sur la théorie générale des systèmes dynamiques. Ann. Inst. Fourier (Grenoble) **6**, 89–115 (1955/56).
3. Sur certains problèmes de topologie algébrique et de topologie générale en dynamique. Bol. Soc. Mat. Mexicana **5**, 199–202 (1960).

REISSIG, R.

1. Kriterien für die Zugehörigkeit dynamischer Systeme zur Klasse *D*. Math. Nachr. **20**, 67–72 (1959).
2. Stabilitätsprobleme in der qualitativen Theorie der Differentialgleichungen. Jber. Deutsch. Math. Verein. **63**, 97–116 (1960).
3. Neue Probleme und Methoden aus der qualitativen Theorie der Differentialgleichungen. Monatsber. Deutsche Akad. Wiss. Berlin **2**, 1–8 (1960).

REISSIG, R.; SANSONE, G.; CONTI, R.

1. Qualitative Theorie nichtlinearer Differentialgleichungen. Pubblicazioni dell'Istituto Nazionale di Alta Matematica. Roma: Cremonese 1963.

REMAGE, R., Jr.

1. On minimal sets in the plane. Proc. Amer. Math. Soc. **13**, 41–47 (1962).

ROHLIN, V. A.

1. Selected topics from the metric theory of dynamical systems (Russian). Uspehi Mat. Nauk **9**, 57–128 (1949). English translation: Amer. Math. Soc. Transl. **49**, 171–240 (1965).
2. New progress in the theory of transformations with invariant measure (Russian). Uspehi Mat. Nauk **15**, 3–26 (1960). English translation: Russian Math. Surveys **15**, 1–22 (1960).
3. Lectures on the theory of entropy of transformations with invariant measure (Russian). Uspehi Mat. Nauk **22**, 3–56 (1967). English translation: Russian Math. Surveys **22**, 1–52 (1967).

ROOS, B. W.

1. Note on generalized dynamical systems. SIAM Rev. **6**, 269–274 (1964).

ROSEAU, M.

1. Vibrations nonlinéaires et théorie de la stabilité (Springer Tracts in Nat. Phil., vol. 8). Berlin-Heidelberg-New York: Springer 1966.

ROXIN, E.

1. Reachable zones in autonomous differential systems. Bol. Soc. Mat. Mexicana **5**, 125–135 (1960).
2. Stability in general control systems. J. Diff. Eqs. **1**, 115–150 (1965).
3. On generalized dynamical systems defined by contingent equations. J. Diff. Eqs. **1**, 188–205 (1965).
4. On stability in control systems. SIAM J. Control **3**, 357–372 (1966).
5. Local definition of generalized control systems. Michigan Math. J. **13**, 91–96 (1966).

ROXIN, E. O.; SPINADEL, V. W.

1. Reachable zones in autonomous differential systems. In: Contributions to Differential Equations, vol. 1. New York: Wiley 1963, pp. 275–315.

SACKER, R. J.

1. A new approach to the perturbation theory of invariant surfaces. Comm. Pure Appl. Mat. **18**, 712–732 (1965).

SAITO, T.

1. On the measure-preserving flow on the torus. J. Math. Soc. Japan 3, 279–286 (1951).
2. On dynamical systems in n-dimensional torus. Funkcial. Ekvac. 7, 91–120 (1965).
3. On the flow outside an isolated minimal set. In: Proceedings U.S.-Japan Seminar on Differential and Functional Equations. New York-Amsterdam: Benjamin 1967, pp. 301–312.
4. Isolated minimal sets. Funkcial. Ekvac. 11, 155–167 (1969).

SANDOR, ST.; WEXLER, D.

1. Sur la stabilité dans les systèmes dynamiques. Rev. Roumaine Math. Pures Appl. 3, 325–328 (1958).

SANSONE, G. (see also REISSIG, R.)

1. Equazioni differenziali nel campo reale. Bologna: Zanichelli 1948.

SANSONE, G.; CONTI, R.

1. Equazioni differenziali non lineari. Roma: Cremonese 1956. English translation, Nonlinear Differential Equations. London: Pergamon Press 1965.

SARD, A.

1. The measure of the critical values of differentiable maps. Bull. Amer. Math. Soc. 48, 883–896 (1942).

SCHÄFFER, J. J. (see MASSERA, J. L.)

SCHAUDER, J.

1. Der Fixpunktsatz in Funktionalräumen. Studia Math. 2, 171–180 (1930).

SCHWARTZ, A. J.

1. A generalization of a Poincaré-Bendixson theorem to closed two-dimensional manifolds. Amer. J. Math. 85, 453–458 (1963); errata p. 753.
2. Flows on the solid torus asymptotic to the boundary. J. Diff. Eqs. 4, 316–326 (1968).

SCHWARTZMAN, S.

1. Asymptotic cycles. Ann. Math. 66, 270–284 (1957).
2. On the existence of strongly recurrent and periodic orbits. Bol. Soc. Mat. Mexicana 5, 181–183 (1960).
3. Global cross sections of compact dynamical systems. Proc. Nat. Acad. Sci. USA 48, 786–791 (1962).

SCHWEIGERT, G. E.

1. A note on the limits of orbits. Bull. Amer. Math. Soc. 46, 963–969 (1940).

SEIBERT, P. (see also AUSLANDER, J.)

1. Prolongations and generalized Liapunov functions. Tech. Rep. 61-7. RIAS, Baltimore, 1961.
2. Zum Problem der Stabilität unter ständig wirkenden Störungen bei dynamischen Systemen. Arch. Math. (Basel) 15, 108–114 (1964).
3. A concept of stability in dynamical systems. In: Topological Dynamics, J. Auslander and W. H. Gottschalk (Editors). New York-Amsterdam: Benjamin 1968, pp. 423–433.

SEIFERT, G.

1. Stability conditions for the existence of almost-periodic solutions of almost-periodic systems. J. Math. Anal. Appl. 10, 409–418 (1965).
2. Recurrence and almost perodicity in ordinary differential equations (Advances in Differential and Integral Equations). SIAM, Philadelphia, 1969.

SELL, G. R. (see also DEYSACH, L. G. and MILLER, R. K.)

1. Stability theory and Liapunov's second method. Arch. Rat. Mech. Anal. 14, 108–126 (1963).

2. A note on the fundamental theory of ordinary differential equations. Bull. Amer. Math. Soc. **70**, 529–535 (1964).
3. On the fundamental theory of ordinary differential equations. J. Diff. Eqs. **1**, 370–392 (1965).
4. Periodic solutions and asymptotic stability. J. Diff. Eqs. **2**, 143–157 (1966).
5. Nonautonomous differential equations and topological dynamics, I, II. Trans. Amer. Math. Soc. **127**, 241–283 (1967).
6. Invariant measures and Poisson stability. In: Topological Dynamics, J. Auslander and W. H. Gottschalk (Editors). New York-Amsterdam: Benjamin 1968, pp. 435–454.

SELL, G. R.; SIBUYA, Y.

1. Behavior of solutions near a critical point. In: Proceedings U.S.-Japan Seminar on Differential and Functional Equations. New York-Amsterdam: Benjamin 1967, pp. 501–506.

SHCHERBAKOV, B. A. (see also BRONSTEIN, I. U.)

1. Classifications of motions stable in the sense of Poisson: pseudorecurrent motions (Russian). Dokl. Akad. Nauk SSSR **146**, 322–324 (1962). English translation: Soviet Math. (Dokl.), pp. 1320–1322.
2. On classes of motions stable in the sense of Poisson. Pseudorecurrent motions (Russian). Izv. Akad. Nauk Moldav. SSR, Ser. Estestven. Tekh. Nauk, no. 1, pp. 58–72 (1963).
3. Decomposition of a set of Poisson-stable motions (Russian). Dokl. Akad. Nauk SSSR **152**, 71–74 (1963).
4. Constituent classes of Poisson-stable motions (Russian). Sibirsk. Mat. Zh. **5**, 1397–1417 (1964).
5. Dynamical systems: review of papers given at the Kishinev seminar on the qualitative theory of differential equations. Differencial'nye Uravnenija **1**, 260–266 (1965).
6. Minimal motions and the structure of minimal sets (Russian). Papers on Algebra and Analysis, Kishinev, 1965, pp. 99–110.
7. On a class of motions stable in Poisson's sense (Russian). Papers on Algebra and Analysis, Kishinev, 1965, pp. 155–160.
8. Recurrent solutions of differential equations and the general theory of dynamical systems. Differencial'nye Uravnenija **3**, 1450–1460 (1967).

SHOLOHOVICH, F. A. (see also BARBASHIN, E. A.)

1. The relationship between a linear dynamical system and a certain differential equation in Banach space (Russian). Dokl. Akad. Nauk SSSR **120**, 43–46 (1958).
2. Linear dynamic systems (Russian). Izv. Vyssh. Uchebn. Zaved., Mat. 1957, no. 1, pp. 249–257.

SHUBIN, M. A.

1. On some properties of generalized ω-limit sets in dynamical systems (Russian). Vestnik Moskov. Univ., no. 3, 58–60 (1966).

SIBIRSKII, K. S.

1. Uniform approximation of points of dynamical limit sets and motions in them (Russian). Dokl. Akad. Nauk SSSR **146**, 307–309 (1962). English translation: Soviet Math. (Dokl.) 3, 1304–1305 (1962).
2. Uniform approximation of points and properties of motions in dynamical limiting sets (Russian). Izv. Akad. Nauk Moldav. SSR, Ser. Estestven. Tekh. Nauk, no. 1, pp. 38–48 (1963).

SIBIRSKII, K. S.; BRONSTEIN, I. U.

1. Partially ordered group dynamical systems (Russian). Uchen. Zap. Kishinev. Univ., no. 54, pp. 33–36 (1960).

SIBIRSKII, K. S.; KRECU, V. I.; BRONSTEIN, I. U.

1. Liapunov stability in partially ordered dynamical systems (Russian). Uchen. Zap. Kishinev. Univ., no. 54, pp. 29–32 (1960).

SIBIRSKII, K. S.; STAKHI, A. M.

1. Limit properties of partially ordered dispersive dynamical systems (Russian). Izv. Akad. Nauk Moldav. SSR, Ser. Estestven. Tekh. Nauk, no. 11, pp. 42–49 (1963).
2. On partial ordering of groups (Russian). Papers on Algebra and Analysis, Kishinev, 1964.

SIBUYA, Y. (see also HARRIS, W. A. Jr. and SELL, G. R.)

SIEGEL, C. L.

1. Note on differential equations on the torus. Ann. Math. **46**, 423–428 (1945).

SINAI, YA. G.

1. Dynamical systems with countably-multiple Lebesque spectrum, I. Izv. Akad. Nauk SSSR, Ser. Mat. **25**, 899–926 (1961). English translation: Amer. Math. Soc. Transl. **39**, 83–110 (1964).
2. Dynamical systems with countably-multiple Lebesque spectrum, II. Izv. Akad. Nauk SSSR, Ser. Mat. **30**, 15–68 (1966). English translation: Amer. Math. Soc. Transl. **68**, 34–88 (1968).

SMALE, ST.

1. Morse inequalities for a dynamical system. Bull. Amer. Math. Soc. **66**, 43–49 (1960).
2. On gradient dynamical systems. Ann. of Math. **74**, 199–206 (1961).
3. On dynamical systems. In: Symposium Internacional de Ecuaciones Diferenciales Ordinarias, Mexico, 1961, pp. 195–198.
4. Generalized Poincaré conjecture in dimensions greater than four. Ann. of Math. **74**, 361–406 (1961).
5. Stable manifolds for differential equations and diffeomorphisms. Ann. Scuola Norm. Sup. Pisa, Ser. 3, **17**, 97–117 (1963).
6. Diffeomorphisms with many critical points. In: Differential and Combinatorial Topology, S. S. Cairns (Editor). Princeton: Princeton Univ. Press 1965, pp. 63–80.
7. Dynamical Systems on n-dimensional manifolds. In: Differential Equations and Dynamical Systems, J. K. Hale and J. P. LaSalle (Editors). New York: Academic Press 1967, pp. 483–486.

SOLNCEV, YU. K.

1. Two examples of dynamical systems determined by infinite systems of differential equations (Russian). Uchen. Zap. Moskov. Gos. Univ., Mat. **155**, 156–167 (1952).

SPINADEL, V. W. (see ROXIN, E. O.)

STAKHI, A. M. (see SIBIRSKII, K. S.)

STALLINGS, J.

1. Polyhedral homotopy sphere. Bull. Amer. Math. Soc. **66**, 485–488 (1960).

STEPANOV, V. V. (see also BEBUTOV, M. V. and NEMYTSKII, V. V.)

1. Lehrbuch der Differentialgleichungen. Berlin: VEB Deutscher Verlag der Wiss. 1956 (German translation of the Russian edition, Gos. Izd. Tekh. Teor. Lit., Moscow 1953).

STERNBERG, SH. (see also BRAUER, F.)

1. On differential equations on the torus. Amer. J. Math. **79**, 397–402 (1957).

2. On Poincaré's last geometrical theorem. Proc. Amer. Math. Soc. **8**, 787–789 (1957).
3. On the structure of local homeomorphisms of euclidean n-space. Amer. J. Math. **80**, 623–631 (1958).

Stokes, A. P. (see Hale, J. K.)

Strauss, A.
1. Continuous dependence of solutions of ordinary differential equations. Amer. Math. Monthly **71**, 649–652 (1964).
2. Liapunov functions and L^P solutions of differential equations. Trans. Amer. Math. Soc. **119**, 37–50 (1965).
3. Liapunov functions and global existence. Bull. Amer. Math. Soc. **71**, 519–520 (1965).
4. On the stability of a perturbed nonlinear equation. Proc. Amer. Math. Soc. **17**, 803–807 (1966).
5. A geometrical introduction to Liapunov's second method. In: Stability Problems of Solutions of Differential Equations, A. Ghizzetti (Editor). Gubbio: Oderisi 1966, pp. 19–27.
6. A note on a global existence result of R. Conti. Boll. Un. Mat. Ital. **22**, 439–441 (1967).
7. Perturbing asymptotically stable differential equations (Advances in Differential and Integral Equations). SIAM, Philadelphia, 1967.

Strauss, A.; Yorke, J. A.
1. Perturbation theorems for ordinary differential equations. J. Diff. Eqs. **3**, 15–30 (1967).
2. On asymptotically autonomous differential equations. Math. Systems Theory 75–82 (1967).
3. Perturbing asymptotically stable differential equations. Bull. Amer. Math. Soc. **74**, 992–996 (1968).
4. On the fundamental theory of differential equations. SIAM Rev. **11**, 236–246 (1969).
5. Perturbing uniform-asymptotically stable nonlinear systems. J. Diff. Eqs. **6**, 452–483 (1969).
6. Perturbing uniformly stable linear systems with and without attraction. SIAM J. Appl. Math. **17**, 725–739 (1969).
7. Identifying perturbations which preserve asymptotic stability. Proc. Amer. Math. Soc. **22**, 513–518 (1969).
8. Linear perturbation of a family of linear systems. Rep. **950**, MRC, Univ. of Wisconsin, Madison, Wisc., October 1968.

Sutti, C. (see Szegö, G. P.)

Szarski, J.
1. Remarque sur un critère d'unicité des intégrales d'une équation différentielle ordinaire. Ann. Polon. Math. **12**, 203–205 (1962).

Szegö, G. P. (see also Bhatia, N. P. and Kuhn, H. W.)
1. Contributions to Liapunov's second method: nonlinear autonomous systems. In: Nonlinear Differential Equations and Nonlinear Mechanics, J. P. La-Salle and S. Lefschetz (Editors). New York: Academic Press 1963, pp. 421–430.
2. Contributions to Liapunov's second method: nonlinear autonomous systems. Basic Eng. Trans. ASME(D) **84**, 571–578 (1962).
3. On a new partial differential equation for the stability analysis of time-invariant control systems. SIAM J. Control **1**, 63–75 (1962).

4. On global stability properties of nonlinear control systems. Rep. Contract NONR-1228(23), Sept. 1964.
5. New theorems on stability and attraction. In: Stability Problems of Solutions of Differential Equations, A. Ghizzetti (Editor). Gubbio: Oderisi 1966.
6. Liapunov's second method. Appl. Mech. Rev. **19**, 833–838 (1966).
7. An application of the extensive theorem to a control problem. Atti Accad. Naz. Lincei Rend., Cl. Sci. Fis. Mat. Natur. **42**, 766–770 (1967).
8. A theorem of Rolle's type in R^n for functions of the class C^1. Pacific J. Math. **27**, 193–195 (1968).
9. Topological properties of weak attractors. In: Topological Dynamics, J. Auslander and W. H. Gottschalk (Editors). New York-Amsterdam: Benjamin 1968, pp. 455–469.

Szegö, G. P.; Arienti, G.; Sutti, C.
1. On the numerical construction of Liapunov functions. Proc. Fourth Int. Congress IFAC, Warszawa, June 16–21, 1969.

Szegö, G. P.; Geiss, G.
1. A remark on "A new partial differential equation for the stability analysis on time-invariant control systems". SIAM J. Control **1**, 369–376 (1963).

Szegö, G. P.; Olech, C.; Cellina, A.
1. On the stability properties of a third-order system. Ann. Mat. Pura Appl., Ser. IV, **78**, 91–104 (1968).

Szegö, G. P.; Treccani, G.
1. Semigruppi di trasformazioni multivoche (Lecture Notes in Mathematics, vol. 101). Berlin-Heidelberg-New York: Springer 1969.
2. Flow without uniqueness near a compact strongly invariant set. Boll. Un. Mat. Ital., Ser. IV, **1**, 113–124 (1969).
3. Noncontinuous Liapunov functions. Ann. Mat. Pura Appl., Ser. IV, **82**, 1–16 (1969).
4. Liapunov functions on the region of unstable weak attraction. In: Seminar on Differential Equations and Dynamical Systems II, J. A. Yorke (Editor) (Lecture Notes in Mathematics, vol. 144). Berlin-Heidelberg-New York: Springer 1970.

Szmydt, Z.
1. Sur la structure de l'ensemble engendré par les intégrales tendant vers le point singulier du système d'équations différentielles. Bull. Acad. Polon. Sci., Cl. III, **1**, 223 (1953).
2. Sur les systèmes d'équations différentielles dont toutes les solutions sont bornées. Amer. Pol. Math. **2**, 234–236 (1955).
3. Remarque sur la méthode topologique de T. Ważewski. Colloq. Math. **18**, 23–27 (1967).

Szmydtowna, Z.
1. Sur l'allure asymptotique des intégrales des équations différentielles ordinaires. Ann. Soc. Polon. Math. **24**, 17–34 (1951).

Ta, Li
1. Die Stabilitätsfrage bei Differenzengleichungen. Acta Math. **63**, 99–141 (1934).

Taam, C. T.
1. Asymptotic relations between systems of differential equations. Pacific J. Math. **6**, 373–388 (1956).

Tonelli, L.
1. Opera scelte (Selected Works), published by UMI, S. Cinquini (Editor), vol. 3. Roma: Cremonese 1960/62.

Treccani, G. (see Szegö, G. P.)

TRJITZINSKY, W. J.

1. Problèmes dans la théorie des systèmes dynamiques. Acta Math. **95**, 191–289 (1956).
2. Aspects topologiques de la théorie des fonctions réelles et quelques conséquences dynamiques. Ann. Mat. Pura Appl. **42**, 51–117 (1956).

TURAN, P.

1. On the instability of systems of differential equations. Acta Math. Acad. Sci. Hungar. **6**, 257–270 (1955).
2. On the property of the stable or conditionally stable solutions of systems of nonlinear differential equations. Ann. Mat. Pura Appl. **48**, 333–340 (1959).

TUROWICZ, A.

1. Sur les trajectoires et les quasitrajectoires des systèmes de commande nonlinéaires. Bull. Acad. Polon. Sci. **10**, 529–531 (1962).
2. Sur les zones d'émission des trajectoires et des quasitrajectoires des systèmes de commande nonlinéaires. Bull. Acad. Polon. Sci. **11**, 47–50 (1963).

URA, T.

1. Sur les courbes définies à la surface du tore par des équations admettant un invariant intégral. Ann. Sci. École Norm. Sup. **69**, 259–275 (1952).
2. Sur les courbes définies par les équations différentielles dans l'espace à *m* dimensions. Ann. Sci. École Norm. Sup. **70**, 287–360 (1953).
3. Sur les periodes fondamentales de solutions periodiques. Comm. Math. Univ. St. Paul **4**, 113–130 (1955).
4. Sur le courant extérieur à une région invariante; prolongements d'une caractéristique et l'ordre de stabilité. Funkcial. Ekvac. **2**, 143–200 (1959).
5. On the flow outside a closed invariant set: stability, relative stability and saddle sets. In: Contributions to Differential Equations, vol. 3. New York: Wiley 1964, pp. 249–294.
6. Local isomorphisms and local parallelizability of dynamical systems. In: Topological Dynamics, J. Auslander and W. H. Gottschalk (Editors). New York-Amsterdam: Benjamin 1968, pp. 493–506.

URA, T.; HIRASAWA, Y.

1. Sur les points singuliers des équations différentielles admettant un invariant intégral. Proc. Japan Acad. **30**, 726–730 (1954).

URA, T.; KIMURA, I.

1. Sur le courant extérieur à une région invariante: Théorème de Bendixson. Comm. Math. Univ. St. Paul **8**, 23–39 (1960).
2. Stability in topological dynamics. Proc. Japan Acad. **40**, 703–706 (1964).

URABE, M.

1. Nonlinear Autonomous Oscillations. New York: Academic Press **1967**.

UTZ, W. R. (see FOLAND, N. E.)

VINOGRAD, R. E.

1. On the limiting behavior of unbounded integral curves (Russian). Dokl. Akad. Nauk SSSR **66**, 5–8 (1949).
2. On the limit behavior of an unbounded integral curve (Russian). Uchen. Zap. Moskov. Gos. Univ. **155**, Mat., no. 5, pp. 94–136 (1952).

VINOKUROV, V. R.

1. On the definition of a dynamic limit point in general dynamical systems (Russian). Izv. Vyssh. Uchebn. Zaved., Mat., **1964**, no. 3 (40), pp. 36–38.

VRKOC, I. (see also KURZWEIL, J.)

1. On the inverse theorem of Chetaev. Czechoslovak Math. J. **5** (**80**), 451–461 (1955).

2. Integral stability (Russian). Czechoslovak Math. J. **9** (84), 71–129 (1959).

VRUBLEVSKAYA, I. N.

1. On trajectories and limiting sets of dynamical systems (Russian). Dokl. Akad. Nauk SSSR **97**, 9–12 (1954).
2. Some criteria of equivalence of trajectories and semitrajectories of dynamical systems (Russian). Dokl. Akad. Nauk SSSR **97**, 197–200 (1954).
3. On geometric equivalence of the trajectories of dynamical systems (Russian). Mat. Sbornik **42**, 361–424 (1957).

WALTER, W.

1. Bemerkung zu verschiedenen Eindeutigkeitskriterien für gewöhnliche Differentialgleichungen. Math. Z. **84**, 222–227 (1964).

WAŻEWSKI, T.

1. Sur un principe topologique de l'examen de l'allure asymptotique des intégrales des équations différentielles ordinaires. Ann. Soc. Polon. Math. **20**, 279–313 (1947).
2. Sur les intégrales d'un systeme d'équations différentielles ordinaires. Ann. Soc. Polon. Math. **21**, 277–297 (1948).

WAŻEWSKI, T.; ZAREMBA, S.

1. Sur les ensembles de condensation des caractéristiques d'un système d'équations différentielles ordinaires. Ann. Soc. Polon. Math. **15**, 24–33 (1936).

WEXLER, D. (see also SANDOR, ST.)

1. Stability theorems for a system of stationary differential equations. Rev. Roumaine Math. Pures Appl. **3**, 131–138 (1956).

WILSON, F. W., Jr.

1. Smoothing derivatives of functions and applications. Div. Appl. Math., Brown Univ., Tech. Rep. 66-3, June 1966.
2. The structure of the level surfaces of a Lyapunov function. J. Diff. Eqs. **3**, 323–329 (1967).

WINTNER, A. (see also HARTMAN, P.)

1. The nonlocal existence problem of ordinary differential equations. Amer. J. Math. **67**, 277–284 (1945).
2. On the convergence of successive approximations. Amer. J. Math. **68**, 13–19 (1946).
3. Asymptotic equilibria. Amer. J. Math. **68**, 125–132 (1946).
4. The infinities in the nonlocal existence problem of ordinary differential equations. Amer. J. Math. **68**, 173–178 (1946).
5. On the local uniqueness of the initial value problem of the differential equation $d^n x/dt^n = f(t, x)$. Boll. Un. Mat. Ital. (3) **11**, 496–498 (1956).

WHITNEY, H.

1. Regular families of curves, I, II. Proc. Nat. Acad. Sci. USA **18**, 275–278, 340–342 (1932).
2. Regular families of curves. Ann. of Math. **34**, 244–270 (1933).
3. On regular families of curves. Bull. Amer. Math. Soc. **47**, 145–147 (1941).

WHYBURN, G. T.

1. Analytic Topology (Amer. Math. Soc. Colloquium Publications, vol. 38). New York 1942.

WU, TA-SUN

1. Continuous flows with closed orbits. Duke Math. J. **31**, 463–469 (1964).
2. Proximal relations in topological dynamics. Proc. Amer. Math. Soc. **16**, 513–514 (1965).
3. Left almost periodicity does not imply right almost periodicity. Bull. Amer. Math. Soc. **72**, 314–316 (1966).

YAMABE, H. (see MARKUS, L.)

YORKE, J. A. (see also BHATIA, N. P.; JONES, G. S. and STRAUSS, A.)

1. Asymptotic properties of solutions using the second derivative of a Liapunov function. Ph. D. Thesis, Univ. of Maryland, College Park, Md., June **1966**.
2. Invariance for ordinary differential equations. Math. Systems Theory **1**, 353—372 (1967).
3. Extending Liapunov's second method to non-Lipschitz Liapunov functions. In: Seminar on Differential Equations and Dynamical Systems, G. S. Jones (Editor) (Lecture Notes in Mathematics, vol. 60). Berlin-Heidelberg-New York: Springer **1968**, pp. 33—38.
4. An extension of Chetaev's instability theorem using invariant sets and an example. In: Seminar on Differential Equations and Dynamical Systems, G. S. Jones (Editor) (Lecture Notes in Mathematics, vol. 60). Berlin-Heidelberg-New York: Springer **1968**, pp. 107—114.
5. Liapunov functions and the existence of solutions tending to zero. In: Seminar on Differential Equations and Dynamical Systems, G. S. Jones (Editor) (Lecture Notes in Mathematics, vol. 60). Berlin-Heidelberg-New York: Springer **1968**, pp. 51—58.
6. Space of solutions. In: Mathematical Systems Theory and Economics II, H. W. Kuhn and G. P. Szegö (Editors) (Lecture Notes in Operations Research and Mathematical Economics, vol. 12). Berlin-Heidelberg-New York: Springer **1969**, pp. 383—603.
7. A theorem on Liapunov functions using $\ddot{V}$. Tech. Note BN-596, IFDAM, Univ. of Maryland, College Park, Md., Feb. **1969**.

YOSHIZAWA, T.

1. The Stability Theory by Liapunov's Second Method. Math. Soc. Japan, Tokyo, **1966**.
2. On the stability of solutions of a system of differential equations. Mem. Coll. Sci. Univ. Kyoto, Ser. A, **29**, 27—33 (1955).
3. Note on the solutions of a system of differential equations. Mem. Coll. Sci. Univ. Kyoto, Ser. A, **29**, 249—273 (1955).
4. On the equiasymptotic stability in the large. Mem. Coll. Sci. Univ. Kyoto, Ser. A, **32**, 171—180 (1959).
5. Liapunov's functions and boundedness of solutions. Funkcial. Ekvac. **2**, 95—142 (1959).
6. Stability and boundedness of systems. Arch. Rat. Mech. Anal. **6**, 409—421 (1960).
7. Asymptotic behavior of solutions of nonautonomous systems near sets. J. Math. Kyoto Univ. **1**, 303—323 (1962).
8. Stability of sets and perturbed systems. Funkcial. Ekvac. **5**, 31—69 (1963).
9. Asymptotic stability of solutions of an almost periodic system of functional-differential equations. Rend. Circ. Mat. Palermo, Ser. II, **13**, 1—13 (1964).
10. Eventual properties and quasi-asymptotic stability of a non-compact set. Funkcial. Ekvac. **8**, 79—90 (1966).

YOUNG, G. (see HOCKING, J.)

ZAREMBA, M. S. C. (see also WAŻEWSKI, T.)

1. Sur les équations au paratingent. Bull. Sci. Math. I, **60**, 139—160 (1936).

ZEEMAN, C.

1. The generalized Poincaré conjecture. Bull. Amer. Math. Soc. **67**, 270 (1961).

ZHIDKOV, N. P.

1. Certain properties of discrete dynamical systems (Russian). Uchen. Zap. Moskov. Gos. Univ., no. **163**, Mat., **6**, **31–59 (1952)**.

ZIPPIN, L. (see also MONTGOMERY, D.)

1. Transformation groups. In: Lectures in Topology. Ann Arbor: Univ. of Michigan Press **1941**, pp. **191–221**.

ZUBOV, V. I.

1. The Methods of Liapunov and their Applications. Leningrad **1964**.
2. Mathematical Methods for the Investigation of Systems of Automatic Control (Russian). Leningrad: Gos. Sojus. Izd. Sudostroit. Promsyl. **1959**, p. **324**. English translation, Oxford: Pergamon Press **1964**.
3. Some sufficient criteria for stability of a nonlinear system of differential equations (Russian). Prikl. Mat. Mekh. **17**, **506–508 (1953)**.
4. On the theory of A. M. Liapunov's second method (Russian). Dokl. Akad. Nauk SSSR **99**, **341–344 (1954)**.
5. Questions of the theory of Liapunov's second method, construction of a general solution in the region of asymptotic stability (Russian). Prikl. Mat. Mekh. **19**, **179–210 (1955)**.
6. On the theory of A. M. Liapunov's second method (Russian). Dokl. Akad. Nauk SSSR **100**, **857–859 (1955)**.
7. An investigation of the stability problem of systems of equations with homogeneous right hand members (Russian). Dokl. Akad. Nauk SSSR **114**, **942–944 (1957)**.
8. Conditions for asymptotic stability in the case of nonstationary motions and estimate of the rate of decrease of the general solution (Russian). Vestnik Leningrad. Univ., Ser. Mat. Mekh. Astron. **12**, **110–129 (1957)**.
9. On a method of investigating the stability of a null-solution in doubtful cases (Russian). Prikl. Mat. Mekh. **22**, **46–49 (1958)**.
10. On stability conditions in a finite time interval and on the computation of the length of that interval (Russian). Bull. Inst. Politechn. Iaşi (N.S.) **4**, **59–74 (1958)**.
11. Some problems in stability of motion (Russian). Mat. Sbornik **48 (90)**, **149–190 (1959)**.
12. On the theory of recurrent functions (Russian). Sibirsk. Mat. Zh. **3**, **532–560 (1962)**.

Author Index

Subject Index

Druck: Strauss Offsetdruck, Mörlenbach
Verarbeitung: Schäffer, Grünstadt